Praise for

Rainwater Harvesting
for Drylands and Beyond
Volume 1

"Brad Lancaster has done it again. In revising his excellent book, he has given us a window into the world through the lens of water. Water connects all things. And Brad shows us water as a practical way of considering context and connection. From a world of water as commodity, he takes us to a world of water as moving, enriching exchanges, the stuff of life. A native friend got a job with his local water company, and rather than an engineering job, he saw it as a sacred trust. This is the shift that Brad leads us carefully through. Water is wealth and health—let's treat it that way, and dance our way from scarcity to abundance.

— Joel Glanzberg, author of the *The Permaculture Mind*; tracker;
and teacher / designer of regenerative living systems; www.PatternMind.org

"In a time of escalating resource scarcity and global conflict, this essential book helps us regain control of our water by showing us how to enhance our water and energy supply with simple, fun, and effective strategies at home and beyond."

— Maude Barlow, author of *Blue Covenant*;
Senior Advisor on Water to the President of the United Nations General Assembly

"Brad Lancaster clearly defines the differences between the path to scarcity and the path to abundance, both revolving around the wise use of water while avoiding the consequences of careless use. Throughout the book, alternatives are plainly described with illustrations that get to the point. I have worked in the field with Brad. He is unabashedly committed to the parallel causes of water and energy conservation. He asks incisive questions, searches for answers, tests solutions, documents findings, and happily shares his conclusions with all who care to listen. Clearly Mr. Lancaster is an agent for change, a true innovator, providing simple but powerful solutions to difficult questions facing society in both urban and rural situations."

—Bill Zeedyk, Zeedyk Ecological Consulting, LLC; co-author of *Let the Water Do the Work*

"Lancaster's book on rainwater harvesting is fantastic and an abundant guidebook for a more sane approach to our most precious resource. I highly recommend it."

— Jason F. McLennan, CEO, International Living Future Institute

Awards

Rainwater Harvesting for Drylands and Beyond, Volume 1 awarded:
- Book of the Year pick by the Pima County Public Library
- Finalist *Foreword* Magazine's Book of the Year
- Silver Medal in the Independent Publisher Book Awards
- Best Gardening/Agriculture Book from Arizona/Glyph Book Awards
- Best First Book from Arizona/Glyph Book Awards

For inside cover figure captions see last page of book.

"This is one of those ideas that you just need to think about for a moment to understand its importance—and here we have all the techniques you'll ever need patiently laid out!"

— Bill McKibben, fellow of the American Academy of Arts and Sciences;
author *Eaarth: Making a Life on a Tough New Planet* and *Oil and Honey*

"A very good book gets even better. The case studies inspire, and the advice is detailed. This book is not just for people in arid lands; rainwater harvesting provides free delivery of clean water, reduces flooding, and improves stormwater quality in wetter lands too. Homeowners, renters, builders, students, planners, and policy makers—read this book!

— David Bainbridge, co-author of *Passive Solar Architecture*;
author of *A Guide for Desert and Dryland Restoration: New Hope for Arid Lands*;
http://works.bepress.com/david_a_bainbridge/

"This book is a must read for all of us who care about water. The personal, heartfelt stories of Brad's own processes and successes motivate us to try, while the clear connections he makes to the global environmental crisis we face deepens the importance of this work."

— Laura Allen, founding member of Greywater Action; www.GreywaterAction.org

"Its said that water is the crisis du jour of the early 21st century as energy was of the late 20th century.

This is true but more importantly the two are intimately related. 20% of our energy use is for pumping water and large percentage of our energy production is based on hydropower much of which is wasted for air conditioning that could be accomplished by proper shading and natural ventilation.

Appropriate technology (finding the right scale to apply technology and design) is the key to addressing this situation and no one does this better than Brad Lancaster. Brad shows how to passively harvest sun, shade, and water using appropriate design principles available to us all."

— Ken Haggard, architect, San Luis Sustainability Group; co-author of *Passive Solar Architecture*;
ISES Passive Solar Architecture Pocket Reference; author of *Fractal Architecture*

"I listen to the music of the water flowing in the river, and I wonder. "Will this beautiful melody remain forever?" Brad Lancaster gives me the answer: YES, if we act now, making the connection between water and life, living cisterns of vegetation and a better climate, and rainwater harvesting and energy saving (reducing toxic emissions). This book is an indispensable tool for all of those who want to minimize their footprint and contribute to greening the planet, especially for those, like me, who live in the Drylands. ¡We all should read it!"

— Alejandra Caballero, Director of Proyecto San Isidro,
a Mexican Training Center in Land Restoration and Sustainable Living; www.ProyectoSanIsidro.com.mx

"Brad Lancaster offers simple, time-tested solutions to making better use of the water falling on properties. The tools and strategies presented have the potential to help homeowners replace nearly all their landscape water use with water derived from on-site sources: rainwater, stormwater runoff, and greywater."

— *Water Engineering Australia*

"Water running off our gardens, streets and farmland is the #1 source of ocean pollution—and ocean users and precious marine life suffer. Meanwhile, many regions lack clean drinking water. Plus, transporting and cleaning water is energy-intensive, contributing to climate change. But we can sponge up that runoff in soil—naturally watering plants, filtering pollutants and recharging groundwater. Brad's methods to do this apply everywhere along the coast. Put them to use on your site, step-by-step. Not a DIY type? Show the book to your landscaper. And share it with your city and water district to spread the knowledge."

— *Paul Herzog, Ocean Friendly Gardens Program Coordinator,*
Surfrider Foundation; www.OceanFriendlyGarden.org

"Brad Lancaster is the country's Messiah of water harvesting, teaching a gospel of catching water that falls from the sky or runs down the street. With the passion of a preacher, he shows us how the water we save ennobles our lives and lifts up our spirits. As we each do our part to make the world a better place, our joy and fulfillment comes from the fruits of our commitment: verdant trees, glorious shade, ripe fruit and vegetables, and precious water. Water harvesting won't turn an arid landscape into the Garden of Eden, but it sure will make you feel better about living in a water-challenged place. Lancaster has made me a believer; this book may change your life."

— Robert Glennon, Regents' Professor at the University of Arizona;
author of *Unquenchable: America's Water Crisis and What To Do About It*

"This book and the thinking behind it should be part of the basic education of civil engineers, architects, landscape architects, and planners everywhere. As a civil engineer working for a progressive municipal water utility in an arid climate, I can see if a majority of our citizens followed these practices, many of our current and future challenges would be alleviated. The positive side benefits in terms of erosion-control, creation of bird habitat, and natural cooling would be exceptional."

— Patricia Eisenberg, P.E., Past president, Arizona Society of Civil Engineers

"Brad Lancaster's *Rainwater Harvesting for Drylands and Beyond* is an important book. Its teachings should not just be applied to drylands. It's about using hydrological cycles to create and support sustainable landscapes, and the lessons are universal and useful wherever you live. This book is where to start with environmental restoration. His story of "The Man Who Farms Water" in Africa is a microcosm and metaphor for the brilliant use of Nature's operating instructions. Most highly recommended!"

— John Todd, Ph.D., Research Professor and Distinguished Lecturer, The Rubenstein School of Environment and Natural Resources, The University of Vermont; President, Ocean Arks International

"*Rainwater Harvesting for Drylands and Beyond, Volume 1* is more than a text on how to harvest rainwater. It is a way of life that gives back nourishment to our earth rather than what has become the standard of continually taking. This way of life has become almost a religion for Brad and as he demonstrates, it should be the same for all of us."

— Heather Kinkade-Levario, R.L.A., President of the American Rainwater Catchment Systems Association (ARCSA); author of *Design for Water*

"The world needs more practical visionaries like Brad Lancaster! Blending his own knowledge, experience and wisdom with the collected wisdom and best practices of water and sun harvesters from around the world, Brad gives us access to a wealth of critically needed tools for rethinking our relationship with the gifts of water and sunshine falling for free from the sky. This man more than walks his talk; he lives, breathes, eats, and drinks it!"

— David Eisenberg, Director of the Development Center for Appropriate Technology;
co-author of the *Straw Bale House Book*; lead author of "Code, Regulatory and Systemic Barriers Affecting Living Building Projects; two-term member of the Board of Directors of the U.S. Green Building Council

"This important and timely water-harvesting book reads like a conversation with a trusted friend. As such, it is an effective how-to and why-for manual for living within our means in our shared watersheds. Heartfelt thanks, Brad, for spotlighting the route to abundance in these arid climes!"

— Barbara Clark, project manager, Teran Watershed Project, Cascabel AZ

"I've been looking for a book like this for a long time. That is, one that will tell me, armed with a shovel and not a great deal of mechanical fussing or back-straining, to begin saving rainwater immediately. And in the process, save money, salve my conscience, and, best of all, come out ahead with a greener life."

— Peter Wild, U.S. Water News

see p. 278 for more testimonials

THE REGENERATIVE FUND

Once production costs are recovered, 10% of the profits from each book sold will go to the "Regenerative Fund." This money will help fund projects, research, presentations, and publications that further promote integrated water harvesting and other regenerative living strategies. The fund will also be used to help distribute these resources to libraries and schools.

Rainwater Harvesting for Drylands and Beyond
Volume 1
Second Edition

GUIDING PRINCIPLES TO WELCOME RAIN
INTO YOUR LIFE AND LANDSCAPE

Brad Lancaster

Illustrated by
Joe Marshall
Silvia Rayces
Ann Audrey
Roxanne Swentzell
Gavin Troy
Kay Sather
Jill Lorenzini
Carol Heffern

Tucson, Arizona
www.HarvestingRainwater.com

Published by:
Rainsource Press
813 N. 9th Ave.
Tucson, AZ 85705
U.S.A.
www.HarvestingRainwater.com

First Printing, 1st Edition 2006
Second Printing, 1st Edition 2006
Third Printing, 1st Edition 2008, Revised
Fourth Printing, 1st Edition 2008, Revised
Fifth Printing, 1st Edition 2010, Revised
First Printing, 2nd Edition 2013, Revised and expanded

Printed and bound in the United States of America on acid- and
chlorine-free, *100% post-consumer* recycled paper

Cover design: Kay Sather
Front cover art: Gavin Troy
Back cover illustration: Joe Marshall, www.planetnameddesire.com
Back cover photo: Kimi Eisele
Book design: Teri Reindl Bingham
Illustrations: Joe Marshall, Silvia Rayces, Ann Audrey,
Roxanne Swentzell, Gavin Troy, Kay Sather, Jill Lorenzini, Carol Heffern
Photographs: Unless otherwise noted, all photographs are by Brad Lancaster

Lancaster, Brad
    Rainwater Harvesting for Drylands and Beyond, Volume 1: Guiding
    Principles to Welcome Rain Into Your Life and Landscape/Brad Lancaster.
    p.   cm.
    Includes bibliographical references and index.
    ISBN 978-0-9772464-3-4
    1. Water harvesting. 2. Garden & landscape design. 3. Nature & ecology.
    4. Sustainable living & green design. 6. Water supply & land use.
    7. Passive solar design. 8. Energy efficiency. 9. Permaculture &
    regenerative design.

Library of Congress Control Number (LCCN): 2013906071

To

My parents Stew and Diana Lancaster,
and my grandparents Herb and Martha Lancaster and
Frits and Jean van't Hoogerhuijs for having always
encouraged me to pursue my wild ideas and dreams

Mr. Zephaniah Phiri Maseko and family
for enabling me to see the whole

All the stewards of the earth
who teach by the example they live

The Eight Principles of
Successful Rainwater Harvesting

1. Begin with long and thoughtful observation.

Use all your senses to see where the water flows and how. What is working, what is not? Build on what works.

2. Start at the top (highpoint) of your watershed and work your way down.

Water travels downhill, so collect water at your high points for more immediate infiltration and easy gravity-fed distribution. Start at the top where there is less volume and velocity of water.

3. Start small and simple.

Work at the human scale so you can build and repair everything. Many small strategies are far more effective than one big one when you are trying to infiltrate water into the soil.

4. Spread and infiltrate the flow of water.

Rather than having water erosively run off the land's surface, encourage it to stick around, "walk" around, and infiltrate into the soil. Slow it, spread it, sink it.

5. Always plan an overflow route, and manage that overflow as a resource.

Always have an overflow route for the water in times of extra heavy rains, and where possible, use that overflow as a resource.

6. Maximize living and organic groundcover.

Create a living sponge so the harvested water is used to create more resources, while the soil's ability to infiltrate and hold water steadily improves.

7. Maximize beneficial relationships and efficiency by "stacking functions."

Get your water-harvesting strategies to do more than hold water. Berms can double as high and dry raised paths. Plantings can be placed to cool buildings. Vegetation can be selected to provide food.

8. Continually reassess your system: the "feedback loop."

Observe how your work affects the site—beginning again with the first principle. Make any needed changes, using the principles to guide you.

Contents

List of Illustrations

List of Boxed and
Tabled Information

Acknowledgments

This book never would have been written, let alone published, without the incredible help and motivation from many water harvesters, friends, family, neighbors, professionals, teachers, and students. I am indebted to you all. You are the community that supports and nourishes me, and part of a larger force I see striving to enhance the quality of life for everyone. Here, I list a few of the many that helped me, though my heartfelt thanks go out to you all.

Thank you Mr. Zephaniah Phiri Maseko and family for your inspirational life and stories; and for updates and elaborations on the Phiris from Ken Wilson of the Christensen Fund; Brock Dolman of the Water Institute; and Mary Witoshynsky, author of *The Water Harvester*. For your teachings, input, encouragement, guiding examples, and ideas, thanks to Tim Murphy, Vicki Marvick, Ben Haggard, Bob and Pamela Mang, Carol Sanford, and Joel Glanzberg of Regenesis Group; Bill Zeedyk of Zeedyk Ecological Consulting; Craig Sponholtz of Dryland Solutions; Van Clothier of Stream Dynamics; Barbara Rose of Bean Tree Farm; Chris Meuli; Mary Zemach; Marco Barrantes and Michelle Matthews of La Loma Development; Jeremiah Kidd of San Isidoro Permaculture; Reese Baker of RainCatcher; Nate Downey and Melissa McDonald of Santa Fe Permaculture; Richard Jennings of Earthwrights Designs; everyone at Watershed Management Group, Inc.; Art Ludwig of Oasis Design; Laura Allen and crew at Greywater Action; Russ Buhrow; Bob Dixon;

Dan and Karen Howell; Dave Tagget; Steve Kemble and Carol Escott; Bert Lopez; Doug Pushard of HarvestH2O.com; Billy Kniffen of ARCSA; Jim and Karen Brooks of Soilutions; Meg Keoppen; Jeff Blau; Toby Hemenway; David Omick and Pearl Mast of Omick.net; Rocky Brittain; Ken Haggard and David Bainbridge, authors of *Passive Solar Architecture*; the staff of the Botswana Permaculture Trust; the staff of the Fambidzanai Permaculture Center; and Justin Willie and the other staff of the Black Mesa Permaculture Project. Thank you Bill Mollison and David Holmgren for creating a "big umbrella" framework within which I can strive for and articulate integrated systems.

Much gratitude to all peer reviewers of the draft manuscripts for your time and wise editing, comments, and suggestions. To David Aguirre, Tony Novelli, and Steve Malone for keeping my computer going, and Josh Schachter for the generous use of your slide scanner and computer. Thank you to Joe Marshall, Silvia Rayces, Ann Audrey, Gavin Troy, Roxanne Swentzell, Jill Lorenzini, and Kay Sather for your beautiful art that brings ideas to life and vision within the book, Carol Heffern for your sun and shadow path videos and images, and Andrew Brown for your "Free Water" video of my work. And to Gary Paul Nabhan for your glowing foreword and visionary work.

Then there is that core of noble, amazing souls that I used endlessly as sounding boards for the book's concept, content, and style: Eileen Alduenda and Ann

Audrey—without your incredible organizational skills, editing, understanding, management, and support this book never could have existed—really you are co-authors. Anastasia Rabin and Gabrielle Pietrangelo, you kept my passion alight, endured never-ending rainbook conversations, and kept the project true to its intent. Brock Dolman, your word play, enthusiasm, and spirit of fun gave life to abundant waterspread. Kevin Moore Koch of Technicians for Sustainability, your friendship, support, and perspective pulled my writing and me out of serious bogs. David Confer, you've added invaluable data, diverse perspectives, and steady support, and along with Wayne Moody, have provided the opportunity for many of this book's ideas to be realized.

Thanks to editors Eileen Alduenda, Ann Audrey, Shay Salomon, Frank McGee, Tom Brightman, Matt Weber, Kimi Eisele, Megan Hartman and Craig Sponholtz (who also helped write sections of this book), Derek Roff, Jan-Willem Jansens, Avery Anderson, Sarah Iams, Dave Schaller, Jeremy Frey and students, Mac Hudson, Dan and Shelly Dorsey, and most recently, skillfully, and thoroughly, June Fritchman. And thanks to many of you, especially June for helping me refine and divide the book into tastier and more easily digestible portions. Thanks to Kay Sather for editing the illustrations, and Brandy Winters for early illustration work. Teri and Bob Bingham, thank you for your beautiful book design and layout.

Thanks to the many people helping provide key information and research such as the City of Tucson and University of Arizona library staff; Brian Barbaris from the University of Arizona's Department of Atmospheric Sciences; University of Arizona professors Tom Wilson and Jim Riley for patient help with soils and text review; Larry Medlin for passive solar resources; D. Yogi Goswami, Ken Haggard, David Bainbridge, Rachel Aljilani, and the International Solar Energy Society for giving me permission to reprint and adapt images (in appendix 8) from your *Passive Solar Architecture Pocket Reference*; Greg Pelletier, Environmental Engineer at Washington State University Department of Ecology, for a crash-course on using Excel macros, which enabled the creation of a sun-angle calculator from the NOAA Solar Calculator's macros; Daniel Snyder of Westwind Solar Electric; Paul James of Just Water Savers; Mark Ragel of Water Harvesting International; Megan Hartman and LeeAnn Lane, and Brandy Lellou and Valerie Strassberg of Nature's Voice Our Choice, for water-energy-carbon (W-E-C) nexus research; Josh Landess and Tanya Kelly for help thinking through the development of the W-E-C charts; Art Diem, Environmental Engineer at U.S. EPA, for helping navigate the eGrid Excel file; Beorn Courtney for research and work helping legalize water harvesting in Colorado; Virginia Welford and Christina Bickelmann of the Arizona Department of Water Resources for water use and conservation data; Hans Hutchinson for local well level research; Henry Jacobson for curb cut innovations; and Frank Sousa for Tucson stormwater data and helping Ann Audrey make Tucson's water-harvesting guide happen.

Thanks everyone who invested in my vision for this book and helped raise the capital for its printing through your preorders. You had to endure delay after delay. I hope your expectations and needs have been met or exceeded.

Big appreciation to Ian Johnson for his incredible work setting up my information-packed website www.HarvestingRainwater.com, and to Megan Hartman for helping me maintain and evolve it.

I am indebted to all who provided me with testimonials, and want you to know that it was your great work in the world that lured me to ask for your thoughts.

Thank you, Mom and Dad, for consistently encouraging me in the pursuit of my passions. Thank you, brother Rodd, and later sister-in-law Chi, for going with, and helping me realize the vision, for our site. Thank you to all my neighbors and activists/organizers in neighborhoods throughout the globe who have stepped up to work with others and become agents of positive change where you live.

Finally, I am indebted to the indigenous dryland cultures who knew—and know— how to sustainably, and even regeneratively, live with the land. May that knowledge never be forgotten.

Foreword

By Gary Paul Nabhan

Although rainwater harvesting has been accomplished by humans in virtually every drought-vulnerable region of the world for millennia, our society seems to have some collective amnesia about the utility, efficiency, sustainability, and beauty of these time-tried practices. Fortunately, this book and Brad's lifelong passion for practical, ecological, and aesthetically pleasing solutions to our water woes may cure us of that amnesia just when we most desperately need to remember such solutions are readily at hand. From where I write this in Northern Arizona, nine out of every ten trees outside my window are dead, due to the worst drought in fourteen hundred years, and the artificial reservoir known as Lake Powell is projected to go dry within six more years. And yet, those of my neighbors who harvest water off their roofs, parking lots, or slopes (as we do) have never had to haul in water during the last six years of subnormal precipitation, and elderly Hopi farmers have still produced crops every year in the floodwater (ak-chin) fields. At a time when surface- and ground-water is becoming increasingly privatized, fought over, and transferred between watersheds and aquifers as if it were but one more globalized commodity, Brad demonstrates a diversity of strategies that can quench our thirst, sustain local food production, and keep peace among neighboring cultures. Because struggles for access to water are likely to be one of the most frequent causes for warfare and social

unrest over the next half century on every continent, Brad should be nominated for the Nobel Prize for Peace for offering the world so many elegant means of avoiding such struggles through local harvesting of both water and traditional ecological knowledge.

Like many arid land ecologists scattered around the world, I was first inspired to consider the supreme importance of water harvesting for desert cultures by reading Michael Evenari's classic, *The Negev—The Challenge of a Desert*, about Israeli Jewish attempts to learn from their ancient neighbors, the Nabateans, who drew upon diverse runoff catchments and storage practices to make their prehistoric civilization flourish at Petra, the Negev, and Sinai. With the likes of arroyo-of-consciousness journalist Chuck Bowden and straw-bale movement founder Matts Myhrman, I sought out older treatises and surviving practitioners of O'odham (Papago) ak-chin farming in the Sonoran Desert. We found that there was much to learn from our desert neighbors about the harvesting of both water and nutrients; Brad has continued and extended our earlier, haphazard efforts of rescuing such knowledge from Native American elders. But Brad has also gone two steps further than many of us. He has essentially accomplished a worldwide survey of water-harvesting practices, humbling his predecessors by compiling a dizzyingly diverse portfolio of strategies, techniques, and technologies. He has then tried and fine-tuned every one of these strategies, so that he now has firsthand knowledge of how they function, and at what cost. His own desert abode is like a walk-through encyclopedia of water-harvesting techniques gleaned from cultures and innovators from around the world.

There is both quantitatively-informed precision and beauty in what Brad has implemented, and this combination is a rarity in our modern world. Technological fixes have grown increasingly ugly, but as you can see from the drawings and photos in this masterwork, Brad's designs sing to us as they solve our water shortages.

A half century ago, Thomas Merton prophesized that "some day, they will even try to sell you the rain," warning us that the privatization and corporate control of our hydrological destiny could become our doom. What Brad's genius safeguards for us is "water democracy," and I predict that this concept will become a keystone of environmental justice throughout the desert regions of the world, if not everywhere. We will no longer think of desert living as "lacking," or "limited," but celebrate the abundance before us. With tongue in cheek, we may even offer our sympathies to those who live in soggy, "drought-deficient" places, who may never be able to share the joy with us of harvesting our own fresh, delicious water, just as horned lizards have done off their very own backs since they first emerged on this dry planet. Blessings to you Brother Brad, the Patron Saint of Water Democracy.

Gary Paul Nabhan is a Kellogg Endowed Chair in Borderland Food and Water Security at the University of Arizona and author of the 2013 book Growing Food in a Hotter, Drier Land *from Chelsea Green Publishing. He farms an experimental orchard with harvested rainwater in Patagonia, Arizona.*

Preface to the Second Edition

People, farms, and even whole communities may periodically run out of fresh water, but planet Earth does not. Why? Because its living systems use water over and over through endless cycles in such a way that the quality of water—and the life it sustains—is not degraded. Instead it is maintained and even *enhanced* as it moves through myriad cycles. This increases what is possible, and leads to more life rather than less.

This book challenges you to do the same where you live. Push the limits. Strive to make your site's beneficial cycles endless, along with their potential. Multiple strategies, tools, and examples are given to help you do this.

This second edition will greatly broaden your site's array of possibilities by helping you integrate your awareness, harvesting, and cycling of water with other *on*-site resources such as sun, wind, shade, and community.

New and expanded aspects of this edition include:

• 100-plus pages of new information, over 120 new images, and revision of over 40 illustrations, to provide more than 280 total images enabling you to *see* in a new way

• Updated case studies of real-life examples including descriptions of their on-going *evolution*, to help inform your site design (chapters 1 and 5)

• The One-Page Place Assessment, a tool to deepen your awareness of—and connection with—the place you live to help you amplify its regenerative potential (chapter 2)

• Tools for identifying, *integrating*, and thus maximizing, the harvest and enhancement of *multiple* free on-site resources, including water (chapter 4 and appendices 5, 7, 8, and 9)

• A completely renovated approach to seeing, understanding, and responding to natural flows of water and sediment, so you can work *with*, not against, these flows (appendix 1)

• A harvest calendar of my region's native, wild, and cultivated foods that you can use as a template for your area to help diversify and augment potential harvests there (appendix 4)

• Charts illustrating the water-energy-carbon nexus of sample households, and how a heightened awareness of this nexus informs actions that dramatically *reduce* your—and your community's—consumption of water and energy, along with emissions of carbon and other contaminants. Better still, this informs actions that can dramatically *increase* the quality and quantity of non-polluting, locally-renewable water and power sources (appendix 9)

Harvesting *on*-site water—primarily rainwater, but also stormwater, greywater (from household drains), and condensate—remains the core of this book. This edition also shows you how to combine water harvesting with harvests of other on-site resources in a way that generates and regenerates whole new resources.

This is all simple, fun, powerful, and often low cost or free. Best of all, we don't have to do all the work when we and our systems are *integrated with*—or are in *relationship with*—our planet's natural and perpetual conditions and processes. These include the force of gravity and materials moved by it; the sun's differential heating of the Earth; air and ocean currents that are set in motion by this heating and the spin of the planet; the storms and rainwater these currents bring as the hydrologic cycle recirculates the Earth's water; the solar-powered photosynthetic plants that grow tall while drinking, filtering, and releasing cooled water through leaves and fruits; the storm clouds regenerating from the cooled moisture—evapotranspired through the plants—which then condenses

as raindrops around tiny air-borne particles of organic matter also generated by the vegetation; the sheltering shadows these plants cast; the leaf-drop-eating, fertility-building soil microbes that feed the plants, as the plants' biomass and sugars feed the microbes; and the faunal life forms—including us—who are sustained by, and in turn have evolved to contribute to, this amazing self-renewing abundance.

Tapping this potential creates and enhances *living* legacies, which enrich those who are here now and will be cherished by those who inherit them. Everyone benefits in the present and across time.

This compelling vision is what generated this book, fed by years of my on-going hands-on learning, and the copious experience that others have shared with me. In turn, I share this vision, learning, and experience with you to raise the springboard from which we can launch to higher and more dynamic levels of awareness, understanding, caring, ability, action, abundance, and joy.

Introduction

Catch rain where rain falls.

—East Indian proverb

love the rain! I love to drink it, sing in it, dance in it, bathe in it. Of course that's only natural; our bodies are more than 70% water. You and I and everyone else—we're walkin', talkin', *rain.*

Rain is the embodiment of life. It infuses water into our springs, rivers, and aquifers. It cools us, greens the land, and nourishes the plants that feed us. It cleans the air, washes salts from the soil, and makes the animals sing.

Yet the world's supply of fresh water is finite. Less than one half of one percent of all the water on Earth is fresh and available. The rest is seawater, or frozen. Our supply is renewed only through precipitation, a precious gift from the sky that falls as droplets, hail, or snowflakes, and then flows over the landscape as run-off. In this book, I refer to the gift as "rainwater." And the gift is ripe for harvesting.

Rainwater harvesting captures precipitation and uses it as close as possible to where it falls. The process mimics intact and healthy ecosystems, which naturally infiltrate rainwater into the soil and cycle it through myriad life forms. Instead of sealing and dehydrating the landscape with impervious pavement and convex shapes that drain the gift away, as most modern cities, suburbs, and home landscapes do, harvesting accepts rain and allows it to follow its natural path to productivity.

This book provides you with a simple series of integrated strategies for creating water-harvesting "nets" which allow rainwater to permeate and enhance our landscapes, gardens, yards, parks, farms, and ranches. Small-scale strategies are the most effective and the least expensive, so they are emphasized here. They're also the safest and easiest to accomplish. *They can empower you to become more water self-sufficient, and resilient in times of changing climates or a break in the water or power line.*

The benefits are many. By harvesting rainwater within the soil and vegetation—*in* the land, or in cisterns that will later irrigate the land, we can decrease erosion, reduce flooding, minimize water pollution, and prevent mosquito breeding (within water standing on top of the soil for more than three days). The process also generates an impressive array of resources: It can provide drinking water, generate high quality irrigation water, support vegetation as living air conditioners and filters, lower utility bills, enhance soil fertility, grow food and beauty, increase local water resources, reduce demand for groundwater, boost wildlife habitat, and endow us and our community with skills of self-reliance and cooperation!

MY RAINWATER-HARVESTING EVOLUTION

In 1994, my brother Rodd and I began harvesting water in our backyard by digging, then mulching a basin around a single drought-stressed sour orange tree. We graded the soil around the basin so runoff from the surrounding area, and the neighbor's roof would drain to the tree. The results amazed us. After a single rain, the tree burst out with new leaves, a dreamy show of fragrant blossoms, and an abundant crop of fruit that was soon converted into tasty marmalade and "orangeade" by family, friends, and neighbors. Ten years later, we've since kept our irrigation of that tree to just three supplemental waterings per year. Yet we live within the Sonoran Desert where annual rainfall averages just over 11 inches (282 mm), and most folks water their citrus trees at least once a week.

With the citrus tree flourishing, we decided to mimic its success and make rainwater the primary water source for all our outdoor needs. Using methods described in chapters 3, 4, and 5, and more in depth in volume 2, we created and planted undulating water-harvesting earthworks throughout our once barren yard. The rain then gently soaked into the soil, soil erosion ceased, and verdant life began sprouting everywhere. We planted shade trees that grew tall around the house, lowering summer temperatures enough for us to eliminate our evaporative cooler (improved insulation, painting the house's exterior white, and passive ventilation also helped). We then boosted the growth of these trees still further using greywater recycled from the drains of our home's sinks, shower, and washing machine. Our daily municipal water use dropped from the Tucson residential average of 112 gallons (424 liters) per person per day[1] to less than 20 gallons (75 liters) per person per day, and our water and electric bills plummeted. This earned us five visits from workers at both the water and electric utilities because they were sure our meters were broken.

We wanted to do more. Every time it rained our street turned into a river, fed by runoff from neighborhood roofs, yards, and pavement. We redirected that runoff to 19 young native trees we planted in the barren public right-of-way adjacent to our property. These low-water-use trees now sing with nesting songbirds and offer a beautiful shaded canopy for pedestrians, bicyclists, and motorists. Water that once flowed away now supports trees that filter pollutants carried in the road's runoff as they shade and cool the street (see the chapter on reducing hardscape and cre-

ating permeable paving in volume 2 for more details). Mosquito populations have plunged because water no longer stands in puddles, but is instead soaked up by spongy mulch and taken up by plants.

Our lot was once hot, barren and eroded, with a house that could only be made comfortable by paying to mechanically alter its climate. Now our yard is an oasis producing 15 to 25% of our food, and after growing trees and installing solar panels to power fans, we no longer pay a cent to heat and cool our home (keep in mind we are also the type that will put on a sweater before firing up a wood stove). We've switched from contributing to neighborhood flooding to contributing to neighborhood flood control, and our landscape enhances local water resources instead of depleting them. On our 1/8-acre (0.05-ha) lot and surrounding right-of-way we currently harvest annually over 95,000 gallons (359,600 liters) of rainwater within two 1,300-gallon (4,920-liter) and two 1,000-gallon (3,780-liter) tanks, the soil, and vegetation. Meanwhile, we use less than 20,000 gallons (75,700 liters) of municipal water for our domestic needs, neighbors using the community washing machine on our site, and landscape irrigation in dry spells. Four-fifths of the water we now use comes from our own yard, not from city supply.

When friends and neighbors drop by they see the potential of water harvesting and learn how to do it themselves. Many then go home and spread the "seeds" by setting up work parties and creating their own rain-fed oases. That, in essence, is my vision: Harvest rainwater within our own yards and neighborhoods, encourage emulation, enhance rather than deplete our water resources, and improve the lives of everyone in our community.

This has become my passion and my profession. Fueled by what I've learned from hands-on experience, I've taught countless workshops on rainwater harvesting and permaculture—an integrated system of sustainable design. I've designed and consulted on self-sustainable water-harvesting strategies and systems for many backyard gardeners, neighborhoods, city projects, land restoration endeavors, and major housing developments. And all these projects and teachings are based on the principles I give you in this book. It is my hope to plant more water-harvesting "seeds."

If every neighborhood in my hometown, and yours, harvested rainwater in an integrated way we could greatly reduce the need for our concrete-clad, water-draining, multi-million dollar flood control infrastructure. Our yards and public right-of-ways would become a new tree-lined, water-harvesting *greenfrastructure*, no longer requiring us to spend billions extracting and importing water from other communities to supplement our drained and dwindling supply.

The economics speak for themselves. By valuing and harvesting the ignored resource of rain, groundwater levels can stabilize and even rise again, failing springs and creeks can come back to life, native plants can recolonize wasteland, and ultimately the global hydrologic cycle can benefit, while we simultaneously reduce our cost of living! Real life examples of these scenarios permeate this book, encouraging us to think globally as we act locally.

How do we get there? We become aware, apply our awareness, and throw down the welcome mat to invite rainwater into our lives and landscape.

WHO THIS BOOK IS FOR

This book, along with volumes 2 and 3, is for anyone who wants to harvest rainwater in a safe, productive, sustainable way. You can be the expert and steward of your land, whether you live on an urban or rural site, big or small. This book explains what water harvesting is, how to do it, and how to apply it to the unique conditions of your site. The aim is to realize the maximum effectiveness for the least effort and cost. You'll be guided in the design of new water-harvesting landscapes or in the retrofit of ones that exist.

This book will also help you convey water-harvesting ideas to the landscape designers and maintenance workers who may be helping you at your site. Planners and designers will discover how to devise more efficient strategies and integrated environments appropriate for dryland communities as well as those with abundant water. Landscapers and gardeners will learn how to create and maintain water-harvesting earthworks. Activists will learn how water-harvesting projects can bring people together, create a sense of place, and empower the community.

Box I.1. Drylands: A Definition

Drylands are typically defined as areas of the world where potential average yearly moisture loss (evapotranspiration) exceeds average yearly moisture gain (precipitation). Evapotranspiration is the combined measurement of water loss to evaporation and transpiration.[2] Transpiration is the loss of moisture from plants to the air via the stomata within their leaves.

More than 6.1 billion hectares, 47.2% of the Earth's land surface, is dryland. A fifth of the world's population lives in dryland habitat.[3] Normal dry seasons can last six months or more. Droughts can last for years.

Dryland-appropriate strategies are emphasized throughout this book, because this is where the need is the greatest (see box I.1). Many are borrowed from, or based on, traditions that have allowed people to survive and thrive in arid environments for thousands of years (appendix 2). Yet the principles are universally applicable; wet and dry climates are both susceptible to drought and flooding. Rainwater harvesting reduces the impacts of dry seasons, droughts, and floods. By optimizing the capture of the rain we buffer our lands from changing climates and climatic extremes, while making our lands more resilient.

My goal is to enable you to appreciate the value of rainwater and begin to use it as your primary water source—if not for the entire household, at least for your landscape. You'll not only get the most from rainfall, no matter how scarce, but from other water sources as well. An integrated landscape harvests all water and builds up such resources as topsoil, organic matter, and nutrients. It acts as a concave, life-giving sponge rather than a convex, eroding burial mound, which drains water and other resources away. (See figure I.1.)

HOW TO USE THIS BOOK AND VOLUMES 2 AND 3

From the start, I've intended *Rainwater Harvesting for Drylands and Beyond* to be an all-in-one source on how to conceptualize, design, and implement

Fig. I.1A. A home and landscape draining resources.
Arrows denote runoff flow.

Fig. I.1B. A home and landscape harvesting resources. Arrows denote runoff flow.
Dotted lines denote greywater pipe. Solar panels and solar hot water heater added to roof

integrated and sustainable rainwater-harvesting systems. And it has grown and grown with ongoing research, experience, insights, and exposure to the great work of others. *The result was the resource I've always wanted!* But it took the form of a massive single volume too intimidating in size to the uninitiated and too large to easily carry while observing a site, brainstorming design ideas, or implementing the plan. So, I've divided the book up into three user-friendly, more portable volumes. I strongly recommend everyone read volume 1, since it puts all three volumes in context and lays down the foundation of how to conceptualize a truly efficient and productive integrated system that can do far more than just harvest rainwater. Volumes 2 and 3 then expand on this by elaborating on how to employ the specific techniques that flesh out and realize the general strategies presented in volume 1. Volume 2 focuses on earthworks passively harvesting rainwater and greywater within the landscape. Volume 3 focuses on roof catchment and cistern systems. In addition, my website, www.HarvestingRainwater.com offers continually expanded resources, research, and stories of those successfully harvesting and enhancing local on-site assets such as water, sun, shade, soil, food, and community.

Here's a more detailed breakdown:

VOLUME 1

The Introduction makes the case for harvesting rainwater and shifting to a paradigm of more sustainable water management.

The following chapters in this volume then lay out the steps for creating an integrated water-harvesting system:

Chapter 1 is intended to help you conceptualize the basic water-harvesting principles that will enable you to create a system that maximizes safety, efficiency, and productivity. *This chapter is the core of the book and the heart of successful water harvesting.*

In chapter 2, you will walk your watershed and assess your site's water resources.

Chapters 3 and 4 are intended to be an overview of the kinds of techniques you can use.

Chapter 3 is a discussion which will determine which of the water-harvesting strategies (earth-

works, cisterns, or both) would be best for your site and needs. It also provides an overview of and illustrations of various earthworks techniques, and some illustrations and discussion of tanks.

Chapter 4 discusses integrating other on-site resources into your system such as sun, wind, and shade to get more than rainwater for your harvesting efforts.

Chapter 5 describes how my family and neighborhood have implemented many of the strategies persented throughout this book.

There are several appendices. Appendix 1 shows patterns of water and sediment flow with their potential water-harvesting response. Appendix 2, by Joel Glanzberg, is about traditional Native American water-harvesting techniques in the Southwest. Appendix 3 gives you many water-harvesting calculations. Appendix 4 provides a list of example plants and their water requirements, and while this list is specifically for Tucson, Arizona, other readers may find it useful. Appendix 5 provides sample worksheets for figuring on-site water resources, water budgets, etc., and is intended as a structure to write down your observations and calculations for future reference. Appendix 6 in this book is a distilled list of rainwater harvesting resources. A longer version of appendix 6, divided into sections, is at www.HarvestingRainwater.com. Appendix 7 provides information about sun angles and paths. This information is to key to integrated design, and enables you to passively heat, cool, and power your home for free. Appendix 8 illustrates how to harvest and deflect the wind along with optimal window placement maximizing passive ventilation. Appendix 9 shows you how everything is connected with the water-energy-carbon nexus, and how integrated water harvesting also saves energy and money, while dramatically reducing emissions of carbon and other pollutants.

There are also reference notes, a glossary, and an index.

VOLUME 2: EARTHWORKS

In this volume, you will learn how to select, place, and construct your chosen water-harvesting earthworks. It presents detailed how-to information and variations of all the earthworks, including chapters on mulch, vegetation, and greywater recycling so you can customize the techniques to the unique requirements of your site.

VOLUME 3: ROOF CATCHMENT AND CISTERN SYSTEMS

Here, you will learn to select, size, design, build or buy, and install your chosen roof catchment and cistern systems. Principles unique to cistern systems are presented along with numerous tank options, and design strategies that enable your tank to do more than harvest water.

Real life stories of people creating and living with water-harvesting landscapes and systems frame all three volumes. We have honed our skills through countless hours of hands-on design, implementation, maintenance, and living with our systems. The scale and context of some of the systems presented may seem too large, too small, too urban, or too rural to apply to your site, but keep in mind that if you grasp how the principles and ethics have been realized in the various systems, you can adapt them to the scale and context your site requires.

THE VALUE OF RAINWATER

Don't pray for rain, if you can't take care of what you get.

—R. E. Dixon (1937) Superintendent, Texas Agricultural Experiment Station, Spur, Texas

So, you want to harvest rainwater—right on! Let's celebrate the value of rainwater and the many water resources it supports, because how we value our water resources directly relates to how we perceive, utilize, and *manage* them.

Precipitation (rain, hail, sleet, and snowfall) is the primary source of fresh water within our planet's hydrologic cycle. This precipitation, or "rain," supplies all secondary sources of water, including groundwater and surface water in creeks, rivers, and lakes. If consistently pumped or drained faster than they are replenished, these secondary sources eventually cease to exist.

Fig. I.2. Pure rainwater

Precipitation is naturally distilled through evaporation prior to cloud formation (fig. I.2), **and thus is one of our purest sources of water.**[4] Rainwater has about 100 times less total dissolved solids (TDS) than ground and surface water in my hometown![5]

Rain is considered *soft* due to the lack of calcium carbonate or magnesium in solution, and is excellent for cooking, washing, and saving energy. Much of our ground and surface water is *hard* due to the calcium and magnesium compounds that dissolve as water runs through or over soil. These compounds deposit on or in cookware, pipes, and water heaters forming white "scale" that inhibits heat conduction and shortens pipe and appliance life. Using rainwater instead saves energy and maintenance costs, and can prolong the life of water heaters and pipes.[6] Rainwater use also reduces detergent and soap requirements, and eliminates soap scum, hardness deposits, and the need for a water softener (sometimes required with well water systems),[7] besides being a natural hair conditioner.

Rainwater is a natural fertilizer. According to cooperative extension agent John Begeman, rain contains sulfur—important in the formation of plant amino acids, and it contains beneficial microorganisms and mineral nutrients collected from dust in the air—important for plant growth. Rainwater also contains nitrogen, which triggers the greening of plants. During storms, lightning strikes enable atmospheric nitrogen to combine with hydrogen or oxygen to form ammonium and nitrate, two forms of nitrogen that go into solution in atmospheric moisture and can be used by plants.[8]

Rainwater has the lowest salt content of natural fresh water sources so it is a superior water source for plants. Calcium, magnesium, potassium, and sodium salts are abundant in the earth's crust. Soils with high salt concentrations inhibit plant growth by reducing vegetation's ability to take up water and conduct photosynthesis.[9] Soils high in sodium have a tendency to disperse—or lose their structure—resulting in poor water infiltration, and soil crusting, which restricts root penetration and impedes seedling emergence.[10] As David Cleveland and Daniela Soleri write in *Food From Dryland Gardens*, "Salty soils occur naturally in arid areas where not enough rain falls to wash soluble salts down and out of the root zone. Irrigation [with surface or groundwater] makes the situation worse, since surface water and groundwater contain more salt than rainwater. Salt tends to build up in the soil as water is continually added through irrigation."[11] As long as the soil drains and enough rainwater is applied, rainwater can dilute these salts and flush them out of the root zone.[12]

Rainwater comes to us free of charge. It falls from the sky and we don't pay to pump it nor do we pay a utility company to deliver it (fig. I.3).

Yet, current management of household and community water resources does not reflect the true value of rain. Rather than treating it as our primary renewable source of fresh water we typically treat rainwater as a nuisance, diverting it to the storm drain, drainage ditch, or pollutant-laden street. In its place we invest vast resources acquiring lower-quality, secondary sources of ground and surface water. Such contemporary water management contrasts sharply with rainwater-harvesting traditions.

Fig. I.3. Rain is always free.

RAINWATER HARVESTING THROUGH LAND AND TIME

Around the globe, traditions and historic evidence of rainwater harvesting illustrate its importance as a primary water source. According to John Gould and Erik Nissen-Petersen, authors of *Rainwater Catchment Systems for Domestic Supply*, the origins of rainwater collection may extend as far back in human history as the use of fire as evidenced by the traditional practices of the hunter-gatherer Kalahari Bushmen (the San Peoples) collecting, storing, and burying rainwater in ostrich eggs to be recovered months or years later.

Roof runoff was the main source of water for many Phoenician and Carthaginian settlements from the sixth century B.C. into Roman times, when harvested rainwater became the primary water source for whole cities. As far back as 2,000 years, rain-fed cisterns provided domestic water throughout North Africa, the Mediterranean, the Middle East, and Thailand. There is a 4,000-year-old tradition of rainwater-collection systems for domestic supply and agriculture throughout the Indian subcontinent, and water harvesting in China may have extended back 6,000 years. Rooftop collection and storage of rainwater was the principal source of water in Venice, Italy from 300 to 1600 A.D. Aztec ground catchment systems were in use by 300 A.D. Native Americans in the Southwest desert used a variety of techniques (see appendix 2). Island cultures still rely on rainwater in parts of Japan, the Caribbean, and Polynesia. The tradition of harvesting rainwater in cisterns at isolated homesteads and farms continues today in the U.S., Canada, Australia, and New Zealand.[13]

THE SHIFT AWAY FROM RAINWATER

We've moved away from these traditions over the past 150 years as new technologies have enabled us to access, pump, and transport huge volumes of groundwater and surface water: secondary water sources in the hydrologic cycle. These secondary supplies seemed infinite so we kept taking more. In 1930 there were 170 irrigation wells tapping the Ogallala aquifer that stretches 1,300 kilometers from the Texas panhandle to South Dakota; by 1959 there were over 42,000.[14] As Charles Bowden writes, "By the sixties the High Plains had 5,500,000 acres under irrigation and men were working through the night to direct the flow from the ceaseless pumps."[15]

Surface water and groundwater—secondary water sources in the hydrologic cycle—appeared to be more convenient, profitable, and dependable than rain—the primary source. Surface water and groundwater became the "primary" water resources in our modern water management system. Waste became more common than conservation. We came to see rain as a source of flooding that needed to be drained away. This appeared to work for a while, but the reality of this hydro-illiteracy has hit.

SCARCITY OR ABUNDANCE

While the hydrologic cycle continuously recycles earth's water to produce renewed fresh rain, the rate at which fresh water is produced does not meet our ever-growing demand. In the face of this demand, our planet's fresh water resources are finite. Current consumption rates are lowering groundwater levels and depleting surface water flows the world over. According to the Blue Gold Report, global water consumption is doubling every 20 years—more than twice the rate of human population growth. If current trends persist, by 2025 the demand for fresh water will be 56% more than is currently available.[16] The Ogallala aquifer is being depleted eight times faster than nature can replenish it.[17]

We have reached a turning point in our water use and management. As my friend Brock Dolman says, "We can choose to be 'scared in the city' because of water scarcity, or we can choose water *abundance*—the fine and thriving condition in which 'our buns can dance'!"

While I focus primarily on rainwater, the way we value ALL water resources shapes our future. In box I.2, I lay out the tenets of two contrasting paths of water use (see also figure I.4). Read on and ask

Fig. I.4A. A landscape on the wasteful path to *scarcity*. Rain, runoff, and topsoil are quickly drained off the landscape to the street where the sediment-laden water contributes to downstream flooding and contamination. The landscape is dependent upon municipal/well water irrigation and imported fertilizer.

Fig. I.4B. A landscape on the stewardship path to *abundance*. Rain, runoff, leaf drop, and topsoil are harvested and utilized within the landscape contributing to flood control and enhanced water quality. The system is self-irrigating with rain and self-fertilizing with harvested organic matter.

Box I.2. The Scarcity Path versus the Abundance Path

THE WASTEFUL PATH TO SCARCITY

- Water scarcity is the condition in which our local water supply cannot continue to meet demand because our "fresh water bank account" is being drained.

VALUES

- We don't value and appreciate water.
- We treat rainwater as a problem—a substance we must get rid of.
- We think of groundwater and surface water as infinitely available—substances we can afford to mismanage and waste.
- We think that we as humans are separate and independent from nature.

CHARACTERISTICS

- As individuals and communities, we do not take responsibility for managing our own water accounts.
- We continually draw on groundwater savings.
- We don't make new water deposits.

RESULTS

- This extractive relationship with our natural resources leads to their degradation and depletion.

THE STEWARDSHIP PATH TO ABUNDANCE

- Water abundance is the condition in which we adjust our patterns of water management and use until our locally available water supply meets and ultimately exceeds our needs.

VALUES

- We value all water, recognizing it as the basis of our living biological system.
- We treat rainwater as the foundation of the life-sustaining hydrologic cycle.
- We treat groundwater and surface water as reservoirs that naturally accumulate and concentrate rainwater, and do not waste these.
- We understand and celebrate that we as humans are part of the earth's natural system, which sustains us all.

CHARACTERISTICS

- We as individuals and as communities thoughtfully manage our own water accounts.
- We withdraw our groundwater savings only in times of true need.
- We continually make water deposits.

RESULTS

- The abundance path contributes to the regeneration of water and other renewable natural resources. We work to enhance the environment by providing the natural "compound interest" of healthy soil, plant, and animal communities that are the source of water, food, shelter, air, and beauty.

---

yourself: What path am I on now? What path do I want to take from now on?

DRAINING VERSUS INFILTRATING WATER

Living the wasteful path to scarcity

We **drain** our communities by diverting our rainwater *away from* rather than infiltrating it *into* our landscapes, waterways, and aquifers. We replace living nets of pervious vegetation and topsoil with *im*pervious asphalt, concrete, and buildings, inducing rainwater to rush across the land and drain out of the system.

We create landscapes of burial-like mounds (convex shapes), which drain rather than retain water, topsoil, and organic matter. We place plants on top of these mounds, and pump water to them through an irrigation system, while rainwater drains away from the vegetation. Care packages of purchased fertilizer are applied to replace the lost topsoil and fertility. It is a system reminiscent of a hospitalized patient on an intravenous drip.

We direct roof runoff to streets and storm drains via gutters, downspouts, and landscaped river cobble "stream beds" that eject water quickly from yards. (See figure I.5A.)

Box I.3. Draining Facts

- A report prepared by American Rivers states that the rapid expansion of paved-over and developed land in communities all across the U.S. is making the effects of drought worse. Development in Atlanta, Georgia and surrounding counties contributes to a yearly loss of rainwater infiltration ranging from 57 to 133 billion gallons. If managed on site, this rainwater—which could support annual household needs of 1.5 to 3.6 million people—would filter through the soil to recharge aquifers, and increase underground flows to replenish rivers, streams, and lakes.[18,19]

- Twenty-five percent of the land within incorporated Tucson is covered with impervious cover such as asphalt, concrete, or buildings.[20] In higher density cities such as Los Angeles, California, over 60% of the land surface is covered with pavement.[21]

Living the stewardship path to abundance

We **infiltrate** rainwater into our soils and vegetation as close as possible to where it falls. We replace impervious surfaces with water-harvesting earthworks and tanks, and with spongy water-retaining mulch and vegetation to intercept and use runoff from sealed surfaces.

We construct bowl-like landscapes (concave shapes) to passively harvest rainwater, build topsoil, accumulate mulch, and reduce or eliminate the need for irrigation and fertilizer. This deposits the primary water source—rain—within our local soils, and reduces the need to use secondary surface water and groundwater resources.

We harvest stormwater runoff from streets into our landscapes. Streets then become passive irrigators of beautiful shade trees lining the streets and walkways. This inexpensive *greenfrastructure* reduces the need for conventional, costly, concrete-clad storm drains. (See figure I.5B.)

Fig. I.5A. A landscape on life support *draining* its resources away. Note the mounded planted areas. Rainwater available to the landscape is *reduced* by up to 50% due to excessive runoff loss.

Fig. I.5B. A sustainable landscape *infiltrating* and *harvesting* on-site resources. Note the sunken, mulched planted areas and native vegetation. Rainwater available to the landscape is potentially *tripled* due to *all* rain falling on landscape infiltrating, in addition to runoff from roof, plus runoff from ground-level hardscapes.

OVER-EXTRACTION VERSUS CONSERVATION OF WATER

Living the Wasteful Path to Scarcity

We **over-extract** our local water sources by pumping wells and diverting water from rivers and springs faster than rainfall can naturally replenish them.

We reduce natural groundwater recharge—particularly in areas with shallow groundwater tables—by paving over surfaces and causing rapid rainfall runoff.[23]

As we drain more of our rainwater "deposits" away, we simultaneously pump and consume more of our ancient groundwater savings account. As a result, rivers dry up, water tables drop, pumping costs increase, riparian trees die, and water quality declines. (See figure I.6A.)

Living the stewardship path to abundance

We **conserve** our fresh water resources by utilizing rainwater, recycling all water, and by mulching, using low-flow appliances, installing greywater systems, and practicing integrated design to reduce fresh water needs.

We allow natural groundwater recharge to occur by maintaining soil- and vegetation-covered landscapes and healthy watersheds and waterways.

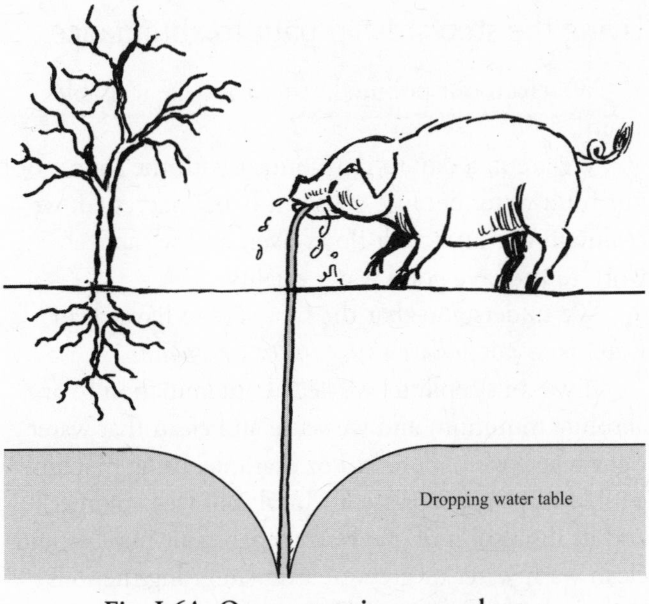

Dropping water table

Fig. I.6A. Over-extracting groundwater

Fig. I.6B. Harvested rainwater conserving municipal/well water

We make deposits to our water account by allowing water to recharge and accumulate in aquifers. We use ancient groundwater supplies only during times of drought and only to meet compelling needs.

Rainwater becomes our primary water source and our groundwater "savings accounts" are enriched and reserved for times of need.

We strive to live in sustainable balance with our local water resources by living within the constraints of our site's rainwater resources/budget. (See figure I.6B.)

POLLUTING VERSUS CLEANING WATER

Living the wasteful path to scarcity

We **pollute** our finite fresh water sources.

We are disconnected from our source of water and our effects on its quality. We see water flowing from taps, delivered by massive central distribution systems, but we aren't confronted with the path water took to get to our tap.

Rain falls through the atmosphere, runs across the land surface, and infiltrates through our soils, so contamination of our air, land surface, and soil contaminates our water. We pollute our environment with sewage, pesticides, herbicides, burning of fossil fuels, chemicals dumped down drains, dripping automobile fluids, and countless other sources. (See figure I.7A.)

Living the stewardship path to abundance

We **clean** our polluted water to make it usable again.

We are in a day-to-day contact with the source of our fresh water because we see it being harvested, we maintain the surfaces it flows over, and we actively work to preserve good water quality.

We understand that the best way to have clean water is to *not pollute water in the first place.*

If we must pollute, we keep contamination to an absolute minimum and we reuse and clean that water right where we pollute it. For example, using biocompatible soaps appropriate for local soils (see volume 2 and its discussion of greywater) enables us to reuse and clean wash water or greywater on-site using the soils and plants within our landscapes.

Box I.7. Additional Water Conservation Strategies in the Home

See the Water Saver House website www.h2ouse.org for water conservation strategies around the home; see appendix 9 for how these strategies also conserve energy and reduce pollution.

WASHING MACHINES

Replacing a standard washing machine that uses 30 to 50 gallons (114–189 liters) per load of wash with a new *Energy Star*™ certified washer that uses 10 gallons (38 liters) per load of wash can reduce water consumption by 30–60% and reduce energy consumption by 50% per load.[33,34] For an average American household that's a savings of nearly 7,000 gallons (26,500 liters) a year.[35] Install a greywater system for your washer to recycle the wash water within your landscape.

PLUMBING LEAKS

Ten percent of a home's water consumption can be due to leaks. Older irrigation systems are prone to leaks, wasting over 50% to 75% of the water consumed.[36] So, regularly inspect for leaks and repair them promptly. Many municipal water companies have a free program to test for leaks.

EVAPORATIVE COOLERS, AIR CONDITIONERS, AND CONSUMPTION OF ENERGY

An evaporative cooler in Phoenix, Arizona, consumes an average 65 gallons (246 liters) per day.[37] Air conditioners don't use water on-site, but the water used to power these and other electrical appliances can be substantial if the power comes from a thermoelectric power plant (nuclear, coal, oil, natural gas, or geothermal). Passive cooling strategies found in chapter 4 can greatly reduce the need for such mechanical cooling and associated energy consumption.

OUTDOOR MISTING SYSTEMS

Outdoor misting systems use as much as 2,160 gallons (8,175 liters) per month to cool 1,000 square feet (92.9 m²) of patio. That's equal to over three times the average summer water use for a homeowner in Tucson, Arizona, yet a study has found that misters reduce temperatures by only 7°F (3.8°C).[38] Low-water-use native shade trees use less water than misting systems while cooling temperatures up to 20°F (11.1°C).[39,40]

POOLS

Multiply the surface area of a swimming pool by the local evaporation rate to determine how much water will evaporate each year. In Tucson a 400-square-foot (37.1-m²) pool will lose 16,000 gallons (60,600 liters) of water per year to evaporation, almost the full volume of the pool.[41] Pool covers can reduce pool water use by nearly 30%.[42] If you don't use such covers in the swimming season—at least use them in the off season. Using a community pool rather than constructing and maintaining your own can save all the water, time, money, and chemicals it takes to keep a home pool functional.

HUMAN HABITS

Consciously reduce your personal water use to reduce household water demand. For example, do not run water while brushing teeth or scrubbing hands, do not spray down driveways, patios, or yards with water, and do not run water over frozen foods to hasten thawing. While setting the example by conserving at home, push for additional conservation measures in the commercial, industrial, government, and agricultural sectors.

Box I.8. Pollution Facts

Runoff from numerous widely dispersed sources or *nonpoint-source pollution* accounts for about 60% of all surface-water pollution in the United States.[43] Nonpoint-source pollution consists of pet feces; automobile emissions; sediments and nitrogen from yards, farms, and rangelands; and residual compounds from the general use of paints, plastics, etc.

Over five billion pounds of pesticides are applied throughout the United States every year,[44] much of which ends up in the country's natural water systems.

Nearly 40% of U.S. rivers and streams are too polluted for fishing, swimming, or drinking.[45]

Potable water is what fills and flushes nearly all American toilets, with 6.8 billion gallons being flushed away every day.[46]

Ninety percent of the "developing" world's, or rather the *majority* world's wastewater is still discharged untreated into local rivers and streams.[47]

- **Keep the crap out of our water.** Defecating in the potable water that typically fills our flush toilets costs us billions of dollars, and billions of gallons of water, annually as we try to remove the feces and other contaminants. Toilets account for up to 30% of all indoor potable water use in a typical U.S. residence.[48] Switching from older, more consumptive toilets to more water efficient 1.0–1.6 gallon per flush toilets can cut that indoor water use by 15–25%.[49] Waterless composting toilets use no water and turn the users' wastes into high-quality fertilizer.

  And then there's the annual "Pee Outside Day" in Sigmota, Sweden, when 50% of the water normally used in toilet flushing is saved.[50]

- **Water harvesting is being used to clean up Los Angeles, California.** Non-profit organization Tree People has spearheaded a program encouraging rainwater and greywater harvesting at residential, business, industrial, and public sites in Los Angeles to clean up pollution and reduce the city's dependence on imported water by up to 50%! Water-harvesting yards, commercial landscapes irrigated by greywater, and sunken ball fields are acting as flood controlling *greenfrastructure* promising significant reduction in runoff pollution into Santa Monica and San Pedro Bays, and the simultaneous removal of the 100-year flood threat on the Los Angeles River. The municipal flood control department—renamed the Watershed Management Division—is subsidizing this work since it reduces the need for expensive, single-purpose flood-control infrastructure.[51]

Fig. I.7A. Polluting our water and watershed at home

Fig. I.7B. Cleaning our water and watershed at home

We invest the community resources needed to purify our polluted water and return it clean to the hydrologic cycle where it can help support all life. (See figure I.7B.)

HOARDING VERSUS CYCLING WATER

Living the wasteful path to scarcity

We **hoard** water as a community by draining, over-extracting, and polluting our *local* water sources,

and instead of changing our policies, lifestyles, and habits to reduce our needs, we divert water from other locations and people.

On an individual level, we watch our local water supply decline even as we buy bottles of imported spring water. We support massive dam and canal projects that divert water from others to support our demands.

Diversions and competition for water are contributing to its commodification, as water shifts from a social resource belonging to all life, to an economic

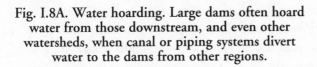

Fig. I.8A. Water hoarding. Large dams often hoard water from those downstream, and even other watersheds, when canal or piping systems divert water to the dams from other regions.

Fig. I.8B. Water cycling along with the harvest of sun, shade, and food

commodity bought, sold, and managed by corporations that profit from the increasing scarcity of water. Water is sold to those thirsty people who have similarly squandered their water, or from whom we have taken it. (See figure I.8A.)

Living the path to abundance

We **cycle** our water as a community by increasing the productivity and potential of our limited fresh water by cycling it—using it again and again. The more life forms, uses, and resources through which fresh water *cleanly* cycles, the more life forms, uses, and resources the water can generate and support.

We stop draining, over-extracting, and polluting our local water sources by changing our policies, lifestyles, and habits to reduce water needs.

We take personal responsibility for how we treat and consume water, recognizing the negative impacts of buying imported bottled water and supporting massive water diversion schemes, and choosing instead to avoid these hoarding behaviors.

We work to prevent and reverse commodification of water by assisting people locally, nationally, and internationally to harvest rainwater, make use of grey-

Box I.10. Hoarding Facts

More than half of all accessible fresh water is now diverted for human use.[52]

Only 2% of America's rivers and streams remain free-flowing and undeveloped.[53]

In 1996, the residents of the high desert city of Albuquerque, New Mexico had to decrease their water use by 30%, while Intel Corporation was allowed to increase its use by the same amount. Intel pays four times less than the city's residents for water.[54]

The U.S. Global Water Corporation has signed an agreement with Sitka, Alaska, to export 18 billion gallons per year of glacier water to China where it will be bottled in one of that country's "free trade" zones to exploit cheap labor. The company brochure tempts investors "to harvest the accelerating opportunity … as traditional sources of water around the world become progressively depleted and degraded."[55]

water, and reduce demand for secondary water sources. Water scarcity attracts market forces; water abundance does not. Water is part of the Commons (box I.12), to which we are all entitled. (See figure I.8B.)

For more information on how to protect the right to clean water for all citizens of the earth (including wildlife) see the resources appendix on my website www.HarvestingRainwater.com.

Box I.11. Cycling Tips

We can consume just 100 gallons (379 liters) cycled through five uses, rather than consuming 500 gallons (1,893 liters) of water in a week for five different uses. And, we can harvest that 100 gallons of water from roof runoff. For example:

- **Bath water**. Harvest rainwater from roofs into a tank and bathe with it in an indoor or outdoor shower. (See chapter 4 for an example, with much more in volume 3.)

- **Irrigation**. Direct that shower greywater to shade trees and 100 gallons of shower water becomes 100 gallons of irrigation water. (See chapter 4, and volume 2, the chapter on greywater.)

- **Cooling**. Place those shade trees on the east and west sides of a building to cool air both outside and inside. This can reduce mechanical cooling that would otherwise consume 100 gallons of water in an evaporative cooler, or 100 gallons in electricity generation for an air conditioner.[56,57,58] (See chapter 4 in this volume for much more on using trees in integrated design.)

- **Food**. Select food-bearing shade trees and each 100 gallons of harvested rainwater offsets the need to use 100 gallons of water to grow food in a distant orchard. (See the plant list in appendix 4 as well as volume 2, the vegetation chapter.)

- **Fertilizer**. Collect fruit and leaves that drop to the ground around the base of the trees to create rich, water-conserving mulch, reducing the need to pump 100 additional gallons per week for irrigation. (See volume 2, the chapter on mulching.)

Box I.12. Water as Commons

I use the term "commons" as defined by Vandana Shiva in her book *Water Wars*, in which she writes, "Water is a commons because it is the ecological basis of all life and because its sustainability and equitable allocation depend on cooperation among community members."[59]

Box I.13. Water as a Human Right

Water is a limited natural resource and a public good fundamental for life and health. The human right to water is indispensable for leading a life in human dignity. It is a prerequisite for the realization of other human rights.

—United Nations, 2002[60]

MY COMMUNITY'S PATH

Today, most of Tucson's 11 plus inches (282 mm) of average annual rainfall pours off roofs, yards, parks, and parking lots creating torrents of street runoff that flow to storm drains (fig. I.9) generating the "need" to import water. Yet from over 4,000 years ago up until the early 1870s, human residents of the area relied entirely on easily obtained rainwater, runoff, surface water, springs, and shallow hand-dug wells. Then Tucson shifted to relying primarily on groundwater extracted with mechanical pumps from natural aquifers beneath the city and surrounding valleys to serve municipal, agricultural, and industrial uses. In the past

Fig. I.9. First Street becomes "Runoff River Street" in a summer storm.

100 years, decades of extracting groundwater at rates exceeding natural recharge have put ground and surface water out of reach, unless one has a mechanical pump to access the water from ever-increasing depths. Overpumping has lowered this groundwater table by more than 200 feet (61 m) in some areas, and it continues to drop an additional 3 to 4 feet (0.9 to 1.2 m) more each year in some areas.[61, 62, 63] Once perennial reaches of the Santa Cruz River and numerous springs have dried up.[64] Cottonwood, willow, and mesquite "bosques" or forests that used to line our waterways have died.[65] Basically we have destroyed and continue to destroy many of the foundational elements of our local hydrology and ecology that enabled the Tucson area to be one of the oldest continuously inhabited areas in North America. (See figure I.10, with its "before" and "after" photos.)

Pollutants from local landfills, businesses, and industry have migrated down to our aquifer, creating several Environmental Protection Agency (EPA) Superfund sites.[66] As we have polluted and depleted our local water supply, we've bought thousands of acres of farmland in surrounding valleys to obtain their groundwater pumping rights for our use. We have spent over 4 billion dollars to construct, and 60 to 80 million dollars a year to operate, the Central Arizona Project (CAP), which diverts water from the Colorado River and pumps it 2,400 feet (731 m) uphill in an evaporation-prone canal over 336 miles (540 km) through the desert to reach our city. Thus the canal and its pumps have become the state's single largest consumer of electricity, and emitter of carbon dioxide (via the coal-fired power plant that provides its electricity).[67] That imported Colorado River water via CAP is now the primary water source for Tucson, but we continue to overpump our groundwater at a rate that exceeds natural recharge.

Projected population growth and increased water use is predicted to outstrip Tucson's "renewable" water supplies by 2025.[68] The Colorado River—designated America's Most Endangered River due to mounting problems with radioactive, human, and toxic waste in the water[69]—has been over-allocated to the point that the southernmost reaches of the river are severely diminished, crippling much of the Colorado River Delta's ecosystem and economy.[70] If the states upstream from Arizona on the Colorado River, and if Mexico below, take the full shares of Colorado River water granted them there will not be enough water left to meet Arizona's needs and fill the CAP canal in drought years.[71] In the meantime, we are importing about 2,000 pounds (907 kg) of salt with every acre-foot of CAP water we pump into the Tucson area.[72] That salt is an additional challenge and can become a contaminant for the already salt-prone alkaline soils of our desert environment.

We can make a shift. A 1/4-acre (0.1-ha) lot in Tucson receives about 67,000 gallons (253,600 liters) of salt-free rain in an average year. The average single-family residence in Tucson (assuming three people) uses about 120,000 gallons (454,200 liters) of water a year, and roughly half of that is for outdoor use. This suggests that most residential outdoor water needs could be met by harvesting the rainwater that falls on the property instead of pumping groundwater—especially if low water-use native plantings are integrated into the landscape design.[73] Reducing Tucson's consumption of groundwater and imported surface water is the key to shifting our city toward sustainable balance with our local water resources.

Tucson's average rainfall actually exceeds our current municipal water use (see box I.14), but most of this rainfall is drained away or lost to evaporation. Harvesting more of that rainwater, coupled with more conservation, brings us to our alternative path.

OUR PATH TO ABUNDANCE

By cycling the water we infiltrate, conserve, and clean within our lives and landscapes, we empower ourselves to do far more with far less. So, more is available for everyone, creating *abundance*. We enhance our own water resources and those of others, especially those downstream and downslope. Rather than *commodifying* fresh water or turning it into a limited-access commodity to be bought, sold, and hoarded, we *communify* it by working together to enhance our local water resources and manage their fair use and equal accessibility. As we enhance our natural resources (our "commons" (box I.12)) within our own lives and throughout our neighborhoods, the

Fig. I.10A. The Santa Cruz River in Tucson, Arizona looking northeast from the base of A-Mountain in 1904. Note the braided running water, densely vegetated watershed, and cottonwood (white), willow, and mesquite trees growing in the floodplain. Credit: Arizona Historical Society/Tucson, AHS Photo # 24868

Fig. I.10B. The same stretch of the Santa Cruz River in 2007. Note the dry, channelized riverbed, and how much of the watershed has been replaced with paving, buildings, or bare, compacted earth. The cottonwood, willows, and most of the mesquites have disappeared with the depleted water table. The river park landscape drains most runoff to the riverbed and is dependent upon a drip irrigation system.

Box I.14. Rain, Rain, Everywhere ... Can We Stop To Think?

According to sustainable development consultant David Confer:

- More free, local rainwater falls on the dryland city of Tucson than the entire community consumes of purchased, imported municipal water in a typical year.

  This was determined by dividing the average annual precipitation falling on the surface area of Tucson, Arizona by its 2010 population (520,116), then dividing again by 365 days per year, to find there is approximately 231 gallons per person (capita) per day (gpcd) (874 liters pcd) of rainwater compared to a (2012) total use (potable and non-potable or reclaimed water) of approximately 177 gpcd (670 lpcd) delivered by municipal water companies for municipal and industrial uses.[74]

- As population densities increase, available rainfall gpcd will decrease. This would eventually become a problem if rain was the source of potable water, but it is not a problem for meeting the water needs of landscapes, because as population density increases, hardscape density also increases. In densely developed areas of Tucson, as much as 3/4 of the land is hardscape consisting of roofs, roads, parking lots, driveways, and sidewalks. If the rainwater that falls on hardscape was directed to and infiltrated within the remaining areas still available for vegetation, the 11.12 inches (282 mm) average annual rainfall is concentrated fourfold, to almost 45 inches (1,143 mm) per year. This approximates the annual rainfall in Columbus, Georgia (48 inches or 1,219 mm). This does not mean we can or should now plant the vegetation of Georgia in Tucson—our rainfall patterns are more erratic, and our evapotranspiration rates are higher than Georgia's. Rather, this illustrates the large volume of local rainwater we currently ignore or expel. The bulk of Tucson's rainfall is currently lost to runoff and evaporation. Nobody really knows the figure, but estimates of up to a 90% loss seem reasonable.[75] (See figure I.11.)

Fig. I.11A. A property losing 75% of its annual rainwater by directing the runoff from the impervious 75% section of its surface area (roof and driveway) to the street. Evaporation adds to the water loss.

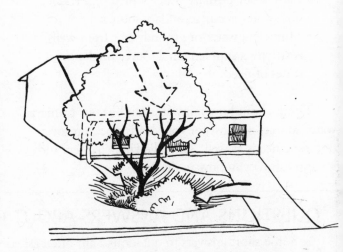

Fig. I.11B. The same property harvesting, and utilizing, 75% more of its annual rainwater by directing the runoff from the impervious roof and driveway into the permeable, sunken, and vegetated quarter of the site.

Note: Sites with less permeable yard space and more impermeable hardscape will likely require a rainwater tank in addition to vegetated basins to handle the runoff.

"community watershed" and the community resources are enhanced many times over!

This must occur in the more water consumptive government, commercial, agriculture, and industry sectors as well as at home, but at home is where it begins because every government official, teacher, student, businessperson, farmer, and industry worker lives in a home. If we realize the potential of water harvesting at home where it is easiest to do so, we can realize it elsewhere, because we will have learned from direct experience, we will be motivated by our success, and we will be living the example we are trying to set.

The main goal of the abundance path is to use less water than nature renewably provides while consistently improving water quality, flow, and dependability and as a result, decreasing groundwater pumping and eliminating the need to import water. The first step is to strive to harvest more rainfall in our landscapes than we use from municipal water sources or private wells. This leads to a more sustainable hierarchy in the household and community management of our water resources in which:

- Rain is our primary water source (fig. I.12);
- Greywater is our secondary source;
- Municipal water or groundwater from wells is strictly a supplemental source used *only* in times of need.

Think about how rainwater can be your primary water source, not just for your landscape, but for domestic needs as well. The following chapters and volumes show you how to do both.

So read on, harvest some rain, and grow abundance. Yet be warned: Once you start putting this information to work, every rainstorm could pump you with so much excitement and wonder that even if it's 3 A.M. when the clouds break you'll be running outside in your underwear to watch your landscape soaking up the water!

Good luck, and may your water-harvesting endeavors be all wet!

Fig. I.12. Rain as our landscape's primary water source

QUESTIONS AND ANSWERS ABOUT RAINWATER HARVESTING

Some short answers to questions I am often asked:

Doesn't harvesting rainwater deplete the water resources of those downstream?

The objective of most water harvesting is to create a "forested hillside" effect in your landscape and community that slows, not stops, the flow of water. This is because a forested hillside that quickly absorbs rainfall, then slowly and consistently releases it from its spongy soils over a period of weeks, months, or years, is much healthier for the stream below it than a denuded hillside that rapidly sheds water, creating sudden, sediment-laden flows downstream that dry up after just hours or days.

Water-harvesting strategies can reduce the amount of surface runoff traveling downstream, but they usually enhance the flow of wells, springs, streams, and rivers, because you are absorbing and cycling water in the landscape rather than quickly shedding it.

This book and especially volume 2 present examples of rainwater-harvesting strategies that turned sporadic flow of streams and rivers into dependable year-round flow, of well levels that rose, and the creation of ephemeral springs. Water resources were enhanced for those downstream as well as for those doing the harvesting.

Is rainwater harvesting legal?

Everywhere I've been and inquired, harvesting rainwater in the soil of one's own property with simple earthworks as advocated in this book is legal. It is important to note that these strategies help rehydrate our soils, and do not divert or impound water from natural channels, streams, and rivers.

On public land, laws can vary and permits are often required. In Tucson, Arizona it used to be illegal to cut the street curb to direct street runoff into public right-of-way street-side water-harvesting basins that passively irrigate street trees shading the street. Now it is legal with a simple permit. Successful local pilot projects; examples from other communities such as Portland, Oregon; public demand; and communicating the many integrated benefits (flood control, improved stormwater quality, reduced groundwater pumping and water use, cooler and more liveable streets/neighborhoods, etc.) helped change the law.

Harvesting roof runoff in a tank(s) is legal almost everywhere, and in places where it wasn't, some laws are similarly changing. For example, this practice used to be illegal throughout the state of Colorado, where it was assumed 100% of the precipitation falling on a site ultimately contributed to stream flows to which many had claimed water rights. These laws did not account for the fact that some of the precipitation was historically consumed by native vegetation and never made it back to the stream.[76] This was proved with a 2007 study based on empirical methods using historical climate data for the study site in Douglas County, Colorado, which found that on average 97% of the precipitation falling on the undeveloped, naturally vegetated test site infiltrated the soil and was consumed by the native vegetation.[77] Even in years of heaviest rainfall no more than 15% of the precipitation would leave the site as runoff.[78, 79] As a result, in 2009 two bills passed, allowing the limited harvest of residential roof and impermeable surface runoff in tanks within approved pilot-project housing developments; or from rooftops where the owner uses, or is legally entitled to, a well.[80]

Ask your authorities about any local, state or national laws or requirements/permits that might affect if, how and where you can harvest water. And if needed, advocate for change.

Do I need a tank or cistern to harvest rainwater?

Not necessarily; you can often easily and effectively harvest rainwater in the soil with simple earthworks presented throughout volume 2 (chapter 3 in this volume provides an overview of some of these). In fact, it is almost always less expensive to harvest the rain in the soil than in a tank, as the soil is already there, and has a far greater storage capacity.

Doesn't rainwater harvesting mean you'll have standing water in which mosquitoes can breed?

Rainwater Harvesting for Drylands and Beyond stresses small-scale strategies that can easily absorb harvested rain into the soil within a couple of hours after a storm. Mosquitoes need water standing for more than three days to complete their life cycle from eggs into adults.[81]

If you are harvesting rainwater in a cistern rather than the soil, you simply ensure that mosquitoes have no access. Techniques in chapter 3 show how to keep sunlight (which encourages algae and bacteria growth), insects, and critters out.

Does it cost a lot to harvest rainwater?

If you harvest rainwater in the soil and you are doing the work yourself, it can be free. If you hire someone to create a rainwater-harvesting landscape for you, it should not cost much more than for a conventional landscape. No additional materials are required; you mainly just need to move more dirt.

If you are installing a cistern the cost will depend on the size and manufacture of the tank, how you plan to plumb it, and how you plan to use the water.

A cost-effective approach is to develop a water budget. Figure how much water can reliably be obtained and sustained on site, then determine what means of harvesting best meets your needs. (See chapter 2 for more.) Passive water-harvesting earthworks (chapter 3 and volume 2) are typically 50 times cheaper than cisterns and can hold far more water. They can be used throughout a site's landscape, and are excellent for passively harvesting dirtier stormwater runoff—such as from streets and walkways—and greywater and air-conditioning condensate. Use active systems (cisterns) to back up your passive system in drought. Only harvest a site's cleanest and most easily harvested water (typically from rooftops) in cisterns, so you'll get the most for your tank investment. Direct cistern overflow to water-harvesting earthworks.

Water-harvesting rebates, grants, and/or credits may be available in your area. Inquire with your local water or conservation department, and enter "financial incentives" in the Search box at www.HarvestingRainwater.com for a few I've listed.

Can I find a competent person to create my own water-harvesting landscape or system?

Rainwater harvesting is starting to become more commonplace, and while some areas have skilled designers and installation crews, many are lacking in such resources. This book is meant to help more people become knowledgeable and skilled at water harvesting. Use it to guide a landscape architect if you hire one, or a crew doing the work on the ground. You will most likely need to do more supervising, but hey, you'll get a better job that way. Be careful whom you hire, check out the quality of their work, show them this book, and see how willing they are to work with you and new ideas. Think about doing things yourself or with the help of friends. You'll learn a lot more and save money. As you learn by doing, you gain the skills that enable you to help others.

Are there rainwater-harvesting building codes?

The International Building Code recommends avoiding the infiltration of water into soils within 10 feet (3 m) of a building's foundation. If you have a basement, you may want to avoid infiltrating water within 20 feet (6 m) of the foundation. Other than this, check with your local building inspector. The International Residence Code (IRC) states no permit is needed for "Water tanks supported directly upon grade if the capacity does not exceed 5,000 gallons (18,927 liters) and the ratio of height to diameter or width does not exceed 2 to 1." Otherwise tanks need a permit. For example, the State of Ohio Department of Health and the State of Virginia Bureau of Sewage and Water Services regulate rainwater cistern systems, though in most counties and municipalities the unwritten code is "CYB"—"Cover Your Butt." That's exactly what the water-harvesting principles and strategies in this text strive to do for you. Follow them all and you won't have any mosquito, flooding, or drowning problems.

Do I need to own property to harvest water and the other local resources highlighted in this book?

No. The first step is to learn how, and you are doing just that by reading this book. But for a more specific list of options type "What Renters Can Do" in the Search box at www.HarvestingRainwater.com. There are more options than we have space for in this question and answer section.

How can I learn more about water harvesting?

Do it! There is no better way than hands-on experience.

You can also look to the Books' Resource Appendices (Volume 1) page at www.HarvestingRainwater.com for a few of the organizations that offer courses in water harvesting and tours of water-harvesting sites. Many other publications and videos are also listed.

The Man Who Farms Water and the Rainwater-Harvesting Guidelines

We forget that the water cycle and the life cycle are one.

—Jacques Cousteau

This chapter is the core of the book and the heart of successful water harvesting. It describes eight guiding principles and three overriding ethics that are the foundation of how to conceptualize, design, and build integrated water-harvesting systems that generate multiple benefits. Use the principles and ethics as an integrated system, while thinking of them as a supportive mantra you can chant or a guiding song you can hum as you play in the rain and experiment with water harvesting.

Used together, these principles and ethics will greatly increase your chance of success, dramatically reduce mistakes, empower you to adapt various strategies to meet your site's specific needs, and allow you to expand the benefits of your work well beyond your site. Learn from doing, but don't go into it blind.

I begin with a story of the man whose life embodies the power of water harvesting, and who made it all click for me. …

THE MAN WHO FARMS RAINWATER

While traveling through southern Africa in the summer of 1995, I heard of a man who was farming water. I set out to find him and soon was packed into a colorful old bus roaring through the countryside of southern Zimbabwe. The scenery was beautiful, with rolling hills of yellow grass on red earth and small thickets of twisting, umbrella-like trees. Nine

hours later we arrived in Zimbabwe's driest region. We crested a pass of low-lying semi-desert vegetation. Below us spread a vast veldt prairie of undulating hills covered with dry grass and capped with barren outcroppings of granite. Trees were sparse. All was covered by a wonderful expanse of clear blue sky, reminding me of the open grasslands of southeastern Arizona. The bus crept down and stopped in Zvishavane, the small rural town where the water farmer lived.

In the morning, I hitched a ride with the local director of CARE International. She took me to a row of single-story houses. One of these was the simple office of the Zvishavane Water Project. There on the porch sat the water farmer, reading a Bible.

As my ride came to a stop he sprung up with a huge smile and warm greetings. Here at last was Mr. Zephaniah Phiri Maseko. When he learned how far I had traveled, he burst into a wonderful laugh. He told me that lately visitors from all over the globe seemed to be dropping in once a week. Nonetheless, for him each was an unexpected surprise. Mr. Phiri jumped into the vehicle and we drove off over worn, eroded dirt roads toward his farm. An endless stream of poetic analogies, laughter, and stories began to pour from his mouth. The best story of all was his own.

In 1964, he was fired from his job on the railway for being politically active against the white-minority-led Rhodesian government. The government told him that he would never work again. Having to support a family of eight, Mr. Phiri turned to the only two things he had—an overgrazed and eroding 7.4-acre (3-hectare) family landholding, and the Bible. He used the Bible as a gardening manual and it inspired his future. Reading Genesis he saw that everything Adam and Eve needed was provided by the Garden of Eden. "So," thought Mr. Phiri, "I must create my own Garden of Eden." Yet he also realized that Adam and Eve had the Tigris and Euphrates Rivers in their region, while he didn't have even an ephemeral creek. "So," he thought, "I must also create my own rivers." He and his family have done both.

The family farm is on the north-northeast-facing slope of a hill providing good winter sun to the site since it is in the Southern Hemisphere. The top of the hill is a large exposed granite dome from which stormwater runoff once freely and erosively flowed. The average annual rainfall is about 22 inches (559 mm). However, as Mr. Phiri points out, this average is based on extremes. Many years are drought years when the land is lucky to receive 12 inches (305 mm) of rain. When Mr. Phiri began, it was very difficult to grow crops successfully, let alone make a profit. There were frequent droughts and he had no money for deep wells, pumps, fuel, and other equipment needed for irrigating with groundwater.

Along with everyone else in the area, Mr. Phiri was dependent on the rains for water. Storms always brought him outside to observe how water flowed across his land. He noticed that soil moisture would linger longer in small depressions and upslope of rocks and plants, than in areas where sheet flow went unchecked (fig. 1.1). He realized he could mimic and enhance areas of his land where this was occurring, and he did so. He then spent ample time watching the effects of his work. Thus began his self-education and work in rainwater harvesting—his "water farming." Over the next 30 years, he created a sustainable system that now provides *all his water needs from rainfall alone* (fig. 1.2).

"You start catchment upstream and heal the young, before the old deep gullies downstream," says Mr. Phiri. Beginning at the top of the watershed, he built unmortared stone walls at random intervals on contour (along lines of equal elevation). These "check dam walls" slow or "check" the flow of storm runoff and disperse the water as it moves through winding paths between the stones. Runoff is then more easily managed because it never gets a chance to build up to more destructive volumes and velocities. Controlled runoff from the granite dome is then directed to unlined reservoirs just below.

Fig. 1.1. More water, soil, seeds, and life gather where their
flow across the land is slowed (here by rocks on contour).

These reservoirs were built with nothing more than hand tools and the sweat of Mr. Phiri and his family. All work on the land was—and is—done on the human scale, so that it can be maintained on the human scale.

The larger of the two reservoirs Mr. Phiri calls his "immigration center." "It is here that I welcome the water to my farm and then direct it to where it will live in the soil," he laughs. The water is directed into the soil as quickly as possible. The reservoirs are located at the highest point in the landscape where soil begins to cover the granite bedrock. (See figure 1.3.)

Above the reservoirs the slope is steep with little soil. At and below the reservoir, the slope is gentle and soil has accumulated. "The soil," Mr. Phiri explains, "is like a tin. The tin should hold all water. Gullies and erosion are like holes in the tin that allow water and organic matter to escape. These must be plugged."

Mr. Phiri's "immigration center" is also a water gauge, for he knows that if it fills three times in a season, enough rain will have infiltrated the soil of his farm to support the bulk of his vegetation for two years. The reservoirs occasionally fill with sand carried in the run-off water. The sand is then used for mixing concrete, or for reinforcing the mass of the reservoir wall. Gravity brings this resource to Mr. Phiri free of charge.

Overflow from the smaller reservoir is directed via a short pipe to an aboveground ferrocement (steel-reinforced concrete) cistern that feeds the family's courtyard garden in dry spells. The family has another cistern, shaded and cooled by a lush food-producing passion vine (fig. 1.4). This cistern collects water from the roof of the house for potable use inside. Aside from these two cisterns, all water-harvesting structures on the farm directly infiltrate water into the soil where the water-harvesting potential

"As Mr. Phiri explains, 'I am digging fruition pits and swales to plant the water so that it can germinate elsewhere.'"

1. Granite dome
2. Unmortared stone walls
3. Reservoir
4. Fence with unmortared stone wall
5. Contour berm/terrace
6. Outdoor wash basin
7. Chickens and turkeys run freely in courtyard
8. Traditional round houses with thatched roofs
9. Main house with vine-covered cistern and ramada
10. Open ferrocement cistern
11. Kraal for cattle and goats
12. Courtyard garden
13. Contour berm
14. Dirt road
15. Thatch grass and thick vegetation
16. Fruition pit in large diversion swale
17. Crops
18. Dense grasses
19. Well and hand pump
20. Donkey-driven pump
21. Open hand-dug well
22. Reeds and sugar cane
23. Dense banana grove

(illustration by Silvia Rayces from a drawing by Brad Lancaster)

Fig. 1.2. Layout of Mr. Phiri's farm

Fig. 1.3. Mr. Phiri in his "immigration center" reservoir

Fig. 1.4. The family house, courtyard,
cistern, and passion vine

Fig. 1.5. A loose-rock check dam that has
healed a once-erosive gully

is the greatest. All greywater (used wash water) from an outdoor washbasin is drained to a covered, unmortared, stone-lined, shallow, underground cistern where the water is quickly percolated into the soil and made available to the roots of surrounding plants.

Across the farm's entire watershed from top to bottom, numerous water-harvesting structures act as nets that collect the flow of surface runoff and quickly infiltrate the water into the soil before it can evaporate. These include check dams (small unmortared stone structures placed within drainages perpendicular to the water's flow; see figure 1.5), vegetation planted on contour, terraces, berm 'n basins (dug out basins and earthen or vegetated berms laid out on contour), and infiltration basins (basins without berms). All these catch water that was once lost to a government-built drainage system.

Many years before, the government had built large drainage swales throughout the region. Unlike most water-harvesting swales or berm 'n basins, these ditches were not placed across the slopes on contour (to retain water), but instead were built so they'd drain water off the land. Vast amounts of unhindered monsoon runoff were caught by the drainage swale, carried away to a central drainage, and shot out to the distant floodplain. The erosion problem was decreased, but drought intensified because this area was being robbed of its sole source of water.

Mr. Phiri turned things around by digging a series of large "fruition pits" (basins about 12 feet long by 3 to 6 feet wide by 4 to 6 feet deep) in the bottoms of all the drainage swales on his land. Now when it rains the pits fill with water and the overflow successively fills one pit after another across his property. Long after rains stop, water remains in the fruition pits percolating into the soil. "You see," giggled Mr. Phiri, "my fruition pits are very fruitful." The fruit of the fruition pits takes the form of thatch grasses, fruit trees, and timber trees, which are planted in and around the pits. This vegetation provides building materials, cash crops, food, erosion control, shade, and windbreaks, all watered strictly by rain and the rising groundwater table underground. As Mr. Phiri explains, "I am digging fruition pits and swales to 'plant' the water so that it can germinate elsewhere." (See figure 1.6.)

"I have then taught the trees my system," continues Mr. Phiri. "They understand it and my language.

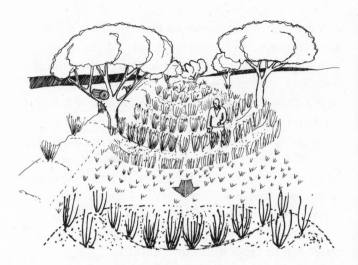

Fig. 1.6. Sketch of Mr. Phiri standing in a
fruition pit full of thatch grass

Fig. 1.7. Mr. Phiri's tree nursery
under the shelter of a mature tree

I put them here and tell them, 'Look, the water is there. Now, go and get it.'" A basin for holding water may be constructed around or beside the trees, but such earthworks are also placed further out from the trees so their roots are encouraged to stretch out and find still more water.

A diverse mix of open-pollinated crops such as basketry reeds, squash, corn, peppers, eggplant, tomatoes, lettuce, spinach, peas, garlic, onion, beans, passion fruit, mango, guava, and paw paws, along with such indigenous crops and trees as matobve, muchakata, munyii, and mutamba, are planted between the swales and contour berms. This diversity gives his family food security; if some crops fail due to drought, disease, or pests, others will survive. Rather than using hybrid and genetically modified (GMO) seed, Mr. Phiri uses open-pollinated varieties to create superior seed stock as he collects, selects, and plants seed grown in his garden from one year to the next. By propagating seed from plants that have prospered off the sporadic rainfall and unique growing conditions of *his* site, each season his seed becomes better suited to his land and climate. This seed saving is another form of water conservation, because Mr. Phiri adapts his seed to live off less water, instead of adapting his farm management to import more water.

Living fertilizer factories pepper the farm in the form of nitrogen-fixing plants. One example, the

edible, leguminous pigeon pea, is also used for animal fodder and mulch. Mr. Phiri has found that soils amended with local organic matter and nitrogen-fixing plants infiltrate and hold water much better than those amended with synthetic fertilizers. As he says, "You apply fertilizer one year but not the next, and the plants die. Apply manure once and plant nitrogen-fixing plants, and the plants continue to do well year after year. Synthetically fertilized soil is bitter."

The abundant food and fruit Mr. Phiri produces is anything but bitter. He's been generous with his abundance, giving away a diverse array of trees to anyone who wants them. Unfortunately, as Mr. Phiri points out, the majority of the trees he gives away die when people don't implement rainwater-harvesting techniques before planting. "The land must harvest water to give to the trees, so before you plant trees you must plant water." Mr. Phiri propagates his trees in old rice and grain bags near one of three hand-dug wells near the bottom of his property (fig. 1.7).

The soil is Mr. Phiri's catchment tank, and it is vast. In times of drought, his distant neighbors' wells go

Fig. 1.8. Mr. Phiri demonstrates how a donkey would power the pump in his lower fields.

Fig. 1.9. Mr. Phiri in his banana grove

Fig. 1.10. Mr. Phiri beside his largest aquaculture reservoir

dry, even those that are deeper than Mr. Phiri's. Yet as Mr. Phiri says, "My wells always have water into which I can dip my fingers." This is due both to the particular hydrologic/geologic conditions of his site and because he is *putting far more water into the soil than he takes out.*

Except for one well, which is lined and has a hand pump for household water use, all are open and lined with unmortared stone. "These wells," explains Mr. Phiri, "are those of an unselfish man. The water comes and goes as it pleases, for you see, in my land it is everywhere." During severe drought, Mr. Phiri uses a donkey-driven pump to draw from these wells to water annual crops in nearby fields (fig. 1.8).

A lush wetland lies below the wells at the lowest point of Mr. Phiri's property. Here, three aquaculture reservoirs are surrounded by a vibrant soil-stabilizing grove of bananas (fig. 1.9), sugarcane, reeds, and grasses. The fish are harvested for food and their manure enriches the water used to irrigate the vegetation. The taller vegetation creates a windbreak around the ponds, reducing water loss to evaporation. The dense, lower-growing grasses filter incoming runoff water, as well as feed his cows when in calf. (See figure 1.10 of the largest reservoir.)

Mr. Phiri has created his Garden of Eden. The rain infiltrates his soil; the reservoirs and vegetation are where it "surfaces." This harvested rain creates the "rivers" of infiltrated moisture his Garden needed to succeed. After 30 years of work his farm continues to grow, and his methods are now starting to be appreciated.

For years Mr. Phiri was the object of scorn since he found himself in opposition to international aid and government programs that pushed groundwater extraction and export crops over rainwater harvesting and local food production and distribution. As a response Mr. Phiri created the Zvishavane Water Project, a nongovernmental organization that is spreading his techniques well beyond his site (see box 1.1). It is having a dramatic effect. He influenced CARE International in his region to the point that it shifted much of its work from giving away imported food, to helping people implement Mr. Phiri's methods of planting the rain and growing their own food.

When I asked Mr. Phiri about the three decades it took him to get his land and his vision to where it is

Box 1.2.
The Rainwater-Harvesting Principles

1. **Begin with long and thoughtful observation.**
2. **Start at the top—or highpoint—of your watershed and work your way down.**
3. **Start small and simple.**
4. **Spread and infiltrate the flow of water.**
5. **Always plan for an overflow route, and manage that overflow water as a resource.**
6. **Maximize living and organic groundcover.**
7. **Maximize beneficial relationships and efficiency by "stacking functions."**
8. **Continually reassess your system: the "feedback loop."**

Principles 2, 4, 5, and 6 are based on those developed and promoted by PELUM—the Participatory Ecological Land-Use Management association of east and southern Africa. Principles 1, 3, 7, and 8 are based on my own experiences and the insights gained from Mr. Zephaniah Phiri Maseko and other water harvesters.

today he answered, "It's a slow process, but that's *life*. Slowly implement these projects, and as you begin to rhyme with nature, soon other lives will start to rhyme with yours."

We then walked back up toward the house and stopped midway. Mr. Phiri's eyes were full of excitement and joy as he pointed across the fence. His neighbor was in the government's diversion swale, digging fruition pits on the adjoining property. "Look," cried Mr. Phiri, "he is starting to rhyme!"

My visit with Mr. Phiri made clear to me that we all have the choice and power to be either the problem or the solution. He told me of a local school where the teachers were striking and threatening to leave due to lack of water and harsh conditions in dusty, hot, wind-blown classrooms. Students were in no condition to learn, being malnourished without school meals and with little food at home. Mr. Phiri listened to the complaints of the teachers then asked them not to run from their problems. He told the teachers, "to look upon wherever they found themselves as home, to set their roots into the ground, and to work to nourish and improve their lives together." Mr. Phiri then made them an offer: If they would stay he would teach them and their students how to turn things around by harvesting the rainfall to grow food, shelter, and beauty. He also warned that if the teachers ran from the situation, they would take their problems with them. Half did leave. The other half stayed, set their roots, and worked with Mr. Phiri and the students. Together they turned the bleak school grounds into lush gardens where lunches are grown on-site and vegetation passively cools buildings and blocks the wind. There is no longer reason to strike or leave, but reason to celebrate.

Years later some of the teachers who left returned. With tears in their eyes they thanked Mr. Phiri for being a man of his word. They also told him his prediction had come true. They had moved on to schools in new settlements in lush lands, but within a few years they had so misused and degraded the land that conditions became as bad as those from which they had run. Mr. Phiri responded by repeating his original offer. The teachers could go back to the schools in the new settlements and heal the scars.

Mr. Phiri turned to me with a huge smile and said, "Remember, children are our flowers; give them rain and they will grow and bloom."

THE EIGHT RAINWATER-HARVESTING PRINCIPLES

Mr. Phiri's story is a wonderful example of a successful, integrated rainwater-harvesting system (see chapter 5 for an urban example—that of my brother and myself). Keep in mind that the specific techniques used on his site are not applicable everywhere. There is no one standardized design for rainwater harvesting. Every piece of land, the plants and animals upon it, and those who steward it, are unique. Each site must be approached with its own distinctive characteristics in mind. However, there are eight rainwater-harvesting principles that are applicable to all sites, and should always be followed. Each is valuable on its own, but you get the full benefit only if all are used together.

RAINWATER-HARVESTING PRINCIPLE ONE

Begin with Long and Thoughtful Observation

Mr. Phiri was not taught by experts or at schools. He learned from long and thoughtful observation of his land (fig. 1.11)—something everyone can do. When he began, his land was dry, eroded, and unproductive, but he was attentive to, and mimicked, the aspects of his land that were working—including rocks and plants found in informal "rows" perpendicular to the slope where they slowed the rainwater and infiltrated it into the soil. Mr. Phiri mimicked this by tucking his water-harvesting structures perpendicular to the slope around existing vegetation, placed at locations where they suited the needs of his family and land. He then spent ample time watching the effects of his work. As Mr. Phiri says, "I enjoy harvesting water. Really, you know, when the rains fall and I see water running, I am running! Sometimes you will find me being very wet!"

Fig. 1.11. Long and thoughtful observation

To observe your site, sit in the dust and dance in the rain, through all the seasons. Sit down, sit quietly, and turn yourself into a sponge. Listen with all your senses—sight, smell, touch, hearing, taste, and your feelings.

Observe all that is happening. Are there lush green areas where moisture naturally collects? Do you see bare spots where water and soil drain away? Is there running water? Is it polluted? If so, by what? Do trees grow straight, or are they bent—perhaps by strong prevailing winds? Is the soil underfoot washed-out and hard-packed, or soft with accumulated organic matter? Can you hear the songs of birds and insects? Some, such as house finches and dragonflies, are never far from water. Where is the life? The resources? The erosion? Where is water coming from? Where is it going? How much water is here?

Relax and be aware. After you've taken plenty of time to observe, contemplate why things are as they are. Why is there erosion? Why does this plant grow here? Why is there more moisture there?

Try to understand the site as a whole, not as separate pieces. Imagine what would happen if you changed something. How would that alter the dynamics of the site's water flow, wildlife paths, prevailing winds, and solar exposure? How would things improve? How would they get worse?

If you listen, the land will tell you things you need to know, and what you need to investigate more deeply. Devoting time to observation and posing and answering questions are a good ways to get to know PLACE.

Once you connect to a place, it begins to show you its resources and challenges and helps guide your plans. Without understanding your site, you might install water-harvesting earthworks and plant fruit trees in an exposed, wind-dried area far from home where runoff is lacking and fruit-eating wildlife abounds. With better site understanding, you would locate these water-harvesting earthworks and trees where ample runoff, shelter from afternoon sun and prevailing winds, and a greywater-producing home are nearby, resulting in far more productive trees conveniently located for easy picking by people and less fruit-eating by wildlife. The outcome is determined by how well you understand your site, and how well you put things together.

It costs nothing to observe, think, and plan. You could pay dearly in the long run if you don't. Try to make your mistakes in your head or on paper before doing anything on the land. Keep imagining different scenarios until you settle on the one you consider best.

RAINWATER-HARVESTING PRINCIPLE TWO

Start at the Top—or Highpoint—of Your Watershed and Work Your Way Down[1]

When you're feeling ready to create water-harvesting structures on your land, start at the "top"—or highest elevation—of your on-site watershed (fig. 1.12). You could also state this rule as "start at the beginning"—the beginning of the water's flow over your buildings and land.

To begin with, form a mental image of your *watershed*. A watershed, or *catchment area* as it is sometimes called, is the total area of a landscape draining or contributing water to a particular site or drainage. The watershed for an erosive rill cut on a bare slope might be no more than 20 square feet (1.9 m²). The watershed for a river may be millions of acres or hectares covering mountains, hills, valleys, mesas, and drainages. Such large watersheds are made up of many small "subwatersheds." These subwatersheds are a patchwork quilt of small areas of land and buildings, often on the scale of a residential lot, a small parking lot, a commercial site, or a field. You will most likely be focused on the subwatersheds that comprise your home and workplace. These subwatersheds directly affect the larger community watershed, and if they are well managed have the potential to enhance the community watershed! Once you've identified the subwatershed of your site, you can begin to practice the art of *waterspread*, emphasizing the gentle harvesting, spreading, and infiltrating of water throughout a watershed rather than the rapid shedding or draining of water out of it.

Next, consider runoff. When more rain falls than surfaces can absorb, water pools and then begins to flow over roofs, roads, and soils on its way downslope. This surface flow of water is called *runoff*: water running off the land. Generally, the further you are

Fig. 1.12. Start at the top.

downslope the greater the runoff volume will have accumulated. The steeper the slope the greater the speed of water you'll have to deal with. Our goal is to turn this *runoff* into *soak-in*: water that no longer runs off the land, but infiltrates into the soil instead.

You may or may not have access to lands upslope of your property line, so begin water harvesting at the top of your "*watershed of influence*"—basically the sub-watershed composed of the area where you have the greatest say. This could mean a hilltop, the highpoint of your property line, the top of a cooperative uphill-neighbor's land, or the roof of your house.

If you begin to harvest water high in the watershed and work your way down, you'll make everything easier in the long run because:

• The volume of runoff you will be dealing with at any one time will be less than if you started lower in the watershed, and will be less likely to get out of control and become destructive. As a result, you can manage it better and construct a water-harvesting system in which most of the water will infiltrate before it runs off the land.

• You can use many modest-sized water-harvesting structures, each retaining an easily managed volume of water. Vegetation in modest-sized water-harvesting structures will get watered without getting flooded.

• The rain will infiltrate more evenly into the soil throughout the landscape, not just at the bottom.

• Water you harvest high in the watershed can be moved around the site more easily than water harvested low in the watershed. Gravity is a free and ever-present energy source that does not break down; use it to your advantage.

RAINWATER-HARVESTING PRINCIPLE THREE

Start Small and Simple

Small is beautiful, and perhaps more importantly when it comes to water harvesting, it is less expensive, easier, and more effective than starting big. Mr. Phiri and his family built everything by hand, spent almost nothing on materials, and did all the maintenance themselves. They could do this because everything was done on a human scale, and kept technically and mechanically simple to reduce the need for maintenance, and enable them to do that maintenance. (See figure 1.13.)

Fig. 1.13. Start small and simple, perhaps by planting a low-water-use native shade tree in a water-harvesting basin to shade the east or west side of your home. Direct roof runoff to the basin to increase water harvested and shade grown.

Small-scale trials of various techniques will quickly show you what works and what doesn't work on your unique site. You'll avoid large-scale mistakes. If a small-scale mistake is made, it will teach you, not break you. Starting small lets you and your friends do the work at your own pace, though of course you can hire folks to help with the work. Either way, don't start by creating an expensive and elaborate system that might not be right for your landscape, lifestyle, and means. Keep in mind that dozens, hundreds, or even thousands of tiny water harvesting "sponges" are usually far easier to create and far more effective than one big dam, because they capture more water and spread it more evenly throughout the land.

RAINWATER-HARVESTING PRINCIPLE FOUR

Spread and Infiltrate the Flow of Water[2]

Spread out the flow of water so it can *slow down* and *infiltrate* into the soil. Make water stroll, not run, through the landscape. This is the act of "waterspread" within the watershed.

Aside from one cistern holding water for a courtyard garden and another capturing roof runoff for household potable water, all Mr. Phiri's water-harvesting strategies direct the rain into the soil. He uses multiple techniques to spread harvested water over as much porous surface area as possible to give the water

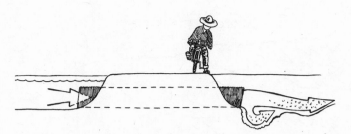

Fig. 1.14. Culvert acting as an erosive shotgun barrel. Note the undercut bed of the drainageway on the downstream side.

maximum potential to infiltrate *into* his land. Once it has infiltrated, water gently travels *through* the soil, not destructively over it. As he says, "I plant water as I plant crops. So this farm is not just a grain plantation. It is really a water plantation."

In contrast, channelization can be compared to a shotgun barrel for water; it typically straightens and constricts water flow by sealing and smoothing the banks and sometimes the bed of a waterway, often with concrete. It's like the hardening of arteries in the body, and it's bad for the health of the system. Channelization increases the velocity of water flow through and downstream of the channelized area, reducing infiltration of water into the soil and sometimes deepening the channel.

A culvert (large pipe) placed in a drainage typically forces water flow through a smaller diameter orifice than the full width and depth of the natural drainage. In a large rain event, water backs up on the upstream side of the culvert, builds up pressure, and speeds through the culvert like it was the barrel of a shotgun. Resulting erosion can often be seen just downstream of the culverts. (See figure 1.14.) Yet, we can reduce erosion and enrich the landscape if we slow down the flow of water, spread it out, and allow it to infiltrate. In the second volume of *Rainwater Harvesting for Drylands and Beyond*, the chapter on check dams provides examples in drainages.

Figures 1.15A and 1.15B illustrate how a landscape can act as either a *drain* or a *net*, respectively. Water flows from the highest point or source of a watershed, to the bottom of the watershed or *sink* where the water and other resources leave the land for good (see figure 1.16 for Source/Sink).

The "drain" example of figure 1.15A shows water, soil, and organic matter quickly draining

---

Box 1.5. Small Dams Yield More Water Than Large Dams

A study by the Central Soil and Water Conservation Research and Training Institute in Dehra Dun, India, found that increasing the size of a dam's catchment from 2.47 acres (1 ha) to about 4.94 acres (2 ha) reduces water yield per hectare by as much as 20 percent.[3] As the Centre for Science and the Environment states, "In a drought-prone area where water is scarce, 10 tiny dams with a catchment of 1 ha each will collect much more water than one larger dam with a catchment of 10 ha."[4] The tiny dams don't need costly water distribution systems either, as they are already located throughout a watershed. Also, small dams displace far fewer people and cause less environmental damage than large dams.

In another example, tiny "dams" yield even more water than small dams. While studying 4,000 year old water-harvesting strategies in the Negev Desert, which enabled ancient people to provide food and water with a scant 4 inches (105 mm) of annual rainfall, Israeli scientist Michael Evenari found that *small watersheds harvest far more water than large watersheds.*

Summarizing Evenari's findings, the book *Making Water Everybody's Business* states, "While a 1 hectare watershed in the Negev yielded as much as 95 cubic meters of water per hectare per year, a 345 ha watershed yielded only 24 cubic meters of water/ha/year. In other words, as much as 75% of the water that could be collected [*in the larger watershed*] was lost [*to evaporation and the soil*]."[5] The loss was even higher during a drought year. According to Evenari "...during drought years with less than 2 inches (50 mm) of rainfall, watersheds larger than 123.5 acres (50 ha) will not produce any appreciable water yield, while small natural watersheds will yield 4,400–8,800 gallons (20–40 cubic meters) per hectare, and microcatchments smaller than 0.24 of an acre (0.1 hectare) [*will yield*] as much as 17,597–21,997 gallons (80–100 cubic meters) per hectare."[6]

---

out of the system *causing* erosion and downstream flooding. Upstream areas are left dry while downstream areas require expensive stormwater management. The system degenerates—or breaks down—over time.

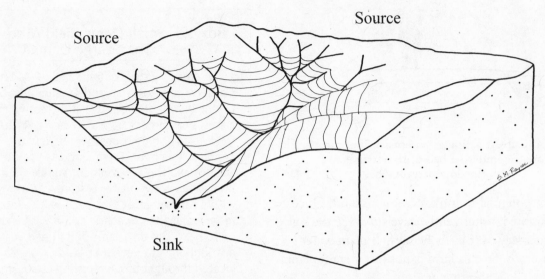

Fig. 1.15A. A bare landscape acting as a drain. The water's Source (beginning of its flow) is at the top; its Sink (departure or end of water flow) is at a lower elevation.

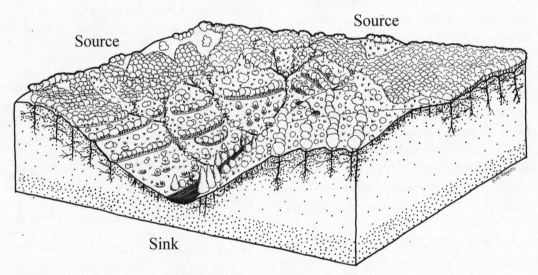

Fig. 1.15B. The same landscape with "nets" of vegetation and water-harvesting earthworks slowing and infiltrating the water

The "net" example (fig. 1.15B) shows the same area with the landscape altered to capture, slow, and spread the flow of water, soil, and organic matter, over and through the entire landscape. This *reduces* erosion, flooding, drought, and monetary costs of storm drain infrastructure while *improving* soil fertility, water infiltration, vegetative production, and ecosystem stability—by growing shade, food, shelter, wildlife habitat,

and erosion control. This system starts to regenerate—or build and take care of itself.

The goal in water harvesting is to create a series of "nets" across our watershed. Like Mr. Phiri, we should direct runoff *into* the soil by spreading and sinking its flow. Still, there will always be storms so big that more water will flow across the site than

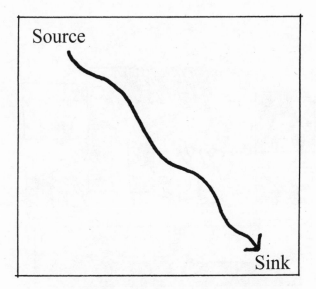

Fig. 1.16A. Source and Sink. A fairly quick and linear downward flow drains the landscape.

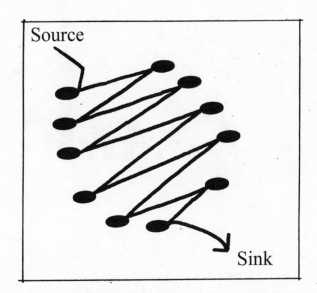

Fig. 1.16B. Source and Sink. The zig-zag increases the time of flow, distance traveled, and ground infiltration from Source to Sink.

the land or tanks can retain, leading us to the next principle.

RAINWATER-HARVESTING PRINCIPLE FIVE

Always Plan for an Overflow Route, and Manage That Overflow Water as Resource[7]

Overflow should not be treated as a problem or a waste. Instead, design the overflow route so that surplus water becomes a *resource* (fig. 1.17). Mr. Phiri converted the government-built drainage swales cut across his land into a water-harvesting project by digging "fruition pits"—or stepped infiltration basins—within the bottom of the large swales. Any excess water overflows from one fruition pit to the next and ultimately drains away down the big swales. All of Mr. Phiri's water-harvesting structures have planned overflow routes. In huge storms extra runoff is directed from one harvesting structure to the next until it reaches the bottom of his site where it is released onto the natural floodplain below.

No matter how well you design your system *always* plan for overflow in very large storm events. Overflow spillways should be stabilized using heavy tightly packed rock, or well-rooted vegetation so they

hold up to large flows. Overflow from tanks and cisterns must be safely routed out of the tank and away from the tank's foundation. Overflow should be directed to a useful location such as a vegetated infiltration basin that passively irrigates a native shade tree that in turn shades the tank, provides food, and creates wildlife habitat. The need to manage overflow applies to all cisterns and water-harvesting earthworks. As the Boy Scouts say, "Be Prepared." Make sure that when your system overflows, it overflows where you want it to, and in a controlled manner.

Fig. 1.17. Cistern overflow water directed to, filling, and then overflowing one earthwork to fill another and another

Fig. 1.18. Planting vegetative groundcover, spreading organic mulch, and planting seed to help permeate, protect, and build soil with roots, leafy cover, mulch, and accumulating leaf drop

Fig. 1.19. A water-harvesting system providing multiple functions of water, passive cooling with shade, stormwater control within earthworks and cistern, wildlife habitat, and food production

Be sure your site has a final overflow outlet at the bottom of your watershed. Ideally this would direct water into a natural vegetated wash or creek, but in the urban environment you may have to settle for a street or storm drain.

RAINWATER-HARVESTING PRINCIPLE SIX

Maximize Living and Organic Groundcover[8]

Rather than infiltrating, water often flows off flat or mounded bare dirt surfaces, or pools for days inside bowl-shaped surfaces and evaporates or supports mosquito breeding. This is because bare dirt is prone to compaction and the surface tends to seal up, both conditions which reduce the ability of rainwater to infiltrate below the surface. In contrast, covering dirt with organic groundcover such as mulch and plantings increases infiltration into the soil. Growing plants set down roots and drop leaves to generate mulch. Earthworms and other soil life convert the leaf drop into more soil, riddled with their holes. I've found the roots of these plants, coupled with surface mulch and the associated soil life, grow to create a living sponge that can more than double stormwater infiltration rates in previously bare basins, reducing evaporation, runoff, erosion, and mosquito breeding! (See figure 1.18.)

Mr. Phiri's site is a living vegetation-covered welcome mat that helps water infiltrate into the soil and pumps soil moisture back to the surface through roots. The vegetation literally brings harvested water to "fruition," transforming it into fruits, vegetables, and grains for people and livestock, hot afternoon shade for home and fields; a dense mat of roots and leaves to stabilize spillways and control erosion; lumber and thatch for building; fiber for clothes; medicinal herbs; and leaf drop that breaks down and fertilizes the soil.

Native vegetation—indigenous plants found within 25 miles (40 km) of your site and within an elevation range of 500 feet (152 m) above or below your site— is generally best adapted to local rainfall patterns and growing conditions, and these plants often make great groundcovers.

RAINWATER-HARVESTING PRINCIPLE SEVEN

Maximize Beneficial Relationships and Efficiency by "Stacking Functions"

Mr. Phiri looks well beyond water infiltration and strives to improve his *whole* site, not just one aspect of it. He does this by designing and placing his water-harvesting structures in relationship to the overall landscape so they perform multiple beneficial functions—he is "stacking functions." By stacking functions, Mr. Phiri gets far more efficiency and productivity for the same amount of effort. The vegetation selected to harvest rainwater also produces food, dust control, shelter, wildlife habitat, and windbreaks. These windbreaks reduce evaporation of water from fields and ponds. Fish raised in his ponds feed the family and fertilize the water used in the fields. Contour berms create raised footpaths. Check dams stabilize path and road crossings over drainages. (See figure 1.19.)

Often, existing strategies designed to perform one primary function can be adapted to perform additional functions. For example, the government's drainage swales were originally designed only to reduce erosion and flooding, which they did, but they also drained away the area's sole source of water—an irreplaceable resource. Mr. Phiri added fruition pits to harvest water within the swales and lined his fruition pits with multi-use plants, creating windbreaks, stabilizing the pits, and generating self-seeding crops that flourish on passively harvested water.

Each site has its own unique potential for stacking functions. For example, when designing rainwater cisterns into a site, they can double as privacy walls, pillars supporting porches, property fences, retaining walls, afternoon sunscreens, and more. Get the Domino Theory working for you. You know you're doing well when you devise a strategy to solve one problem that simultaneously solves many other problems and creates more resources.

RAINWATER-HARVESTING PRINCIPLE EIGHT

Continually Reassess Your System: The "Feedback Loop"

Continual reassessment is the key to long-term maintenance of a water-harvesting system (fig. 1.20).

Mr. Phiri had a great idea: Grow water-harvesting structures by placing plants on contour. He "stacked

Fig. 1.20. Long and thoughtful observation again. How is the land responding to your work? What still needs to be addressed?

functions" by seeking out plant species that produced crops as they harvested water and reduced erosion. He quickly settled on hardy sisal plants (*Agave spp.*) that use little water, require almost no maintenance, produce large amounts of biomass to hold back water and soil, and produced fiber to use on site or sell.

Mr. Phiri thought long and hard about this strategy. He began high in his watershed where the sisal contour berms helped spread runoff and infiltrate it into the soil. He built rock-stabilized spillways for overflow. His system quickly maximized groundcover, harvested water, stabilized soil, and produced sisal fibers. The only thing he forgot to do was to start small.

Sisal plants covered his land and slowed runoff. Everything was great until winter arrived in a drought year. Grass was sparse, and plants appeared yellow and dead in their dormancy, but the evergreen sisal stood out verdant and lush: Mr Phiri's livestock went right for it. The long, strong fibers of the plants bound up in the intestines of the animals, killing them. Mr. Phiri was devastated. He had not foreseen these

Fig. 1.21. Mr. Phiri beside his remaining
stand of sisal planted on contour

consequences. There are always consequences that we
cannot or do not predict. Mr. Phiri subsequently spent
many hard days removing all but one small stand of
sisal. He left this stand as a reminder and teacher—and
keeps the livestock away from it (fig. 1.21).

Following *all* the principles *together* can decrease
your mistakes and increase your chances of success.
No matter how good a plan or design is, maintenance
and adaptation will be required over time. When the
design is well thought out in the first place, these
changes are likely to be minor. Mr. Phiri finds himself reinforcing spillways, maintaining berms and
swales, and pruning vegetation for livestock forage
and mulch. Sometimes, as with the sisal, he needs to
change or alter some of his strategies. After the sisal
mistake, Mr. Phiri did not abandon his idea of planting vegetation as water-harvesting structures—he was
just more careful about plant selection. Now he starts
with smaller plantings to see their effects before he
expands to larger areas.

As with Mr. Phiri's site, all landscapes are continually evolving and we need to continually work *with*
them. Go back frequently and observe how your site is
performing, repair elements if needed, and see if there
are ways you could improve on your site plan and techniques. We cannot escape the need for maintenance,
but we can reduce the need for excessive maintenance
by following the eight water-harvesting principles.
Balanced maintenance should not be feared or
neglected; it is an opportunity to learn and to improve.

WATER-HARVESTING ETHICS

Mr. Phiri's site and life provide a wonderful example of embodying water-harvesting principles within
an integrated system. They also embody an ethical
basis that further increases the benefits of his work.
The three ethics of permaculture[9] described below are
realized in Mr. Phiri's work, and are important guides
to me in making decisions about water-harvesting and
integrated-system design.

1. The CARE OF THE EARTH[10] ethic reminds us
   to care for all things living and nonliving, including
   soil, water, air, plants, animals, and entire ecosystems. As Bill Mollison states in *An Introduction to
   Permaculture*, "It implies harmless and rehabilitive
   activities, active conservation, ethical and frugal use
   of resources, and 'right livelihood' (working for useful and beneficial systems)."[11]

2. CARE OF PEOPLE[12] directs us to strive to
   meet our basic needs for air, water, food, shelter,
   education, fulfilling employment, and amiable
   human contact in ways that do not hamper or
   prevent others from doing the same. We do not
   exploit or disregard others for our own gain. Nor
   do we destroy the environment that supports us
   all. Instead, we sustain a basic quality of life that
   improves our environment while enabling
   others to do the same.

3. REINVESTMENT OF SURPLUS TIME,
   MONEY, AND ENERGY[13] to achieve the aims of
   earth and people care encourages us to extend our
   influence and surplus energies to help others attain
   the ethics in their own life and work. This helps us
   all because it strengthens the greater communities
   in which we all live.

Mr. Phiri embodies these ethics: He improves
his land, the earth, and his community by working
with local resources so his land and community can
sustainably regenerate more resources. He eschews
synthetic fertilizers and clear-cutting that provide
short-term gains but pollute and weaken the land in
the long run. He practices infusion rather than

extraction. He gives his land more than he takes—in the form of water. He gives his community more than he takes—in the form of information, trees, and water. He empowers others to do the same. The Zvishavane Water Project was formed by Mr. Phiri to contribute surplus time, energy, and money to spread these ideas. He teaches people how to harvest rain, improve soil, grow food, and build community. And he learned it all from living it.

By following these eight principles and the "care" ethics above, you can thoughtfully and effectively practice water harvesting and create the best techniques and strategies for your unique situation. Use the principles and ethics as a checklist of guidelines while you assess your site, imagine what water-harvesting strategies would work best and where, and as you implement your ideas. As long as all the guidelines are met you'll be on the path to abundance.

You now have the tools to conceptualize and plan an integrated rainwater-harvesting system. Read on to find out how much rain you have to harvest.

2013 UPDATE ON MR. ZEPHANIAH PHIRI

Many thanks to Ken Wilson, Executive Director of the Christensen Fund, a longtime ally and supporter of Mr. Phiri, for his writings and reports from which most of this update was produced.

Mr. Phiri's efforts and all he brings to life continue to evolve and grow. He is now 85 years old and his first wife—and co-generator of his farm—Magrate, has passed away, Yet, though Mr. Phiri says he's "down to one eye and one ear" and his "days are numbered," he has no sense that his work is finished.

Mr. Phiri is joined in his work by some of his children (his youngest daughter just graduated from college), grandchildren, and in particular his second wife and master farmer/demonstrator/leader, Constance. Together they keep increasing the fertile organic matter in their soil, the diversity of life growing from that soil, and the water they plant within it—all of which increases their ability to cope with drought and floods. The diversity of multi-use, woody plants on the farm has increased 25% since 1999. And new innovations

such as gently-sloping diversion swales now direct water from areas of their land that were too wet to those that would otherwise be too dry.

In this hot, dry region of Zimbabwe water arrives rapidly and leaves rapidly. And it doesn't leave alone. Mr. Phiri explains the situation and his response as follows. "When there are thunderstorms, soil and water try to elope together and run away from my land. It is my job to persuade them to settle down here and raise a family."

Settle down and raise a family they have, which over the decades has attracted over 8,000 visitors to Mr. Phiri's farm to see how this common-sense approach has led to ever-increasing productivity and sustainable resilience. Guests have come from over 14 African countries, and 9 other countries in Asia, Europe, and North and South America. However, most important perhaps are the many hundreds of farmers who came by themselves or with local non-government organization (NGO) staff, agricultural extension officers, and other government officials.

This has helped lead to the widespread adoption of the kinds of water-soil management systems that Mr. Phiri, his family, and those who have been inspired by them, have shared with their communities in southern Africa. Tens of thousands of farmers are now taking up these methods across Zimbabwe. According to a 2010/2011 study in Mazvihwa, an area about 62 miles (100 km) from the Phiri home, people have heard this message and taken it to heart, and almost all households are now involved in some form of water harvesting.

Until the last decade most farmers were focused on exploring "modernization" approaches to agriculture—believing hybrid plant varieties, synthetic fertilizer inputs, and extractive water and plant systems from the outside world would make the difference. What Mr. Phiri recommended offered no technological miracle and involved no handout. Instead it relied on farmers developing their own knowledge and investing their own labor to dig and shape their lands to manage water and soil over the long term. Rather poignantly, a large number of men suffering from HIV-AIDS dug water-harvesting earthworks as investments to help their families, even as they slowly died of the disease. These strategies tackled the real constraints of this arid region: the water available for crops. More people came around as the first generation of followers achieved success while external solutions failed to transform the land. This pattern is now seen across the country, with whole communities shifting their farming systems towards water harvesting. Such can be the impact spearheaded by one dedicated, visionary man, and those who have supported him and furthered this work.

Yet Mr. Phiri does not see his approach as a silver bullet, but as a set of principles for approaching soil and water with curiosity, respect and innovation. Copying his practices exactly doesn't work because every place is unique. He's delighted to see so many of his followers doing things differently in order to work with the characteristics of their own land.

Likening himself to a bullfrog, Mr. Phiri says he is constantly croaking about the delights of harvesting local waters. For nine years he did so through his NGO, the Zvishavane Water Project (ZWP). In 1996 he retired from the ZWP, and gave the organization a "natural endowment" by granting a portion of his land to train staff, generate income from crops, and serve as a demonstration plot. Mr. Phiri was selected in 1997 as an Ashoka fellow, and in 2006 he received the National Geographic/Buffet Award for Leadership in African Conservation. While receiving a Lifetime Achievement Award at the University of Zimbabwe in 2010, Mr. Phiri invited the researchers at the college to work with him and document what he's done for the hydrology of his land, its soils, its agricultural productivity, and its biodiversity. He offered to share with them "all the results of forty years of experimentation, except those that I have already eaten." The offer was eagerly accepted.

Fig. 1.22. Mr. Zephaniah Phiri Maseko in 2013.
Credit: Brock Dolman

Fig. 1.23. Constance and grandchild in 2013.
Credit: Brock Dolman

CHAPTER 2

Assessing Your Site's Water Resources and More

*But especially as I drink the last of my water, I believe that we are subjects of the planet's hydrologic process,
too proud to write ourselves into textbooks along with clouds, rivers, and morning dew.
When I walk cross-country, I am nothing but the beast carrying water to its next stop.*

—Craig Childs, *The Secret Knowledge of Water*

It's important to begin harvesting water knowing *how your site fits into larger water flows* and knowing *how much water there is to harvest.*

This chapter builds on the principle of long and thoughtful observation by beginning with a description of the hydrologic cycle. This will help you understand your site's water flow in the context of global hydrologic patterns and interconnections. Next the chapter moves to local watersheds and subwatersheds, where small-scale portions of the hydrologic cycle occur. You will learn how to determine the boundaries of your site's subwatershed, where to concentrate your water-harvesting efforts, how to calculate rainfall and runoff volumes that affect your site, and how other on-site waters such as greywater and air-conditioning condensate can add to your site's harvestable water resources. This site assessment then expands to other resources and challenges unique to your place. The chapter ends with the story of an Arizona couple striving to live within their site's rainwater budget. Appendix 3 (Calculations), appendix 5 (Worksheets), and appendix 9 (Water-Energy-Carbon Nexus) are meant to be adjuncts to this chapter.

THE HYDROLOGIC CYCLE—OUR EARTH'S CIRCULATORY SYSTEM

The following is primarily drawn with permission from water-harvester Ben Haggard's great little book, *Drylands Watershed Restoration*:

Of the world's total water, a small percentage is fresh water. The majority of that is tied up as ice in polar ice caps and glaciers. The remainder is continually recycled in order to support the world's living systems. This recycling is known as the hydrologic cycle.

Water is evaporated from the oceans and precipitated as rain or snow over the continents. This water is absorbed by plants and evapotranspired back into the air. This pumping of water back into the air by plants accounts for much of our atmospheric water. This water forms clouds and rains again … Forests play an important role in maintaining [and retaining] rain in the landscape.

Raindrops form around ice crystals in clouds. These ice crystals require a nucleus for their formation. Dust, tiny bits of leaf, and bacteria are among the particles that initiate rain. A number of natural systems [such as forests] encourage rain by giving off columns of tiny particles that seed the clouds causing drops to form.[1]

Raindrops are soaked up by the living sponges of forests, prairies, and desert thornscrub. These, along with their associated leaf drop, topsoil, and the cavities created by burrowing animals, help hold onto that water and slowly release it.

If forests, grasslands, and other rain seeders and sponges are removed from a landscape, the landscape begins to dry. Rain can become less common and vegetation has trouble reestablishing. Rivers and streams become dry.

Ben Haggard continues:

Rivers and streams generally flow throughout the year; in spite of the fact that rain is a localized and fairly infrequent event in arid settings. Even in rainy climates, rain occurs a relatively small percentage of the time. Rivers have a sustained flow because most of the water is actually stored in the soil where it slowly releases into the drainage. In disturbed watersheds, this slow and sustained release is disrupted. Water runs rapidly off the ground's surface rather than soaking into the ground. This process creates floods followed by drought. To repair such a watershed, infiltration of the water into the ground must be increased.

Living systems create complex interactions with water. Water falls as rain. Trees intercept this water, directing it into the ground where a layer of organic material deposited by the trees absorbs and holds it. Some of the water flows slowly through the ground where it supports the growth of forests and the sustained flow of rivers. Rivers act as transportation networks, allowing nutrients from the forests to wash downstream and fish and other animals to swim upstream, importing phosphate and other minerals into the forests. Water in the landscape also attracts and supports wildlife, the active planters, fertilizers, and maintainers of the forest. The forests breathe water back into the air, where it condenses around particles also released by the forests. The clouds form and the entire process repeats itself.[2]
(*Reprinted with permission from* Dryland Watershed Restoration—Introductory Workshop Activities *by Ben Haggard, copyright 1994, Center for the Study of Community*)

Each of us depends on and is a part of the hydrologic cycle. As water moves through the global cycle, so it moves through the watersheds of our communities, the subwatersheds of our individual sites, and our own bodies, which are over 70% water. We can slow, cycle, and enhance that flow as we improve our lives and community by harvesting rainwater. First, we need to identify and thoughtfully observe our watersheds.

WATERSHEDS AND SUBWATERSHEDS— DETERMINING YOUR PIECE OF THE HYDROLOGIC CYCLE

A *watershed* is the total area of land from which water, sediments, and dissolved materials flow by gravity to a particular end point. At the largest scale this endpoint might be a river, lake, or ocean. A watershed is a geographic entity clearly defined by high points or ridgelines that split the flow of water, creating the boundaries of each watershed. Watersheds are made up of many smaller *subwatersheds*, each defined by lower elevation ridgelines that further split the flow of water and direct portions to particular endpoints. These subwatersheds are made up of a collection of still smaller subwatersheds. If you trace the boundaries of all these subwatersheds, the pattern you'll see looks like pieces of jigsaw puzzle forming an interconnected whole. The terms "watershed" and "subwatershed" are relative terms that can refer to a variety of scales of water drainage areas.

On a large scale, your land will almost surely be part of a regional watershed that drains thousands of square miles of land, creating streams and rivers. My Tucson home is a part of the Santa Cruz River watershed, which covers approximately 8,600 square miles in southern Arizona and northern Mexico. Within this regional watershed, water drains toward me from up-slope areas, away from me to downslope areas, and ultimately flows to the Santa Cruz River. The Santa Cruz River watershed is in turn a subwatershed of the larger Gila River watershed, which is in turn a subwatershed of the still larger Colorado River watershed.

The land area of a city flowing toward one regional watershed consists of smaller subwatersheds throughout the city, broken up into many smaller neighborhood-sized subwatersheds, made up of

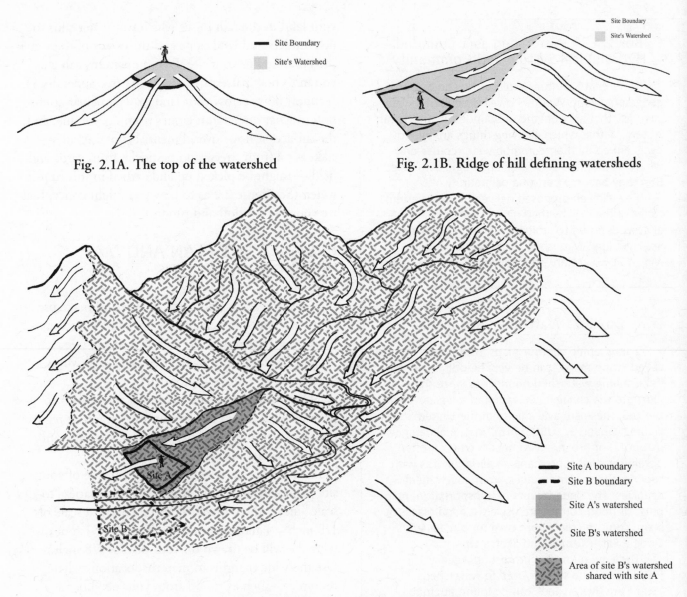

Fig. 2.1A. The top of the watershed

Fig. 2.1B. Ridge of hill defining watersheds

Site Boundary
Site's Watershed

Site Boundary
Site's Watershed

Site A
Site B

Site A boundary
Site B boundary

Site A's watershed

Site B's watershed

Area of site B's watershed
shared with site A

Fig. 2.1C. Watersheds and subwatersheds, the larger picture

property-sized subwatersheds, made up of still smaller subwatersheds consisting of residents' roofs, yards, patios, and driveways. Each small urban-scale subwatershed directs flowing water toward a different urban endpoint. Urban landforms, buildings, and parking lots act as the dividing "ridges" between these tiny watersheds. Pitched roofs divide the flow of water between front and back yards. Parking lots act as gradually sloped fields. Roads act as linear ridges if raised, or as drainageways if built as a lower element in the landscape.

If your site is at the very top of a hill it is also at the top of a watershed, because all runoff water will be draining off the hill away from your site (fig. 2.1A). If your site is at the bottom of the hill, your site's watershed will be that part of the hill's slope that drains or sheds water toward your property. Most likely, the water on the other side of the hill will drain to a different endpoint, so it will be part of a different watershed (fig. 2.1B). However, if runoff from the other side of the hill does eventually drain toward your site—perhaps via an arroyo or wash curving around the hill—then it too is part of your site's watershed. In addition to this local hill, there may be other areas of land that drain toward your site. If so, the watershed affecting your site is even larger (fig. 2.1C).

IDENTIFY YOUR SITE'S WATERSHED AND OBSERVE ITS WATER FLOW

To assess your site's water resources, first define the boundaries of your property and the watershed directly affecting your site. A topographic map will give you a general idea of your watershed's "ridgelines"—the tops of slopes that determine if water is flowing toward or away from your location (see box 2.2). You can walk your land in the rain to see which way water runs to help understand land slope and the extent of the watershed draining to your site. Erosion patterns can clue you in to flow patterns when it's dry (see appendix 1). If runoff flows across your land, pay particular attention to what direction it comes from, its volume, and the surfaces it flows over. Potential contaminants—such as oil from streets and pesticides from yards and fields—might be picked up and carried in this runoff water. (See figure 2.2 as to how you might conceptualize your site's runoff and runon.)

CREATE A SITE PLAN AND MAP YOUR OBSERVATIONS

Creating a site plan helps you see and make use of site resources and challenges; integrate your water-harvesting system with the rest of your site (see chapter 4); and place and size water-harvesting earthworks, vegetation, and tanks appropriately.

Use the worksheets provided in appendix 5 to prompt ideas. You can use your own graph or regular paper to map out your site (or perhaps create your own water-harvesting journal).

Leave wide margins around the outside of your site plan, and draw your property's boundaries "to-scale" inside these margins. If you choose a scale of 1/8 inch = 1 foot, a measured distance of 1 foot on your site will be drawn on your plan as 1/8 inch. Use the wide margins to map the locations where resources—such as runoff from your neighbor's yard—flow on, off, or alongside your site. Draw buildings, driveways, patios, existing vegetation, natural waterways, underground and above-ground utility lines (to avoid damaging them and yourself), and other important elements of your site to-scale on the plan. Make multiple copies of your basic site plan on which to draw a number of drafts of your observations and ideas. (Figure 2.3 is a sample site map.)

CALCULATE YOUR SITE'S RAINFALL VOLUME

Once you've defined and mapped the boundaries of your site, use the calculations in box 2.3 to determine average volume of rain falling on your site each

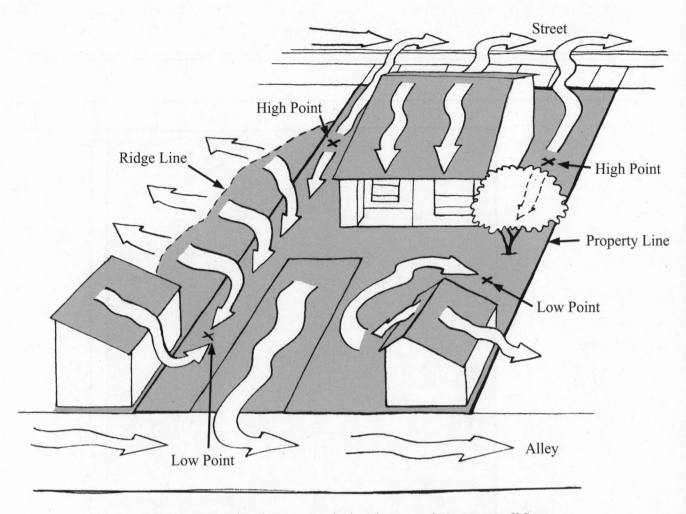

Street

High Point

Ridge Line

High Point

Property Line

Low Point

Low Point

Alley

Fig. 2.2. An urban home watershed with arrows depicting runoff flow

year. This is the "income" side of your "water budget." (Again see figure 2.3.)

MAP YOUR SITE'S CATCHMENT SURFACES AND CALCULATE THEIR RUNOFF VOLUMES

Roofs, paved surfaces such as driveways and patios, and compacted earth surfaces such as paths are useful catchment surfaces from which to harvest rainfall. Indicate on your plan any catchment surfaces that drain water *off* your site (for example, a driveway sloping toward the street) and *subtract* this lost runoff volume from your site's calculated average annual rainwater resources. (You can devise strategies to recapture that lost runoff later.)

Indicate on your plan any catchment surfaces draining water *onto* your site from off-site (for example, runoff from your neighbor's yard that drains into your yard). *Add* this bonus runoff (or "*runon*") volume to your site's calculated average annual rainwater resources. See box 2.4 for instructions on calculating runoff volumes. See figure 2.4 for the example site map with runoff volumes and runoff coefficients.

Note: This book and the rainwater-harvesting principles emphasize the harvest and utilization of localized runoff and runon high in the watershed *before* it enters a drainageway. Once the water is in a drainageway, it is not to be diverted from it, although its flow can be slowed to allow for more infiltration as is the case with a one-rock dam.

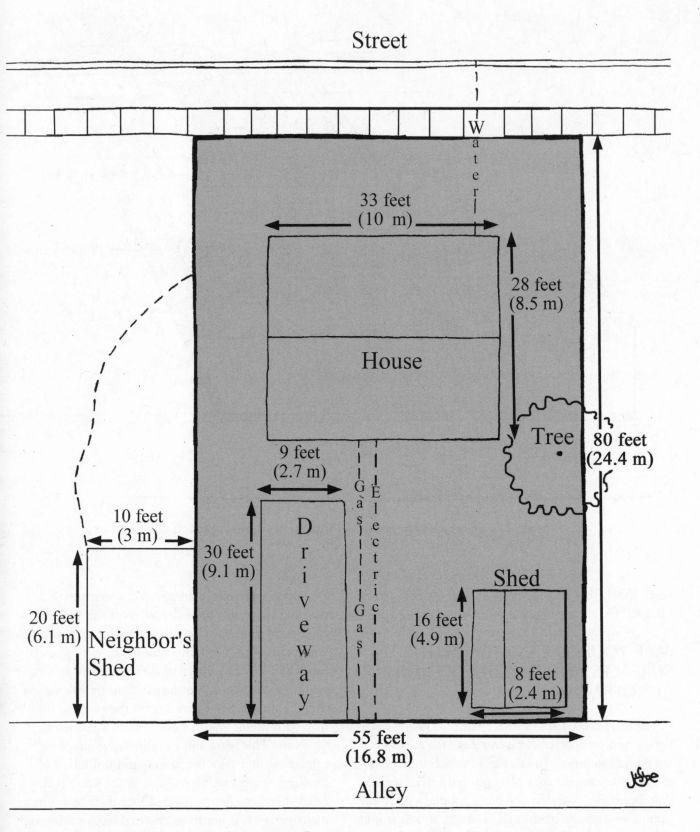

Fig. 2.3. Site map (overhead/plan view) of a 4,400 square foot (409 m²) property. In an average
year of 12 inches (305 mm) of precipitation the site receives 32,912 (124,585 liters) gallons of rainfall "income."

Box 2.3. Calculating Rainfall Volumes

CALCULATING RAINFALL VOLUMES IN ENGLISH UNITS

To calculate the volume of rainfall in *cubic feet* that falls in an average year on a specific *catchment area*, such as your roof, yard, neighborhood, or other subwatershed:

CATCHMENT AREA (in square feet) multiplied by the AVERAGE ANNUAL rainfall
(in feet) equals the TOTAL RAINWATER FALLING ON THAT CATCHMENT
IN AN AVERAGE YEAR (in cubic feet)

(or)

CATCHMENT AREA (ft^2) $\times$ AVG RAINFALL (ft) = TOTAL RAINWATER (ft^3)

If you normally measure annual rainfall in inches, simply divide inches of rain by 12 to get annual rainfall in feet. For example, folks in Phoenix, Arizona get about 7 inches of annual rainfall, so they would divide 7 by 12 to get 0.58 foot of annual rain.

Once you get your answer in cubic feet of annual average rainfall, convert cubic feet to gallons by multiplying your cubic foot figure by 7.48 gallons per cubic foot. The whole calculation looks like this:

CATCHMENT AREA (ft^2) $\times$ RAINFALL (ft) $\times$ 7.48 gal/ft^3 = TOTAL RAINWATER (gal)

For example, if you want to calculate how much rainwater in gallons falls on your 55 foot by 80 foot (4,400 square feet) lot in an normal year where annual rainfall averages 12 inches the calculation would look like this:

4,400 square foot catchment area $\times$ 1 foot of average annual rainfall $\times$ 7.48 gallons
per cubic foot = 32,912 gallons of rain falling on the site in an average year

CALCULATING RAINFALL VOLUMES IN METRIC UNITS

To calculate the volume of rainfall falling on a specific catchment area in liters:

CATCHMENT AREA (in square meters) $\times$ AVERAGE ANNUAL RAINFALL
(in millimeters) = TOTAL RAINWATER FALLING ON A CATCHMENT
AREA IN AN AVERAGE YEAR (in liters)

CALCULATING THE VOLUME OF RAINFALL ON A SPECIFIC CATCHMENT FOR A GIVEN RAIN EVENT IN ENGLISH OR METRIC UNITS

Use the calculations above, but enter the amount of
"rainfall from a given rain" in place of "average annual rainfall."

Note: Appendix 3 "Calculations" provides more detailed information on conversions, constants, and calculations for water harvesting.

Box 2.4. Calculating Runoff Volumes

You can get a ballpark estimate of runoff volume from any sloped surface by multiplying the volume of rain that falls on that surface by its "runoff coefficient"—the average percentage of rainwater that runs off that type of surface. For example, a rooftop with a runoff coefficient of 0.95 estimates that 95% of the rain falling on that roof will run off.

The runoff coefficient for any given surface depends on what the surface is composed of. Rainfall intensity also affects the coefficient: the higher the rainfall intensity, the higher the runoff coefficient. Ranges and averages of various runoff coefficients I use in the southwest U.S. are as follows:

• A roof or impervious paving (such as an asphalt street): 0.80–0.95
• Sonoran Desert uplands (healthy indigenous landscape): range 0.20–0.70, average 0.30–0.50
• Bare earth: range 0.20–0.75, average 0.35–0.55
• Grass/lawn: range 0.05–0.35, average 0.10–0.25
• For gravel use the coefficient of the ground below the gravel.

The runoff coefficient for earthen surfaces is greatly influenced by soil type and vegetation density. Large-grained porous sandy soils tend to have lower runoff coefficients while fine-grained clayey soils allow less water to infiltrate and therefore have higher runoff coefficients. To harvest runoff in earthworks, whatever your soil type, the more vegetation the better, since plants enable more water to infiltrate the soil.

CALCULATING ROOF RUNOFF: AN EXAMPLE IN METRIC UNITS

Determine the size of a roof catchment by measuring only the outside dimensions—or "footprint"—of the roof's edge (if your house has a roof with overhangs the roof's footprint will be larger than the building's footprint). Ignore the roof slope; no more rain falls on a peaked roof than falls on a flat roof with the same footprint. (See figure 2.5).

To calculate the runoff in liters from a metal roof's 9 meter × 10 meter "footprint" (90 square meters) in a climate averaging 304 millimeters of rain a year:

90 square meter roof × 304 millimeters of average annual rainfall = 27,360 liters of rain falling on the roof in an average year.

$$90 \text{ m}^2 \times 304 \text{ mm} = 27,360 \text{ liters/average year}$$

Multiply the above figure by the roof surface's runoff coefficient 0.95*:
27,360 liters × 0.95 = 25,992 liters running off the roof in an average year.

Note*: 5 to 20% of runoff from impervious catchment surfaces such as roof can be lost due to evaporation, wind, overflow of gutters, and minor infiltration into the surface itself. In volume 3, the chapter on cistern components includes a table for runoff coefficients specific to roof type.

CALCULATING YARD RUNOFF: AN EXAMPLE IN ENGLISH UNITS

Let's say we are on a site receiving 18 inches of rain in an average year, and the neighbor has about a 25 foot by 12 foot bare section of his yard that drains onto our example property. The soil is clayey and compacted.

Determine the available rainwater running off that section of the neighbor's yard onto our land by multiplying its catchment area (300 square feet) by the average annual rainfall in feet (1.5) by 7.48 (to convert the answer to gallons):

$$\text{CATCHMENT AREA (ft}^2) \times \text{RAINFALL (ft)} \times 7.48 \text{ gal/ft}^3 = \text{TOTAL RAINWATER (gal)}$$

300 × 1.5 × 7.48 = 3,366 gallons of rain falling on that section of the neighbor's yard in an average year.

Multiply that figure by the soil surface's runoff coefficient of 0.60:

3,366 × 0.60 = 2,019 gallons annually running off the neighbor's compacted yard into ours. Add that to our site's annual rainwater budget.

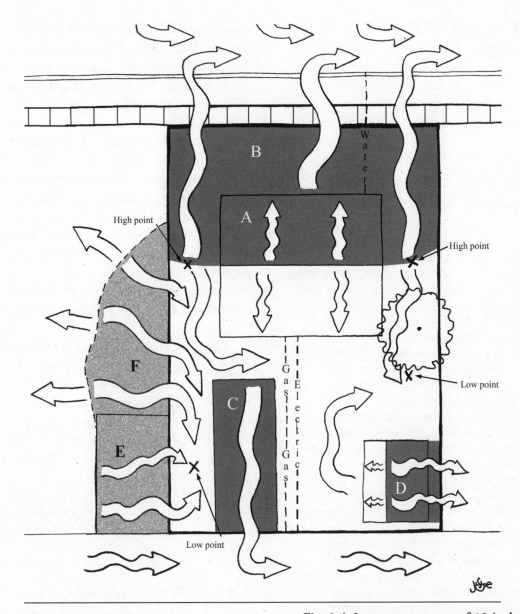

RUNOFF LOST

A. 3,283 gallons (12,424 liters) runoff lost from 462 ft² (42.9 m²) half of metal roof having a 0.95 runoff coefficient.

B. 2,246 gallons (8,502 liters) of runoff lost from 858 ft² (79.7 m²) section of gravel yard having a 0.35 runoff coefficient.

C. 1,818 gallons (6,878 liters) of runoff lost from a 270 ft² (25.1 m²) concrete driveway having a 0.90 runoff coefficient.

D. 539 gallons (2,037 liters) runoff lost from 80 ft² (7.4 m²) section of shed's asphalt shingle roof having a 0.90 runoff coefficient.

RUNON GAINED

E. 1,421 gallons (5,379 liters) runon gained from neighbor's 200 ft² (18.6 m²) metal shed roof having a runoff coefficient of 0.95.

F. 1,571 gallons (5,943 liters) runon gained from 350 ft² (32.5 m²) section of neighbor's compacted dirt yard having a runoff coefficient of 0.60.

Fig. 2.4. In an average year of 12 inches (305 mm) of rainfall this site receives 32,912 gallons (124,572 liters) of rainfall, gains 2,992 gallons (11,325 liters) of runon from the neighbor's yard and shed roof, and loses 7,885 gallons (29,845 liters) of runoff for a total site rainwater budget of 28,019 gallons (106,052 liters). If the landscape were changed to harvest both the runon and runoff, the site's annual rainwater resources could increase up to a total of 35,904 gallons (135,897 liters). Estimated annual runoff and runon volumes off each type of catchment surface are listed with surface material and runoff coefficient. Still more runoff from sidewalk, street, and alley could be harvested within the public right-of-way to grow public street and alley trees (see chapter 8 of volume 2 for strategies on harvesting street runoff). See box 2.4 for calculations of runoff and runon volumes.

ESTIMATE YOUR SITE'S WATER NEEDS

Determine the "expense" side of your water budget by estimating your household and landscape water needs. Your water bill reflects current water use. The user-friendly website www.h2ouse.org provides water use rates for household appliances, and recommended conservation strategies. Estimated water needs of plants can be obtained from the local agricultural extension office (or see, for example, appendix 4 "Example Plant Lists and Water Requirement Calculations for Tucson, Arizona"). Better yet, take a hike to observe native plants that grow naturally in *your* area on rainfall alone. Water needs of plants will vary widely depending upon the plant, its size, and the soils, climate, and microclimate in which it is planted. Keeping that in mind, in Tucson, Arizona, a mature 20-foot (6.1-m) tall and wide native mesquite tree will use about 3,000 gallons (11,356 liters) of water per year, while a mature 16-foot (4.9-m) tall and wide exotic non-native citrus tree will use about 8,000 gallons (30,283 liters) per year.[4]

Compare your site's water needs to the volume of rain falling directly on, or flowing through, your site. How much of your domestic water needs could you meet by harvesting rooftop runoff in one or more tanks? How much vegetation could you support by harvesting rainfall directly in your soil? How can you balance your water budget using harvested rainwater as your primary water source?

Box 2.5. The Example Site's Water Expenses

Refer to figure 2.4.

The estimated annual water "expense" or needs of the water budget for the Tucson, Arizona site from figure 2.4 is 8,000 gallons (30,280 liters) per year for the landscape (one mature citrus tree), and 104,248 gallons (394,622 liters) for the four-person household's interior requirements (washing, bathing, cooking, drinking, toilet, evaporative cooler, with usages based on www.h2ouse.org data for a non-conserving household).

So all the water needs of the current, or a denser, landscape could easily be met by the 35,904 gallons (135,897 liters) of harvestable rainfall and runon, and the landscape's density could be increased still more with the on-site reuse of greywater and harvest of street runoff.

Of the household's current interior water needs 6% could be met by harvesting the home's 6,566 gallons (24,855 liters) of roof runoff. By implementing simple water conservation strategies recommended by h2ouse.org (use of low-flow toilets, faucet aerators, efficient washer, and evaporative cooler without bleed valve) the family could reduce annual interior water needs to 62,684 gallons (237,284 liters), with roof runoff providing 10% of that. An additional 7,300 to 20,600 gallons (27,600 to 78,000 liters) could be conserved with such strategies as planting cooling shade trees to offset cooler use and the use of a composting toilet to eliminate water used for flushing. Lifestyle changes and learning to live with less can lead to still more water conservation.

By combining all these conservation strategies, and expanding the roof surface (perhaps with a covered porch on the east, west, and winter-shade-sides of home) this household could meet *all* its water needs from rain in an average year. Read on and see what strategies would be appropriate for your site and needs. Every site is unique, and you decide how far you want to take it, using strategies found in this and the next two volumes of this series.

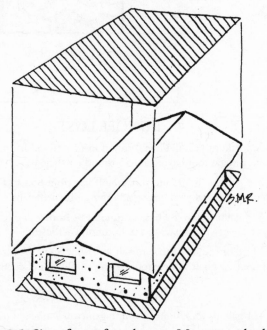

Fig. 2.5. Size of a roof catchment: Measure only the outside dimensions—or "footprint"—the roof's edge. Ignore the roof slope; no more rain falls on a peaked roof than falls on a flat roof with the same footprint.

ESTIMATE YOUR SECONDARY ON-SITE WATER SOURCE—GREYWATER

Greywater is the water that drains from your household sinks, tubs, showers, and washing machine. It does *not* include the water draining from your toilet, which is called *blackwater*. Ideally greywater's original source is harvested rainwater, but more often it is municipal or well water drawn from the tap. Whatever the source, you can turn this household wastewater into a resource by using it to safely and productively irrigate your landscape. The volume of available household greywater depends on how much water goes down your drains. Every household is different, but the information in box 2.6 offers ballpark estimates of typical volumes used. Add up your site's

GREYWATER SOURCES/VOLUMES

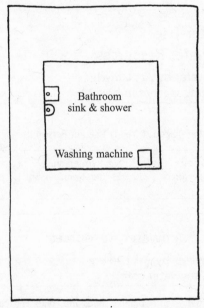

Low water-use washing machine	Low-flow bathroom sink & shower
weekly: 60 gallons 227 liters	weekly: 224 gallons 848 liters
monthly: 261 gallons 988 liters	monthly: 974 gallons 3,687 liters
yearly: 3,132 gallons 11,855 liters	yearly: 11,688 gallons 44,239 liters

**Fig. 2.6. Example of a home's average estimated greywater resources discharged through four residents' use of water. This greywater could be accessed and recycled within the landscape with the installation of a greywater system.
Note: One month equals 4.35 weeks.**

estimated greywater volume from each source (see figure 2.6 for an example). You'll then use this information in chapter 4 to create an integrated conceptual rainwater-harvesting and greywater-harvesting plan for your site. In volume 2 there's a chapter which provides more specific information on greywater-harvesting systems.

ESTIMATE ADDITIONAL ON-SITE "WASTE" WATERS THAT CAN BE USED AS "RESOURCE" WATERS

Reverse osmosis (RO) water filters often discharge more than half the household tapwater they purify. And while the discharge water contains the salts and other total dissolved solids (TDS) that the filter filtered out, it can still be used to irrigate plants.

Evaporative coolers, common in hot dry climates, discharge even saltier water. Their salty bleedoff water can be used to irrigate salt-tolerant vegetation (like four-wing saltbush or *Atriplex canescens*), particularly if the soil is occasionally flushed with salt-free rainwater or condensate.

Condensate is discharged from air conditioners (AC). Like rainwater, it is basically distilled water that is great for watering plants (though I wouldn't drink AC condensate). The hotter and more humid the climate, the more condensate air conditioners will discharge.

See Box 2.9 to estimate these water sources' discharge volumes. See the Energy Costs of Water in appendix 9 for the associated energy consumption and carbon emissions of condensate, greywater, rainwater, and more.

After you have a good idea of your site's rainwater, greywater, and other potentially "wasted" waters that can become harvested waters you are now ready to consider some of the strategies presented in chapter 3 with which you can harvest that water. However, first expand your site assessment to harvesting opportunities and challenges in addition to water. This will better prepare you to take full advantage of strategies in chapter 4 that increase your water harvests exponentially by integrating them with additional free resources such as sun, wind, vegetation, and more.

Box 2.6. Estimates of Household Greywater Resources[5]

Based on figures from *Create an Oasis with Greywater* by Art Ludwig
(see also www.OasisDesign.net and www.Greywater.com)

Fixture	Frequency Used	Volume Used	Weekly Use Per Person	Yearly Use Per Person
Top-loading washing machine	1.5 uses/person/wk	30 gal/use 114 liters/use	45 gal/wk/person 170 liters/wk/person	2,340 gal/yr/person 8,858 liters/yr/person
Bathtub	1.5 uses/person/wk	20 gal/use 76 liters/use	30 gal/wk/person 114 liters/wk/person	1,560 gal/yr/person 5,905 liters/yr/person
Shower	5 uses/person/wk	13 gal/use 49 liters/use	65 gal/wk/person 246 liters/wk/person	3,380 gal/yr/person 12,795 liters/yr/person
Bathroom sink	21 uses/person/wk	0.5 gal/min 1.9 liters/min	10.5 gal/wk/person 40 liters/wk/person	546 gal/yr/person 2,067 liters/yr/person
TOTAL			150.5 gal/wk/person 570 liters/wk/person	7,826 gal/yr/person 29,625 liters/yr/person

Box 2.7. Estimates of Low-Water-Use Front-Loading Washing Machine Greywater Resources

Based on figures from *Create an Oasis with Greywater* by Art Ludwig.
Compare to the data for a high-water-use top-loading washer in box 2.6.

Fixture	Frequency Used	Volume Used	Weekly Use Per Person	Yearly Use Per Person
Front-loading washing machine	1.5 uses/person/wk	10 gal/use 38 liters/use	15 gal/wk/person 57 liters/wk/person	780 gal/yr/person 2,953 liters/year/person

Box 2.8. Estimates of Household Kitchen Sink Drainwater Resources

Based on figures from *Create an Oasis with Greywater* by Art Ludwig
(see also www.OasisDesign.net and www.Greywater.com)

NOTE: Some regulators consider kitchen sink drainwater to be blackwater due to higher levels of organic solids and coliform bacteria than other greywater sources. If those sending the water down the drain are environmentally- and soil-conscious gardeners using appropriate soaps, I prefer Art Ludwig's classification of "dark greywater." For soap recommendations and safe strategies harvesting kitchen sink drainwater on site see Chapter 12 and the Kitchen Resource Drain appendix in *Rainwater Harvesting for Drylands and Beyond, Volume 2*.

Fixture	Frequency Used	Volume Used	Weekly Use Per Person	Yearly Use Per Person
Kitchen Sink	Daily	3 to 15 gal/day 11 to 55 liters/day	21 to 105 gal/wk/person 79 to 397 liters/wk/person	1,095 to 5,475 gal/yr/person 4,015 to 20,725 liters/yr/person

Box 2.9. Estimates of Discharge Volumes from Household Evaporative Coolers, Air Conditioners, and Reverse-Osmosis Filters

Water Source	Example Discharge Rates	Factors Affecting Discharge Volume	Water Quality	Compatible Vegetation
Evaporative-Cooler Bleed-off	Evaporative coolers are most effective in high-temperature, low-relative-humidity conditions. Example discharge in dry conditions of Phoenix, Arizona: Cooler bleed-off discharge is 1 to 5 gallons/hour (3.7 to 18.9 liters/hour)[a] for estimated average single-family residential cooler use.	- Relative humidity - Temperature - Number of hours of run time	Typical bleed-off water has high salts ranging from 375 to 4,043 mg/L, with average 1,580 mg/L.[b] Some evaporative coolers have a "sump dump" discharge, which often contains higher levels of total dissolved solids (TDS) than bleed-off.	Use bleed-off only on salt-tolerant vegetation. If TDS is < 5,000 mg/L, can irrigate salt-tolerant Bermuda and salt grass. If TDS < 2,000 mg/L can use on low-salt-tolerant plants such as citrus trees and almond, apple, peach, and pear trees.[b]
Air-Conditioner Condensate	Example discharge in *dry* season conditions of Phoenix, Arizona[c]: 0.01 to 0.02 gal (0.04 to 0.09 liters) of condensate/hour for estimated average single-family air conditioner, 2-3 ton capacity. Condensate from a large commercial unit can reach 500 gallons (1,895 liters)/day.[c] Example discharge in *humid* conditions of Austin, Texas[d]: 0.1 to 0.2 gallons (0.3 to 0.7 liters) of condensate/hour for estimated average single-family air conditioner, 2-3 ton capacity. Condensate from a large commercial unit can add up to 2,000 gallons (7,580 liters)/day.[d]	- Relative humidity - Temperature - Ton-hours of run time: typical 2 – 3 ton capacity residential air conditioner may run 60% to 70% of the time on hottest, most humid days.[d] - Frequency of air exchange within building. Commercial buildings typically exchange air more frequently than similar sized residential units, so can produce more condensate. - Amount of humidity produced inside the building by various activities	AC condensate is typically high-quality irrigation water produced when humid air condenses on the cold evaporator or coil of an air conditioning unit.	Use salt-free condensate as irrigation source throughout landscape and garden in both humid and dry landscapes. Can use condensate to help leach accumulating salts from primary root zone in top 2 feet (0.6 m) of soil.
Reverse-Osmosis (RO) Water-Filter Discharge	Discharge of 2 to 5 gal (7.58 to 18.95 liters) per gallon of water purified. Assuming 1 gal/person/day (3.79 liters) drinking and cooking water needed, discharge would be 2 to 5 gallons (7.58 to 18.95 liters)/person/day.		RO filter discharge water has about 25% higher concentration of TDS than the tapwater it is filtering.[e] It has no suspended solids.	Use as irrigation source on landscape and garden.

Notes: These waters are not free waters delivered by the natural hydrologic cycle and/or gravity, but high-embodied energy waters pumped to, or on the site, or condensed on a high-energy-use appliance. But if these waters are present on a site we should reuse them through additional beneficial cycles, such as irrigating appropriately-placed, passive-cooling shade trees, rather than wastefully sending these waters down the drain.

And in doing so, we would be wise to simultaneously increase our awareness of the true costs of these waters to inform still better and more efficient practices. See appendix 9 and its table A9.1 and the Energy Costs of Water, Water Costs of Energy, and Carbon Costs of Energy charts for some of these hidden costs.

A ONE-PAGE PLACE ASSESSMENT AND HOW TO USE IT

Want to expand your site assessment? Here is a tool with which to do so. The Watershed, Climate, Sun, Wind, Water, Watergy, and Totem Species information in figure 2.7 constitutes a "starter" Place Assessment, in this case describing Tucson, Arizona. Your One-Page Place Assessment (a.k.a. One-Pager) fleshes out more of the context of the *place* in which your site is situated (not the site itself) by focusing on some of its broad parameters. I recommend you expand on the information in figure 2.7 with more specific on-*site* observation, such as what direction your site's slope(s) face, the conditions of your soils, where your site's cooler and hotter microclimates are, and how your site's wind, water, and wildlife patterns differ from the community's averages. The simple, concise summary of these contextual observations and statistical information in the One-Pager help you integrate the assessment, harvesting, and regeneration of water and other on-site and community resources. This assessment goes well beyond water, opening a door to a greater perception and understanding of where you are, what you can do, and what you should do. The goal is to know and work *with* Place in a way that continuously and sustainably enhances it and you. The One-Page Place Assessment should evolve as your observations, understanding, and roots deepen. The fourth section "Water" directly applies to this chapter, while the other sections apply more directly to chapter 4 (Integrated Design) and the appendices.

See the One-Page Place Assessments page at www.HarvestingRainwater.com/one-page-place-assessments/ for Place Assessments for a variety of communities, and for guidance on how to find the information you need to create one for your community.

Below are some tips on how to use your Place Assessment.

Watershed

What we do, and how we live in a watershed affects all others in that same watershed; and what they do, and how they live affects us. This awareness, along with an awareness of what watershed we live in, can help strengthen our understanding of, connections to, and stewardship of our shared land, water, community, ecologies, and ecosystems. See the blog "Watershed Maps Are Community Maps" on my website for strategies highlighting local watersheds.

Climate

The Climate section of the Place Assessment lists Average High and Low Temperatures for each month of the year, along with Record High and Low Temperatures to determine what plants can survive/thrive in your area based on their heat and cold tolerances. These temperatures also highlight opportunities to take advantage of, or to modify those temperatures within your site's microclimates using simple, low-cost, zero-power-consuming passive heating and cooling strategies.

For example, notice the over *20°F (10°C) difference* between temperatures in September (fall equinox) and March (spring equinox). In fall, Tucson's days are hot and nights are cool, while in spring, days are warm and nights are cold. Retractable awnings, shade cloth, or exterior blinds—when extended over equator-facing windows—can shade out that direct hot sun in summer and fall, while retracting that shade in winter and spring to utilize the sun's free heat and light as needed. See chapter 4—Integrated Design for more.

We can take advantage of the contrast between hot days and cool nights in late spring and early fall, and even summer, by opening screened windows and doors at night for free cooling, then closing them during the day to keep cool air inside. This reduces power consumption for mechanical heating and cooling, thereby reducing water consumption and pollution from power plants. See appendix 9—Water-Energy-Carbon Nexus for more on the water-energy connection.

Tucson winter temperatures can and do drop below freezing—as shown by the Record Low. Frost-intolerant tropical and semi-tropical plants will die unless seasonally protected and placed in warmer microclimates, for example, beneath protective plant canopies open to low angle winter sunlight. In hot months, these tender plants—plus those from higher/cooler elevations and latitudes—often require protection/shade from the midday and afternoon sun.

ONE-PAGE PLACE ASSESSMENT: TUCSON, ARIZONA
LOCATED IN THE SANTA CRUZ RIVER SUBWATERSHED WITHIN THE COLORADO RIVER WATERSHED

CLIMATE

AVERAGE HIGH & LOW TEMPERATURES[1] — 1894 – 2008

	JAN	FEB	MAR	APR	MAY	JUN	JUL	AUG	SEP	OCT	NOV	DEC	ANNUAL
°F HIGH	65.5	68.9	74.3	82.2	90.8	99.8	100.1	97.9	95.2	85.9	74.3	66.2	83.4
°F LOW	37.6	40.2	44.1	49.9	57.6	66.8	73.9	72.4	66.9	54.9	43.8	38.1	53.9
°C HIGH	18.6	20.5	23.5	27.9	32.7	37.7	37.8	36.6	35.1	29.9	23.5	19.0	28.6
°C LOW	3.1	4.6	6.7	9.9	14.2	19.3	23.3	22.4	19.4	12.7	6.6	3.4	12.2

RECORD HIGH[2]	118° F	47.8° C	June 27, 1990	RECORD LOW[1]	6° F	-14.4° C	January 7, 1913

SUN

	MAR 21	JUN 21	SEP 21	DEC 21
DEGREES N or S of DUE EAST THE SUN RISES[3]	0°	29°N	0°	27°S
DEGREES N or S of DUE WEST THE SUN SETS[3]	0°	29°N	0°	27°S
SOLAR-NOON ALTITUDE ANGLE (ABOVE HORIZON)[a,3,4]	58°	81°	58°	34°

LATITUDE 32.2°

ELEVATION 2,555 FT / 779 m

SOLAR-NOON WINTER-SOLSTICE SHADOW RATIO[b]	1 : 1.46	...AND AZIMUTH[c]	0°
9AM & 3PM WINTER-SOLSTICE SHADOW RATIO[b,3]	1 : 2.79	...AND AZIMUTH[c,3]	44°

WIND

MAX SPEED[6] 81 130 MPH kmph

PREVAILING WIND DIRECTION (FROM WHERE)[5] & AVERAGE SPEED[6]

	JAN	FEB	MAR	APR	MAY	JUN	JUL	AUG	SEP	OCT	NOV	DEC	
	ESE	ESE	WSW	WSW	WSW	WSW	SE	ESE	ESE	ESE	ENE	ESE	ANNUAL
MPH	5	5	6	6	6	6	6	5	5	5	5	5	5.4
kmph	8	8	10	10	10	10	10	8	8	8	8	8	8.7

WATER

AVERAGE RAINFALL (GAIN)[1] — 1894 – 2008

	JAN	FEB	MAR	APR	MAY	JUN	JUL	AUG	SEP	OCT	NOV	DEC	ANNUAL
INCHES	0.88	0.83	0.75	0.39	0.18	0.26	2.06	2.15	1.15	0.74	0.77	0.96	11.12
mm	22.4	21.1	19.1	9.9	4.6	6.6	52.3	54.6	29.2	18.8	19.6	24.4	282.4

AVERAGE PAN EVAPORATION (POTENTIAL LOSS)[d,7] — 1894 – 2005

	JAN	FEB	MAR	APR	MAY	JUN	JUL	AUG	SEP	OCT	NOV	DEC	ANNUAL
INCHES	3.25	4.57	6.95	9.88	12.87	14.91	13.17	11.65	10.35	7.81	4.73	3.37	103.51
mm	82.6	116.1	176.5	251.0	326.9	378.7	334.5	295.9	262.9	198.4	120.1	85.6	2,629.2

WETTEST YEAR'S RAIN[2]	26.2 INCHES	666 mm	1983	DRIEST YEAR'S RAIN[1]	5.1 INCHES	129 mm	1924

LONGEST PERIOD WITH NO MEASURABLE PRECIPITATION[8]   RAINFALL INCOME[e] 231 GPCD / 874 lpcd

155 DAYS: DECEMBER 26, 1971 – MAY 29, 1972

AREA[f,9] 226.7 SQ MILES / 587 km²   POPULATION[f,9] 520,116 / 2010   UTILITY-WATER USE[10] 112 GPCD / 424 lpcd

HISTORICAL 30 FT 9.1 m 1946   DEPTH TO GROUNDWATER[8,11] 120 FT 36.6 m 2012 CURRENT

CURRENT GROUNDWATER EXTRACTION > NATURAL GROUNDWATER RECHARGE[h,i,12,13]

WATERGY

% MUNICIPAL ENERGY CONSUMPTION USED TO MOVE & TREAT WATER[14] 44%

TOTEM SPECIES

PLANT: Tumamoc Globeberry[15]   MAMMAL: Mexican Long-Tongued Bat[15]

FISH: Sonora Suckerfish[15]   BIRD: Rufous-Winged Sparrow[15]   REPTILE: Mexican Garter Snake[15]

AMPHIBIAN: Lowland Leopard Frog[15]   MEGAFAUNA: Mexican Gray Wolf,[16] Jaguar,[17] Grizzly Bear (Catalina-Rincon Mtns)[18]

Available online at HarvestingRainwater.com/one-page-place-assessments

Fig. 2.7. "A One-Page Place Assessment" for Tucson, Arizona.
For a downloadable color version (including assessments for many other communities)
see the One-Page Place Assessment page at www.HarvestingRainwater.com. Notes on next page

TUCSON PLACE-ASSESSMENT NOTES

a. Altitude angle (a.k.a., elevation angle) refers to the number of degrees the sun is located above the horizon at the given time and date.

b. The solar-noon winter-solstice shadow ratio is the object's height : length of object's shadow cast on December 21 at noon (the longest noontime shadow of the year). The ratio is 1 : x, where x = 1 ÷ tangent (90 - (latitude + 23.44)).

c. Azimuth is the angle formed between a reference direction (here, due south) to the point on the horizon directly below the sun.

d. Pan evaporation daily measures the depth of water evaporated from a pan of water to determine evaporation rates at the given location.

e. Average annual rain falling on Tucson divided by Tucson's population, then divided again by 365 days per year.

f. City proper.

g. Depths to groundwater vary widely in Tucson-Basin wells. This Tucson Water well (Local ID D-14-13 13CBC) is very close to downtown Tucson, the Santa Cruz River, and their histories. It is within 600 yards (547 m) of a hand-dug well on South Main Street (near El Tiradito or the Wishing Shrine) from which in the 1870s Adam Sanders and Joseph Phy obtained water to sell at 5¢ per bucket. According to *The Lessening Stream: An Environmental History of the Santa Cruz*, by Michael F. Logan (University of Arizona Press, 2002), "The two entrepreneurs filled an iron tank on a wagon from their well and traveled daily through town selling water. Within 25 years municipal water use in Tucson would progress from well water sold by the bucket, to a piped supply tapping the aquifer. When the mains were first opened in September 1882, an almost immediate decline in the water table downstream resulted."

h. Due to rapidly depleting groundwater tables and associated surface water in areas of Arizona with a heavy reliance on mined groundwater, the 1980 Groundwater Management Code identified and designated five such areas as Active Management Areas (AMAs), and mandated that they attain safe-yield, on an AMA-wide basis, by the year 2025. Safe yield, according to the 2010 Arizona Department of Water Resources DRAFT Demand and Supply Assessment of the Tucson Active Management Area, "is a balance between the amount of groundwater pumped from the AMA annually, and the amount of water naturally or artificially recharged. Groundwater withdrawals in excess of natural and artificial recharge leads to an overdraft of the groundwater." All projections from the Assessment predict the Tucson AMA will not attain safe-yield by 2025. None of the projections presented in the Assessment consider the potential benefit of wide promotion and adoption of on-site harvest of on-site waters, coupled with integrated conservation advocated by this book.

i. Groundwater levels continue to drop, but are rising in some parts of the Tucson Active Management Area (AMA) due to reduced groundwater pumping in those areas where purchased CAP water (Colorado River water imported 300+ miles (483+ km) via the Central Arizona Project canal and its pumping stations) is replacing groundwater use, or artificially recharging groundwater. See appendix 9 to compare costs of our water and energy options.

CREDITS: Brad Lancaster, Resource concept, research, content oversight | **Megan Hartman**, Research, resource creation

Sun

To harvest the benefits of sun and shadow, you must understand the annually changing flow of sunlight and shadow where you live, so the "One-Pager" records the location of the sun when it Rises, Sets, and at Noon at different times of year. Use this information to place objects such as trees and above-ground cisterns so they cast cooling shade where you want and need it in hot seasons, without blocking the free heat and light of the winter sun where needed.

To further aid you in that aim, the Winter-Solstice Shadow Ratio tells you the distance an object's shadow will be cast on the winter solstice—the day with the longest shadows of the year, so you won't shade winter-sun-facing windows in winter, or solar panels and water heaters. For example if the ratio is 1:1.46, an object will cast a shadow 1.46 times longer than the object's height—a 10-foot tall tree will cast a 14.6-foot long shadow (see the Maintaining Winter Sun Access section of chapter 4 for illustrations, more explanation, and ratios for other latitudes).

The Azimuth indicates the angle of the sun's location east or west of True South on a given date and time, which helps you figure out the direction shadows will fall. For example, at 3 P.M. on the winter solstice in Tucson, the sun is 44° west of True South, so shadows will be cast 44° east of True North on that date and time (again, see the Maintaining Winter Sun Exposure section of chapter 4 for illustrations and more explanation).

Latitude and Elevation show where your site fits into the global-scale atmospheric circulatory patterns and sun exposure patterns that strongly affect a site's climate. The higher the elevation, the greater the solar radiation, but the cooler the temperature. On average, temperatures drop 3.5° F or 1.94° C per 1,000 feet or 304 m of elevation.[6]

Coupled with the Water and Climate information, knowing your Latitude and Elevation can help you create analog climate assessments. As Dave Boehnlein of Bullock's Permaculture Homestead (www.PermaculturePortal.com) describes, the process entails:

> ... creating a profile for the climate for which we are creating a design. This is based on precipitation, temperatures, seasonality, etc. Then we try to find other places around the world with profiles that match very closely: analog climates. Once we've found these different places we can start to analyze them in terms of native vegetation, agricultural products, medicines, architectural styles, cuisine, and a variety of other factors (especially with regard to traditional or indigenous peoples). As these people usually live in ways that are highly responsive to their environment (e.g., they harvest, rather than drain their local water; and design smart buildings that passively cool themselves instead of poor buildings that require mechanical air conditioning), we can look at what they grow and how they live as a source of inspiration for what types of design features we might use to best approach sustainability.[7]

Wind

By knowing the monthly variations in Prevailing Wind Direction—the direction winds come from— and the Average and Maximum Speed of winds, we can correctly place and design windbreaks, windows, roads, buildings and other features to harvest, divert, or deflect the winds and whatever they carry.

Windbreaks and wind diversions, along with landforms, can concentrate, speed up, and enhance the harvest of wind for pumping or power.

Simple wind-harvesting techniques can slow the speed so wind-borne snow drops within shelterbelts that are irrigated by that snow, while keeping roads and driveways free of snow drifts. Light summer breezes can be diverted through windows for free cooling. Windbreaks can deflect cold winter winds around buildings and can sometimes deflect fires fueled by the wind. Windbreaks also reduce evaporation and evapotranspiration loss from protected plantings, soils, and water bodies and can filter dust and reduce the speed of wind blowing on human settlements. See appendix 8—Wind Harvesting for more.

More complex forms of wind harvesting can pump water or generate power. Windmills need a minimum wind speed of 3-4 mph or 5 to 6.5 km/h to pump water.[8] Wind turbines need an average wind speed of between 10 to 14 mph or 16 to 22 kmph to generate power—the higher the better for higher electricity production. [9]

Water

Average Rainfall is provided by month so you know when to expect rain, and how much you will typically receive over the long term. This helps you plan the size of roofs, hardscapes, and other catchment areas that harvest water in tanks or concentrate water in the soil. It also helps you size the tanks and/or raingardens to hold that rainwater.

Pan evaporation is a measurement of the loss of water from an open water body or pan of water. The potential for evaporation reflects the combined effects of temperature, humidity, rainfall, drought dispersion, solar radiation, and wind on standing water and bare soil. It also indicates the level of stress plants are under, which lose water through the process of evapotranspiration from their leaves. This information can be used to help determine plant water demand, whether and how much you may need to irrigate those plants, and how to reduce those irrigation needs.

If annual pan evaporation rates exceed rainfall rates you are in a dryland environment. If pan evaporation rates exceed rainfall only during certain times of year, you may just experience dryland seasons. In either case, evaporation-reducing strategies such as mulch, windbreaks, shading, and covered water storage areas are very important.

Rainfall can vary tremendously from year-to-year and place-to-place, particularly in drylands environments. The Wettest Year's Rainfall helps you plan for wet extremes so you have enough capacity in soil, tanks, and overflow routes to turn potential floods into harvests. The Driest Year's Rainfall along with the Longest Period with No Measurable Precipitation enables you to plan ahead for storage capacity, backup water supplies, more efficient water cycling and reuse, and other mitigation measures to get you through extreme years. If you are living solely on rainwater, you want enough water in your tanks on day one of that drought to last you through it.

What's the potential of rainwater harvesting in your community? Comparing Rainfall Income to Utility-Water Use in gallons per capita/person per day (GPCD) and in liters per capita/person per day (lpcd) helps you see just that. Note how Rainfall on Tucson's surface area exceeds that of the community's Utility-Water Use. This is the case for most communities.

Historical and Current Depths to Groundwater, and information on whether current Groundwater Extraction *Exceeds* or *Is Less Than* Natural Groundwater Recharge are indicators of the declining or improving health of the community's hydrologic systems. Rainwater, greywater, and stormwater harvesting can help improve the volume and quality of our groundwater.

Other measurements you may want to research and add to your own Assessment are historical and current creek and river flows and various water quality parameters. Start recording now, as these are benchmarks that you can re-measure in the future to determine progress or decline.

For example, total dissolved solids (TDS) or mineral content, is an important water quality parameter in a community's water supply. The TDS of the groundwater delivered by Tucson's water utility from 1978-1990 was 280 mg/L—much of it salt that can harm plants and soil. By 2012, TDS concentrations had increased to 460 mg/L due in large part to the post-1990 importation of Colorado River water to Tucson through the Central Arizona Project (CAP) canal, which provides water having a TDS of 605 mg/L.[10] Coal-burning power stations provide the energy that pump this water uphill to Tucson, so CAP water not only degrades the quality of our water supply, but also pollutes our air (and thus our rainwater), as the costly importation of that river water attempts to offset the community's over pumping of groundwater.

We Tucsonans, like millions or others across the globe, must use a wide array of strategies—such as those described in this book—to better steward our precious water supplies. We must make the best possible use of our local salt-free rainwater, reduce excessive water loss to evaporation, recharge surplus stormwater, raise groundwater levels as we reduce groundwater extraction without depleting the quality and quantity of water available to others, and by doing so—extend the duration of surface water flows. We must also reduce our demand for water-consumptive energy supplies and energy-consumptive low-quality imported waters—the watergy connection.

Watergy

Watergy is the relationship between pumped and treated water and the resulting energy consumption. The City of Tucson expends 44% of its energy consumption moving and treating water. Such energy consumption typically leads to more water consumption and release of water-polluting, air-polluting, and climate-altering emissions from the power plant. See chapter 5 and appendix 9—Water-Energy-Carbon Nexus to learn ways this can be remedied on individual and community scales.

Totem Species

A Totem Species is a threatened indigenous plant, mammal, insect, bird, fish, amphibian, or reptile whose role in your community is much like that of a canary in a coal mine. The decrease or increase of its numbers reflects the declining or improving health of a culture, a watershed community, and the ecosystem it is part of. The change in Totem Species numbers alerts us to our need to help regenerate our mutual ecological habitat and community—or provides evidence of success in this regeneration. In this way, we save and enhance the diversity of life that we all depend on and enjoy. All species listed for Tucson were formerly more common along the water-filtering, water-infiltrating riparian habitats and wetlands of our disappearing perennial and ephemeral water flows. However, Tumamoc globeberry is starting to make a comeback where it has been planted under native food-producing shade trees in street-side stormwater-harvesting basins.

USING THE ONE-PAGE PLACE ASSESSMENT IN SITE DESIGN

The Climate, Sun, Wind, Water, Watergy, and Totem Species described in the Place Assessment interact continuously in your environment to create the physically and energetically complex and synergistic conditions you live within. Understanding the interaction between these elements and others is key to creating a sustainable, or better yet, regeneratively integrated site. Chapter 4 provides more detail on the One-Page Place Assessment elements and their relationships, chapter 5 provides an example of a site capitalizing on these relationships, while appendix 5 provides notes, questions, and worksheets to help you dig deeper into the Place Assessment and relate it back to your site design. Proceed to appendix 5 when you are ready to develop and utilize a Place Assessment for your site.

REAL LIFE EXAMPLE

LIVING WITHIN A DRYLAND HOME'S RAINWATER BUDGET

Matthew Nelson and Mary Sarvak come very close to living within their rainwater budget. You can see much of their annual rainwater supply in the middle of their living room. A 2,500-gallon (9,450-liter) ferrocement cistern rises three feet from a hole in the floor (fig. 2.8). It looks like a *tinaja*—a desert water hole carved into bedrock. Peering into the topless cistern you come face to face with the rainwater that has drained in from the downspout in the middle of their 1,500-square-foot (139.4-m²) roof. That rainwater supply meets all their domestic needs except drinking water. As they use their water, they see its level drop. "You get more conservative as you see the

Fig. 2.8. Matt and Mary beside their indoor cistern

water level fall," says Matt. Though when the water level is high in the middle of the rainy season, they feel free to indulge a little. Talking about their harvested rainwater Mary says, "During the intense heat of July and August, we can walk in the door of our home and splash cool water all over ourselves. Or we can just jump in!"

The system was designed to supply the average water needs of a family of three for four months between rains typical for the site. Their rainfall averages 14 inches (356 mm) a year. A 10-gallon RV pump pressurizes the water and sends it throughout the house to all the sinks, the bathroom, and the washing machine.

Outside, 100% of the landscape's water needs are met by rainwater. When the home was built, extreme care was taken to avoid disturbing or destroying any of the existing native vegetation beyond the footprint of the house. As a result, nothing was spent on landscaping—it already existed. In addition, 98% of the landscape has never needed supplemental irrigation because it was already well established and perfectly adapted to natural rainfall. The 2% of the landscape needing irrigation includes an apple tree planted after the home was built and some potted plants. This irrigation is done entirely with recycled rainwater in the form of greywater from the drains of their washing machine and kitchen sink.

Matt and Mary's rainwater oasis in the Sierrita Mountains of southwest Arizona is located in a boondoggle development from the 1970s that sold home sites to folks sight-unseen. No utilities were installed and few sites were ever developed. At sites that were eventually developed, neighbors run generators around the clock for electricity, and nearly all truck in their water, ignoring rain that falls freely from the sky. Matt and Mary generate most of their electricity from solar panels on their roof, and aside from drinking water they have had to haul in water only once, during a five-month drought.

Designing around the assessed on-site resources of rain and sun is what has enabled Matt and Mary to live with modern comforts while consuming and paying for just a fraction of the off-site resources their neighbors consume. Matt and Mary are also directly connected with their beautiful desert surroundings, the source of their resources, and their need to keep consumption in balance. These are the ideas I encourage others to mimic, and why I feature Matt and Mary here. However, two of the specific techniques used on their site should be altered, rather than duplicated. First, a closed exterior tank would be superior to the open living room tank. That is because, while the open water beneficially cools the home in summer, it also cools it in winter as water pours in from frigid downpours (window and door screens along with

ten mosquito-eating fish take care of the potential mosquito problem associated with the open water). Second, Matt and Mary are concerned that the elastomeric paint on their roof may taint their water. This is the reason they haul drinking water in from town. I recently told them about various elastomeric paints and roof coatings approved for use with potable rainwater catchment systems (see the Roofing section in the Materials and Suppliers section of the Rainwater Harvesting menu tab at www.HarvestingRainwater. com). With the application of one of these coatings (and cooling their bodies in the rainwater-fed shower, rather than plunging into the cistern), they'll be worry-free and rainwater-rich (but winters will still be cold).

Strive to live in balance with your site as Matt and Mary do by first assessing your resources and needs—the first principle of water harvesting: *Long and thoughtful observation*. And as you plan what strategies and techniques you use to harvest the site's resources persistently think of how the system will work as a whole—the last principle of water harvesting: *Continually reassess your system: the "feedback loop."* The better your assessment the better your system will be, and the following chapters give you more tools to do just that.

Overview: Harvesting Water with Earthworks, Tanks, or Both

Humankind—despite its artistic pretensions, its sophistication, and its many accomplishments—owes its existence to a six-inch (fifteen-cm) layer of topsoil and the fact that it rains.

—Author Unknown

Once you've estimated your on-site water resources and needs (from the previous chapter), the next step is to answer the following question: How do you plan to use your water resources? How you answer that question will largely determine the best strategy to harvest that water— be it water-harvesting earthworks, tanks, or both. This chapter compares these strategies, and briefly describes some of their more detailed techniques and applications to help you visualize and decide which are most appropriate for your unique site. (*These strategies are covered in detail in volumes 2 and 3.*) Basic recommendations are then given for harvesting water within a residential landscape. The chapter wraps up with a homesteading couple harvesting water in earthworks and tanks. The next chapter (Integrated Design) then shows you how to integrate these strategies with other aspects of your site so your water-harvesting efforts can capitalize on additional free resources such as the sun's light, power, and winter heat; provide privacy, and summer shelter and cooling; filter out noise, light, and air pollution; while also controlling flooding and erosion. In other words, how to maximize your site's potential!

HOW DO YOU PLAN TO USE YOUR HARVESTED WATER?

LANDSCAPE OR GARDEN USE

If you plan to use your harvested water for landscape or garden use, begin harvesting water in the soil using earthworks. Water-harvesting earthworks, sometimes referred to as *rain gardens*, *green infrastructure*, and/or *low-impact development (LID)* strategies, are simple structures and strategies that change the topography and surface of the soil to collect water and amplify its quick infiltration—in essence, "planting" the water. They increase organic matter, life, and fertility atop and within the soil, further augmenting soil moisture and nutrient availability. Earthworks maximize the density and diversity of "living pumps" of vegetation that utilize water to generate more resources (food, shade, windbreak, wildlife habitat, etc.), multiplying the gains of water planting and nutrient cultivation. Once vegetation is established, landscapes that harvest water in the soil and are planted with low-water-use native plants not only subsist but also thrive on rainfall alone without supplemental irrigation from tank or tap. However, tanks give you the option of applying supplementary irrigation in dry times—especially if a vegetable garden or less hardy non-native vegetation is planted.

POTABLE USE AND WASHING

If you plan to use your harvested water for potable use and washing, begin harvesting water in *tanks*. But do not forget the soil, and continue to direct overflow from the tank, greywater from your house, and the runoff from the general landscape into water-harvesting earthworks. While a cistern can be installed to irrigate a garden or supplement a landscape in dry times, the more water you can effectively harvest and hold in the soil, the less supplemental cistern irrigation will be needed.

MATCH THE QUALITY OF WATER BEING HARVESTED TO THE STRATEGY USED TO HARVEST IT

Your site's highest quality rainwater, typically run-off from clean roof materials such as metal, slate, tile, or elastomeric paints approved for rainwater collection systems, is the most appropriate for storage within cisterns, domestic consumption, or use on vegetable gardens. Stormwater from dirtier surfaces such as earthen slopes, streets, or sidewalks should be directed to trees and shrubs within passive water-harvesting earthworks. Household greywater should be directed to and utilized within mulched basins planted with trees and shrubs. Do not store greywater in tanks.

Keep in mind the question "How do you plan to use your harvested water?" as you read the comparisons of these water-harvesting approaches in box 3.1.

RECOMMENDATIONS FOR HARVESTING WATER FOR A LANDSCAPE AROUND A HOME

I recommend that all my clients *begin* with water-harvesting *earthworks*, which need cost no more than the price of a shovel if you do the work. I stress the importance of placing earthworks and the plants they support in areas where they will provide beneficial functions, such as the east and west sides of a home for passive cooling with shade trees. I stress using a low-water-use native plant palette of indigenous vegetation found within a 25-mile (40-km) radius of the site and 500 feet (150 m) above or below the site's elevation.

Some sites may require defining *native* with a larger radius to bring in more diversity, but start with the small radius to ensure you do not overlook superior local species. Such plants are typically the most beneficial for native wildlife, and create a Sense of Place rooted to our local bioregion. These plants are adapted to local growing conditions and can be cut off from supplemental irrigation once the vegetation is well established, or has grown to a desired height. Such native plant landscapes are beautiful and low maintenance.

I recommend *cisterns to complement* the foundation of water-harvesting earthworks laid throughout the landscape or garden. The cisterns stretch the availability of the rain long into the dry season—especially for more water-needy landscapes and vegetable gardens. The earthworks utilize the overflow water from the tanks, and make all the water used within the landscape or garden go further.

Fruit trees or other water-needy vegetation, if used, should be placed close to the house to create an oasis-effect around the home. The plants are then supported by the rainwater from the roof; greywater from the sinks, showers, tubs, and washing machine; and the care of those living inside the house. *The goal is to select and plant the vegetation in such a way that rainwater will be the landscape's primary source of water, greywater its secondary source, and imported water (municipal or well water) will only be used as a supplementary source in times of need.* Sprawling water-needy vegetation placed far from the house makes achieving this goal far more difficult. The plants will be out of sight and out of mind, neglected and underutilized by you, and potentially too distant for easy irrigation from roof runoff or household greywater. Instead, hardier native vegetation goes to the periphery of the property, and gets planted within water-harvesting earthworks. This hardy vegetation can shelter the less hardy plants, such as the fruit trees, from excessive sun or wind. Hardy native plantings placed along the periphery support more native wildlife, which appreciates less interference from us. (See figure 3.1.) And of course, a low-water-use native landscape placed in water-harvesting earthworks is easy and inexpensive to establish and maintain. (See appendix 4 for an example plant list for Tucson, Arizona, and also volume 2, the chapter on vegetation, for more planting tips.)

Box 3.1. Comparing Earthworks and Tanks

CHARACTERISTICS	EARTHWORKS	TANKS
Water uses	Provides large quantities of high quality rainwater to garden and landscape	Provides water for drinking, washing, fire control, flushing toilets (but first consider water-less composting toilets), and supplemental irrigation. Water quality will vary with catchment surface, tank construction, screening, maintenance, and first flush system. Rainwater has very low hardness.
Water collection areas	Can collect water from roofs, streets, vegetation, bare dirt, greywater drains, air conditioner condensate, etc.	Need a relatively clean collection surface (typically a metal, tiled, or slate roof) located higher than the tank
Water storage capacity	Very large potential to store water in the soil	Storage capacity limited by the size of the tank
Cost	Inexpensive to construct and maintain. Can build with hand tools, though earth-moving equipment can speed up the process	Much more expensive than earthworks to construct and maintain. Cost varies with size, construction material, above- or below-ground placement, self-built or prefab, etc.
Location	Do not locate within 10 feet (3 m) of wall or building foundation. May be difficult to use in very small yards with adjacent large roofs	Can locate within 10 feet (3 m) of wall or building foundation, but you must be able to walk around entire above-ground tank to check for, and repair, leaks. Tanks increase water-storage potential in very small yards.
Time period water is available	Water available for limited periods after rainfall depending on soil type, mulch, climate, and plant uptake	Water is available for extended periods after rainfall.
Maintenance	Earthworks work passively; require some maintenance after large rainfalls	Maintenance required; must turn valve to access water and may need pump to deliver water
Erosion control	Very effective for erosion control	Can assist with erosion control
Greywater collection	Very effective at harvesting greywater from household drains	Not appropriate to harvest greywater in tanks due to water-quality issues. Never store greywater in a rainwater tank.
Water quality impacts to environment	Pollutants in greywater and street runoff intercepted in the soil stay out of regional waterways	Less impact than earthworks to the broad environment
Impacts on urban infrastructure and flooding	Can capture large volumes of water, reducing need for municipal water, stormwater drains, stormwater treatment, and decreasing flooding	Can capture low to moderate volumes of water, reducing demand for municipal water, stormwater drains, stormwater treatment, and decreasing flooding
Groundwater recharge	Can sometimes directly recharge shallow groundwater tables. However, use of rainwater and greywater in earthworks instead of municipal/well water for irrigation reduces groundwater depletion.	Not an efficient use of tank water. However, use of cistern water instead of municipal/well water reduces groundwater depletion.

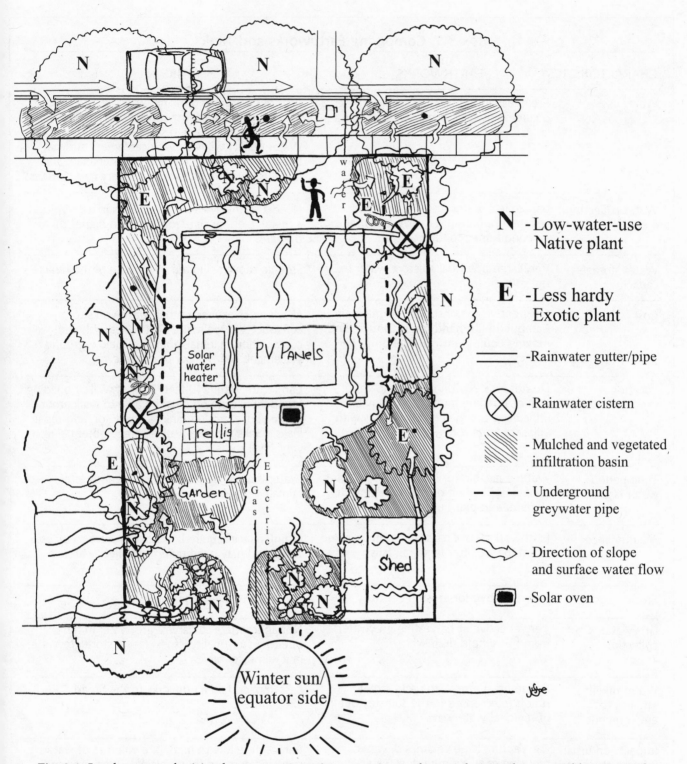

N – Low-water-use Native plant

E – Less hardy Exotic plant

──── – Rainwater gutter/pipe

⊗ – Rainwater cistern

▨ – Mulched and vegetated infiltration basin

– – – – Underground greywater pipe

〰➤ – Direction of slope and surface water flow

▪ – Solar oven

Fig. 3.1. Landscape emphasizing low-water-use native vegetation at the periphery, in less accessible areas, and in areas of less water. Needier exotics are placed close to the home where roof runoff, household greywater, and attention are easy to access. However, a lower maintenance landscape could consist entirely of low-water-use natives. Utilize multi-use, food-producing native, or climate-appropriate vegetation to increase on-site production of resources.

SAMPLING OF STRATEGIES/ TECHNIQUES

Now that you have some idea of how you want to harvest your rainwater—in the soil, tanks, or both, you are ready for specifics! In volume 2 of *Rainwater Harvesting for Drylands and Beyond*, the chapters describe in detail how to harvest rainwater in the soil using various earthwork techniques. The cistern section later in this chapter addresses harvesting rainwater in tanks. The illustrations in this chapter are for you, the reader, to conceptualize earthworks and water storage as you formulate an integrated water-harvesting plan for your site— and to whet your appetite for more in volumes 2 and 3 plus the resources on my website www. HarvestingRainwater.com.

OVERVIEW OF WATER-HARVESTING EARTHWORKS

The techniques outlined below are discussed in detail in volume 2. Additionally, appendix 1 in this volume has more illustrations and tips on how to place and use these structures so the structures work with—rather than against—the natural flows of water and sediment.

Berm 'n basin

A berm 'n basin is a water-harvesting earthwork laid perpendicular to land slope, designed to intercept rainwater running down the slope and infiltrate this water in a localized area, usually the soil and root zone of existing or planned vegetation. It usually consists of two parts: an excavated basin and a raised berm located just downslope of the basin. The berm can be made of earth excavated to form the basin or it can be made from brush, rock, or additional earth. The basin holds water; adding the berm enables you to harvest even more water. Use this strategy on sloped land up to a 20:1, 2.9°, or 5% grade.[1] Size for the maximum stormwater event. This strategy is not appropriate in drainage ways. (See figures 3.2, 3.3, and 3.4.)

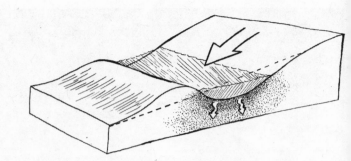

Fig. 3.2. A berm 'n basin holding runoff water and infiltrating it into the soil

Fig. 3.3. A contour berm doubling as a raised path

Fig. 3.4. A series of boomerang berms overflowing one into the other

Fig. 3.5. Terraces stabilized with dry-stacked (mortar-less) urbanite (salvaged pieces of broken concrete sidewalks) by La Loma Development Company in Pasadena, California. This job alone pulled over one million pounds (500,000 kg) of concrete out of the dump and turned it into a beautiful, water-harvesting resource. La Loma's combined urbanite projects from 2005 to 2012 have removed a total of 16 million pounds (7 million kg) or over 4,000 cubic yards (3,000 m³) of concrete from the waste stream.[2] For color photo of another view of this installation see inside back cover.

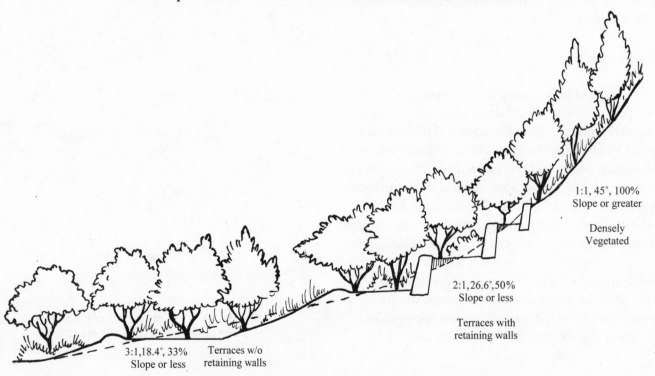

1:1, 45°, 100%
Slope or greater

Densely
Vegetated

2:1, 26.6°, 50%
Slope or less

Terraces with
retaining walls

3:1, 18.4°, 33%
Slope or less

Terraces w/o
retaining walls

Fig. 3.6. Different terracing strategies for different grades of slope

Terrace

A terrace, sometimes called a *bench*, is a relatively flat "shelf" of soil built parallel to the contour of a slope. Unlike berm 'n basins, terraces do not have depressions built into them, so in order to retain rainwater and accumulate organic material, they need to be bordered by berms or low walls. Terraces can be used for gardens, orchards, and other plantings in drylands. Well-built terraces also help control erosion on slopes. Use on sloping land up to a 2:1, 26.6°, or 50% grade. Terraces on slopes exceeding a 3:1 grade likely need a retaining wall for stabilization. Size for the maximum storm event. Terraces are not appropriate in soils prone to waterlogging where rainwater infiltration could lead to saturated subsurface conditions due to the presence of layers—such as clay—that impede the movement of water further down through the soil. (See figures 3.5 and 3.6.)

French drain

A French drain, or dry well, intercepts rainwater into a trench or basin filled with porous materials including gravel, pumice stone, or rough organic matter. These materials have ample air spaces between them that allow water to infiltrate quickly into the drain and percolate into the root zone of the surrounding soil, while creating a stable surface you can walk on. A French drain can be constructed with or without perforated pipe placed horizontally within the porous material that fills the drain. Use on flat to gently sloped land; drains are appropriate where deep subsurface irrigation of landscapes is needed. It is important that the French drain only harvest runoff water that is relatively sediment free, such as from a roof gutter downspout, or edge of paved patio, in order to prevent premature silting up of the structure. French drains are not appropriate in drainage ways, areas with sediment-laden water, or beneath or across roadways. (See figures 3.7 and 3.8.)

Infiltration basin

An infiltration basin or rain garden is a landscaped level-bottomed, relatively shallow depression dug into the earth that intercepts and infiltrates rainfall,

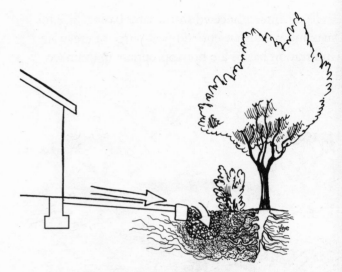

Fig. 3.7. French drain infiltrating intercepted runoff from a roof and patio

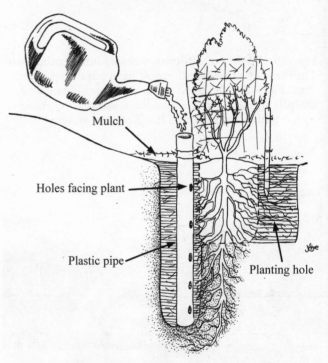

Mulch

Holes facing plant

Plastic pipe

Planting hole

Fig. 3.8. Deep pipe variation of a French drain, in which a plastic pipe directs water deeper into soil, reducing weed growth, and encouraging deeper, more dought-hardy root growth. This can increase plant survival/establishment in basins by up to 65%. After establishment, pull and reuse pipe.

runoff, and/or greywater in the planting basin it creates. This technique works best on flat landscapes where it will have no berm, so all surrounding runoff can drain into it. It can also work on moderate slopes as a terraced basin, water-harvesting tree wells, or interconnected stormwater basins. Size for a maximum storm event and peak surge of greywater. Infiltration basins are not appropriate in drainage ways, areas of shallow groundwater where they might result in standing water, or over septic drain fields. (See figures 3.9 through 3.12.)

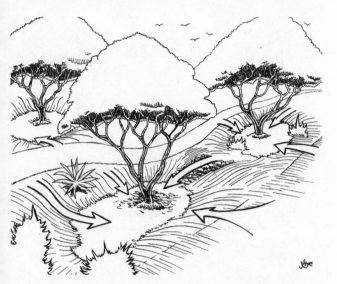

Fig. 3.9. A series of infiltration basins intercepting and infiltrating rainfall and runoff from adjoining street and footpath. Designed right, these basins can act as the sole irrigation system for the associated vegetation, while doubling as a flood control system.

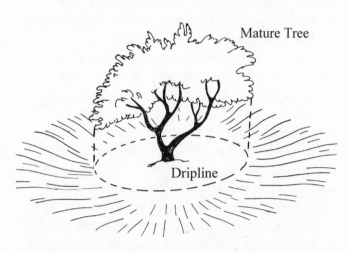

Fig. 3.10. Ideally, the infiltration basin diameter is at least 1.5 times (and up to 3 times) the diameter of the associated *mature* plant's canopy dripline, since roots spread, and *most* water used by plants is drawn from the root zone *outside* the canopy drip line.[3] However, even an undersized basin is better than no basin. Multiple, smaller basins can work just as well as one large basin.

Fig. 3.11A. *Without* sponge. Rainfall collected in newly constructed infiltration basins minutes after a large summer storm. The basins have not yet been mulched or planted. Milagro Cohousing, Tucson, Arizona. Credit: Natalie Hill. For color photo see inside front cover.

Fig. 3.11B. *With* sponge. Same basins mulched and vegetated. Basins are designed to infiltrate water quickly so there are no problems with mosquitoes or anaerobic soils. These basins, with their spongy mulch and soil-burrowing plant roots, infiltrate all water within 20 minutes. For color photo see inside front cover.

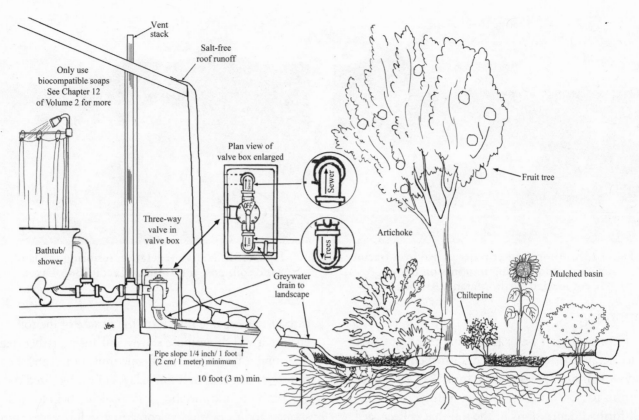

Vent stack

Salt-free roof runoff

Only use biocompatible soaps
See Chapter 12 of Volume 2 for more

Plan view of valve box enlarged

Sewer

OFF

Trees

Fruit tree

Three-way valve in valve box

Bathtub/ shower

Greywater drain to landscape

Artichoke

Chiltepine

Mulched basin

Pipe slope 1/4 inch/ 1 foot (2 cm/ 1 meter) minimum

10 foot (3 m) min.

Fig. 3.12A. Roof runoff and bathtub/shower greywater directed to a well-mulched and vegetated infiltration basin. Note P-trap and vent stack between interior drain and exterior greywater outlet, which prevents potential odor and insect entry into house. A three-way valve (downstream of the P-trap and vent) in a valve box allows for distribution of greywater to either the landscape or sewer (compare to fig. 3.12B). End of greywater pipe discharges a few inches (7.5 cm) above the mulch in the basin to prevent roots growing into pipe and solids from backing up and clogging pipe. Greywater immediately infiltrates beneath the surface of the mulch to be used by plants.

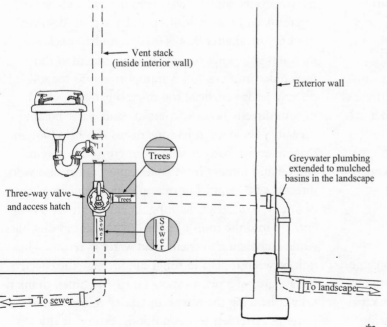

Vent stack (inside interior wall)

Exterior wall

Trees

Greywater plumbing extended to mulched basins in the landscape

Three-way valve and access hatch

Trees

Sewer

Sewer

To sewer

To landscape

Fig. 3.12B. A three-way valve enables you to direct greywater to the landscape or to the sewer, depending on what goes down the drain, and on whether soil can use the water or is too saturated or frozen for water to infiltrate. The three-way valve (with flow destinations marked) should ideally be accessible *in the same room* where the greywater is generated to maximize its convenience, ease, and likelihood of use. However, the three-way valve can still function outside the room (fig. 3.12A); and retrofitting existing plumbing often necessitates this. Optimally, the three-way valve, and its greywater stub out plumbing (directing greywater outside the building), are installed at the time of construction.

Note: the greywater line should exit the building at as high an elevation as possible (even above grade) to allow for the maximum number and longest distance of low-cost, zero-energy-consuming, gravity-fed distribution options within the landscape. See the Greywater chapter in *volume 2* for more.

Fig. 3.13A. An imprinter roller pulled by a tractor creating and seeding imprints on barren, compacted earth, Marana, Arizona

Fig. 3.13B. Native vegetation restoring the land and soil, post imprinting. Credit: Bob Dixon

Imprinting

Imprinting is a water-harvesting technique used to accelerate the revegetation of disturbed or denuded land with annual precipitation from 3 to 14 inches (76 mm to 356 mm) by creating numerous small, well-formed depressions in the soil that collect seed, rainwater, sediment, and plant litter, and provide sheltered microclimates for germinating seed and establishing seedlings. Ideally, each imprint captures enough water to germinate one or more seeds and sustain their growth. Imprints have enough water-storage capacity to increase infiltration to levels above most dryland rainfall rates, thus eliminating nearly all runoff and associated erosion. The imprints are V-shaped and are made with an imprinter roller, though hoof prints, slow-moving deep-knobbed wheels, or bulldozer tracks can create somewhat similar effects. Sites should be at least an acre to justify bringing in equipment. The slope may be up to a 2:1, 26.6°, or 50% grade. Individual imprints are about 4 to 7 inches (10 to 18 cm) deep, 10 inches (25 cm) long, and 8 inches (20 cm) wide. Imprinting is not appropriate in drainage ways. (See figure 3.13.)

Mulching

Mulching is the application of organic or inorganic material onto the surface of the soil, *not* mixed into the soil. Organic mulch consists of porous materials such as compost, aged manure, straw, or wood chips. Placing carbon-rich organic mulch on the surface of the soil ensures it will decompose slowly, enhancing rather than depleting soil nitrogen. Inorganic mulch can consist of a thin layer of gravel or a single layer of cobble-sized rock.

Both organic and inorganic mulches help form a spongy and/or porous welcome mat to lure water into the soil and create a sheltering cover that reduces loss of soil-moisture to evaporation (see figures 3.14 and 3.15). In addition, they protect the soil from the erosive force of falling raindrops and buffer temperature extremes at the soil surface. Note: Never use weed-inhibiting plastic over the soil or beneath mulch. Impervious plastic diverts rainfall away from the soil and can create anaerobic conditions prime for plant diseases.

Organic matter that is used as mulch quickly increases your soil's organic matter content as the mulch decomposes and is transported into the soil. This directly feeds beneficial soil microorganisms, which in turn directly boost soil fertility and water-holding capacity. These soil microorganisms also help bioremediate materials such as soap from greywater systems, and oil and nitrates from street runoff. They also work directly *with* plants. For example, the branching "roots" of one type of beneficial fungi or mycorrhizae, if present, tap into the roots of plants to obtain carbohydrates, while exchanging nutrients and water in return. This link between roots and fungi can increase the effective surface area of a plant's roots up to 700 times, dramatically enhancing the plants' uptake of nutrients and water in extremely wet conditions, extremely dry

Fig. 3.14A. A basin without mulch losing the bulk of the soil moisture to capillary action and evaporation. Note the small stressed tree.

Fig. 3.14B. The basin mulched for improved infiltration and retention of water into the soil

conditions, and everything in between.[4] To do all this, you need organic matter to feed the beneficial microorganisms and using organic mulch provides this.

Finesse things further by giving different plant types their preferred type of organic mulch to attract the soil microorganisms best suited to that plant. Woody, long-lived, perennial plants such as shrubs and trees (and their preferred soil microorganisms) favor woody, longer-lasting mulches like wood chips and bark chips. Annual, short-lived plants such as vegetables and flowers (and their preferred soil microorganisms) favor lighter, more ephemeral mulches like straw. Basically, the best mulch for any plant is that plant's own tissue. So harvest leaf drop and cut up prunings and place them below and beside the plant they came from instead of raking or blowing them away.

Water-harvesting earthworks do a great job of holding super-spongy organic matter, which then enhances the water-harvesting capacity of the earthwork. Organic mulch also delineates basins from paths and limits weed growth. This mulch works best in flat and gently sloped areas where water calmly pools and infiltrates, though it can be placed on steeper slopes when combined with earthworks that slow or stop water runoff. Organic

Mulched Basin

Vertical Mulch in Hole

Fig. 3.15. Vertical mulch variation (mulch-filled hole or trench) encouraging infiltration and retention of water deeper into the root zone of the soil

mulch alone is not appropriate in drainageways, or *flow-through systems*, since it would be washed away.

Rock mulch is much heavier than organic mulch, and thus can be used in areas of higher water flow such as within *flow-through* systems or where there is heavy sheet flow. The rough surface of rock mulch traps sediment and organic matter that help hold moisture, which in turn supports the growth of stabilizing vegetation. Inorganic mulches should be thin enough to let in enough sunlight to grow seedlings that have germinated beneath the mulch. Rock used for mulch should be placed only one rock deep. Gravel should be applied in a shallow layer not more than 1-inch (2.5-cm) deep. Besides allowing plants to grow through them, thin inorganic mulches allow morning dew that condenses on chilled rocks and gravel to trickle down to the soil below.

Sunlight heats inorganic mulch more than it heats organic mulch, so sun-warmed inorganic mulch can be used in colder climates to help initiate the growth of plants earlier in the spring than would occur with organic mulch. However, until plants grow broad enough to shade the inorganic mulch in the hot season, inorganic rock and gravel mulch can have the negative effect of substantially heating the air above the rock.

Reducing hardscape and creating permeable paving

Hardscape reduction is a strategy to minimize the need for pavement through creative planning and design, and the removal of impervious pavement, where possible. By reducing impervious hardscape (and sloping it toward adjoining earthworks), you can increase the adjoining pervious areas to enhance on-site water infiltration of the hardscape's runoff, reduce runoff leaving the site that would otherwise contribute to downstream flooding and contamination, and decrease the heat-island effect caused by excessive, exposed hardscape (see figures 3.16, 3.17 and 4.23 to 4.26).

Permeable paving is a broad term for water-harvesting techniques that use porous paving materials to enable water to pass through the pavement and infiltrate into soil, passively irrigating adjoining plantings, dissipating the heat of the sun, reducing soil compaction, allowing tree roots beneath the paving to breathe, filtering pollutants, and decreasing the need for expensive drainage infrastructure. As a water-harvesting technique, it is most useful on densely

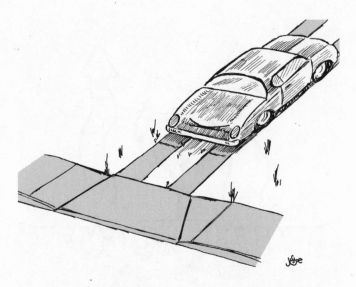

Fig. 3.16. A two-track driveway can reduce impermeable surface area by 60% compared to conventional concrete driveways. Shortening the driveway reduces impermeable surface area still more.

Fig. 3.17. Narrower streets and young native food-producing mesquite trees irrigated by harvested street runoff, Civano, Tucson, Arizona. Arrow denotes water flow.

developed sites with little unpaved surface. Permeable paving is most effective when it only harvests the rainwater that falls directly upon it, without additional runoff from upslope areas. It is not appropriate in drainage ways. Raise permeable paving above the surrounding landscape to prevent settling or pavement displacement due to poor draining subsoil, and prevent sediment-laden stormwater runoff from plugging pores in permeable pavement.

Fig. 3.18. A small yard with hardscape kept permeable by installing recycled sidewalk chunks with ample gaps for water infiltration, Amado residence, Tucson, Arizona. Designed by Blue Agave Landscape Design. For color photo see inside back cover.

Diversion swale

A diversion swale is a gradually sloping drainageway that slowly moves water from one point to another. Like a berm 'n basin, it usually consists of a generally linear basin with the excavated earth placed downslope to form a berm. Unlike a berm 'n basin, which is constructed on-contour to contain and allow water to soak into the earth locally, a diversion swale is built slightly off-contour, allowing a portion of the water to soak into the soil locally while moving surplus water slowly downhill from one place to another, infiltrating water all along the way.

Diversion swales are used to intercept, infiltrate, and redirect both sheet flow and channelized water. Diversion swales can tame the force of water that rushes out from a culvert or roadside bar ditch, transforming the concentrated fast-moving water into a valuable resource by spreading out and calming the flow. Diversion swales can direct runoff to a water-harvesting berm 'n basin, infiltration basin, pond, or other final destination.

Diversion swales are not appropriate in drainageways, and they should not be used in alkaline soils prone to salt buildup and waterlogging. (See figures 3.19 through 3.21.)

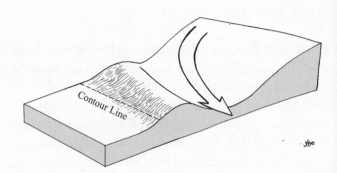

Fig. 3.19. Diversion swale

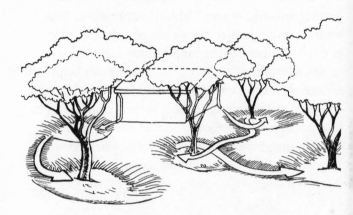

Fig. 3.20. A series of diversion swales as a water-harvesting overflow route from one infiltration basin to another

Fig. 3.21. Rolling dip/diversion berm speed hump directing runoff from the road to diversion swale freely irrigating roadside vegetation

Sheet flow spreader

A sheet flow spreader, otherwise known as a media luna (half moon), is a one-rock high, crescent-shaped, rock mulch structure that is used to slow, spread, and infiltrate sheet flow that would otherwise erosively speed up, concentrate, and run off as more destructive channel flow. Innovator Van Clothier first introduced me to the media luna, and how rock is laid along the land contour of a flat or alluvial-fan-shaped area in the form of a predominantly level arc. The ends of the arc point upslope to capture and focus water flow over the middle of the structure and prevent water from cutting around the ends. This structure works great in locations where water carries lots of sediment. With enough sediment-laden runoff water—and the space to spread it onto—a sheet flow spreader can help induce the formation of an alluvial fan. It will harvest much of the sediment, making the structure stronger, and it will hold more water in that sediment, supporting the growth of soil-stabilizing, water-slowing, water-spreading, water-sinking vegetation (figures 3.22 to 3.23C).

Fig. 3.23A. Sheet flow spreader just after construction, built to spread out the concentrated flow from an upstream culvert. Southern Crescent, Galisteo Basin Preserve, northern New Mexico. Arrows denote water flow.

Fig. 3.23B. Sheet flow spreader two to three months after construction, with sediment-harvesting and soil-stabilizing vegetation growing through the single course of rock.

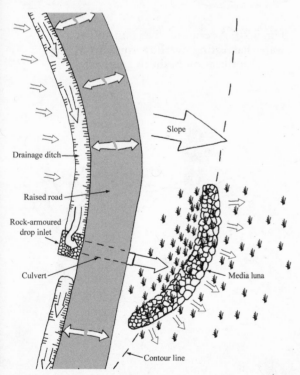

Fig. 3.22. A *sheet flow spreader* or *media luna* slowing, spreading, and sinking flow from a culvert to transform erosive channelized flow into productive sheet flow.

Fig. 3.23C. Media luna two years after construction, concealed by the abundant native grasses it enabled to grow. Photos and work by Craig Sponholtz

One-rock dam

A one-rock dam (ORD) is a low barrier or grade-control structure placed perpendicular to the flow of water within a drainage. It is built just one rock tall and three to six tightly packed rock-courses wide, and typically does not exceed one foot (30 cm) in height. It is built in the shape of a gradually sloped speed hump with the top of those rocks in the downstream row placed at the same level as the bed of the channel. Unlike with a check dam, there is no sudden elevation drop on the downstream side, which would induce the erosive force of falling water. It is purposefully permeable or "leaky." The dam is basically a constructed riffle that *slows*—but does not stop—the flow of water. It helps stabilize the bed of the channel by increasing its roughness, recruiting vegetation within and around the dam, capturing sediment, and *gradually* raising the level of the channel bed over time.

Running water and sediment slow down and back up behind the dam, spreading out over the channel bed before flowing through and over the dam. By slowing and spreading the flow of water, one-rock dams help moisture infiltrate into the soil, reduce downstream

Fig. 3.24. A one-rock dam is easy to build, and is very stable due to its low profile. It is one rock high creating about five parallel rows packed tightly together across the drainage and up part of the banks. Minimum rock size in this illustration is 8 to 12-inch (20 to 30 cm) diameter. Better, tightly fit placement of rock is more important than size of rock, so many rocks act like a few large flattish boulders. Place the structure in the crossover riffle or straight run of a drainage where sediment naturally accumulates.

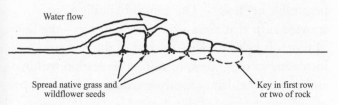

Water flow

Spread native grass and
wildflower seeds

Key in first row
or two of rock

Fig. 3.25A. Key in the first downslope row or two of rocks in a one-rock check dam to anchor the structure, and avoid a waterfall-like drop off from rock to bare soil that would create a scour hole. Seed the structure with native grasses to quicken establishment of anchoring vegetation.

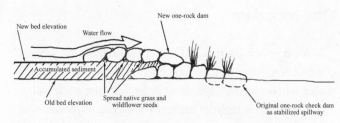

New bed elevation

Water flow

New one-rock dam

Accumulated sediment

Old bed elevation

Spread native grass and
wildflower seeds

Original one-rock check dam
as stabilized spillway

Fig. 3.25B. Another one-rock check dam can be built upon the first after it has naturally back filled with sediment and been stabilized with vegetation. The original one-rock dam then acts as a stabilized spillway for the second.

Fig. 3.26A. One-rock check dam *before* summer rains, Red Windmill Draw, Malpai Ranch, near Douglas, Arizona. One-rock dams are just one rock high and placed in three to five parallel rows packed tightly together across the drainage. Reproduced with permission from *An Introduction to Erosion Control* by Bill Zeedyk and Jan-Willem Jansens. Photo credit: Van Clothier of Stream Dynamics. For color photo see inside front cover.

Fig. 3.26B. Two months later, *after* summer rains, new vegetation has grown upstream of the stabilizing dam. Because the structure is only one rock high, vegetation will easily grow between the rock to further stabilize the structure and slow water flow. Keying the downslope row of rocks deeper into the soil could have minimized the scouring just below the structure. Reproduced with permission from *An Introduction to Erosion Control* by Bill Zeedyk and Jan-Willem Jansens. Photo credit: Van Clothier of Stream Dynamics. For color photo see inside front cover.

flooding by moderating the peak flow of water, retain soil and organic matter upslope of and within the dam, and reduce erosion. A one-rock dam should only be built across a drainage that flows intermittently or ephemerally, not in a perennial drainage. It is often used to help heal a deepening arroyo or gully. These dams can improve roads or paths crossing ephemeral drainages when the dams are built on the downslope side of the crossing. A "paving" of flat-topped rock can be placed on the road or path surface upslope of the dam to keep the crossing from becoming boggy when wet. Position one-rock dams in straight sections of a drainageway, not on curves (see figures 3.24 – 3.26, plus Crossover Riffles in appendix 1).

Rock-mulch rundown

A rock-mulch rundown is an erosion-controlling layer of rock mulch, constructed one-rock high, which is used to armor a sloped, low-energy water-flow location. Its primary purpose is not to harvest water, but to direct flowing water to a less potentially erosive location where water can be more easily and effectively harvested and infiltrated. The rock-mulch rundown works great to direct water from a downspout outlet into a water-harvesting earthwork (fig. 3.27) and as an overflow spillway conveying water from one water-harvesting earthwork to another.

A rock-mulch rundown can also be used to control headcut erosion, but ONLY on low-energy headcuts like those with small catchment areas found at the top of upland rills and gullies where calm sheetflow concentrates into more destructive channelized flow (see figure A1.6 in the Headcut section of appendix 1). The erosive vertical face of the headcut is dug back to a stable angle of repose (no steeper than a 3:1, 19°, 33% slope) then covered with a single layer of tightly fitted rock mulch (fig. 3.28). The rundown *must be lower in the middle than on either side* to ensure that water flows down the middle of the structure as opposed to erosively flowing around it. Since there is no mortar in the structure and it is only one rock high, water can infiltrate between the tightly fitted rocks and sustain the growth of vegetation that helps stabilize the structure and slow water flow. A rock-mulch rundown is NOT used within channels that

Fig. 3.27. Downspout spillway stabilized by a rock-mulch rundown that directs water into the bottom of a mulched infiltration basin. The rundown is constructed much lower in the middle of its downward run than on the sides in order to direct water flow in a contained and stabilized manner.

have moderate- to high-energy flows, such as in channels below headcuts. In those instances check out one-rock dams or rock-lined plunge pools to see if they are appropriate for the site.

Rock-lined plunge pool

A rock-lined plunge pool, otherwise known as a Zuni bowl (named by Bill Zeedyk after he observed the structures built by the people of the Zuni Pueblo), is a structure built at headcuts to control headcut erosion. It consists of an arc-shaped rock-mulch rundown leading into a constructed plunge pool where the pooled water will dissipate the energy of the water falling over the rundown. The plunge pool is built by placing rows of rocks around the bowl of the headcut up to half the height of the original headcut. Water collects in the rock-lined pool, then pours over the

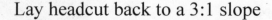

Lay headcut back to a 3:1 slope

Fig. 3.28A. Side or elevation view of a headcut where sheet flow becomes channelized flow.
Dotted line represents the more gradual slope to which the headcut needs to be dug back, filled,
and compacted before the rock-mulch rundown is laid down.

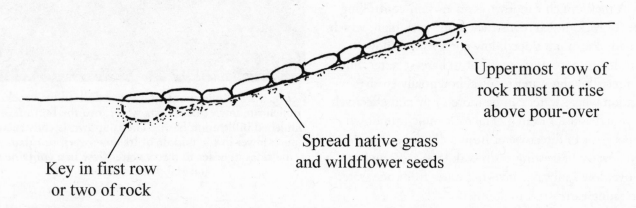

Uppermost row of
rock must not rise
above pour-over

Spread native grass
and wildflower seeds

Key in first row
or two of rock

Fig. 3.28B. Side or elevation view of the rock-mulch rundown. To anchor the structure, key-in the first downslope
row or two, typically using the structure's largest rocks. Ensure the downslope edge of the top of the rock is level
with the bed of the channel to avoid creating an erosive waterfall-like drop. Just before laying the rest of the rock
one course high, broadcast native grass seed over the section of the slope you'll lay rock on to speed the growth
of anchoring vegetation between the rock. The uppermost row of rock must not rise above the headcut pour-over
location, as this would cause water to flow around—rather than down—the structure. Lay all rock tightly
together and fill any gaps with smaller rock and gravel. Adapted from *Erosion Control Field Guide*
by Craig Sponholtz and Avery C. Anderson, Quivira Coalition, 2012

lower end. The pooled water hydrates the soil above, below, and within the structure to sustain the growth of more stabilizing, sediment-accumulating vegetation between the structure's rocks. Like a rock-mulch rundown, a Zuni bowl can be used to stabilize or prevent a headcut from forming at the point where sheet flow is being concentrated into channelized flow. Unlike a rock-mulch rundown, it can also be used within an existing channel or drainage to stop a headcut that is

cutting the channel deeper. (See figures 3.29A and B, plus more in the Headcuts and Stepped Pools sections of appendix 1.) Note that a one-rock dam is typically built downstream of a rock-lined plunge pool at a distance 4 to 6 times the height of the headcut the rock-lined plunge pool is stabilizing. This creates a secondary shallow pool of water held by the one-rock dam that further dissipates stream energy.

ELEVATION/SIDE VIEW

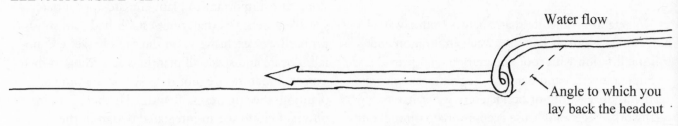

Water flow

Angle to which you
lay back the headcut

Fig. 3.29A. Side or elevation view of a headcut. Arrow represents water flow over headcut. Dotted line represents
more gradual slope the headcut needs to be dug back to before laying rock.

PLAN/OVERHEAD VIEW

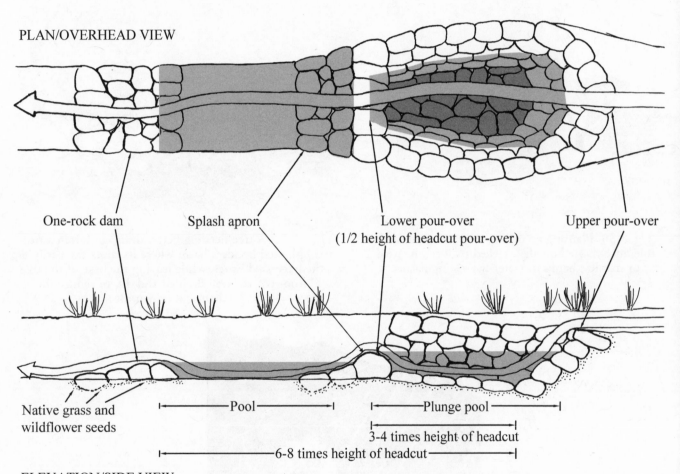

One-rock dam

Splash apron

Lower pour-over
(1/2 height of headcut pour-over)

Upper pour-over

Native grass and
wildflower seeds

Pool

Plunge pool

3-4 times height of headcut

6-8 times height of headcut

ELEVATION/SIDE VIEW

Fig. 3.29B. Overhead and side or elevation view of a rock-lined plunge pool or Zuni bowl with one-rock dam
below. Arrow represents water flow. Downstream from the base of the sloped-back headcut, measure and mark a
spot at a distance equaling 3 to 4 times the height of the headcut. Place the largest rocks of the structure here to
create the downslope side of a bowl, making the depth of the bowl about half the height of the headcut. Just
downslope of the bowl, construct a splash apron about two rows of rock wide to diffuse the water spilling out of
the bowl. To prevent erosive waterfalls, key the lowest splash apron rock into soil so the top of the downslope-side
of rock is level with the bed of the channel (as with one-rock dams and rock-mulch rundowns). To keep the structure
stable, never lay rocks more than one course high. All rocks that are stabilizing the rundown to the bowl should
be angled into the slope, as you would with a dry-stacked rock retaining wall. The uppermost course of rock should
be highest at the downslope ends of the arc and lowest in the middle of the headcut pour-over location to
ensure water flows over—not around—the structure. Adapted from *Erosion Control Field Guide*

Vegetation

Vegetation is the life that emerges from a water-harvesting system. It increases water infiltration and soil stabilization with root penetration, and provides multiple resources and benefits. Vegetation is encouraged and used within or beside every water-harvesting earthwork. Vegetation's use is appropriate throughout all kinds of watersheds, with no limit to the applicable slope. It is important to plant climate-appropriate species at densities that, once established, can subsist primarily, or exclusively, on the site's harvested rainfall. Locate and space all plantings according to their expected mature size, not their size at the time of planting. (See figures 3.30 and 3.31; there is more on vegetation's use in integrated design in the next chapter.)

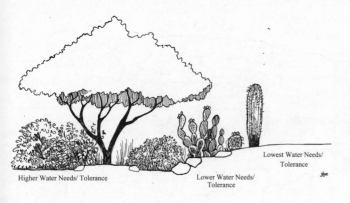

Higher Water Needs/ Tolerance

Lower Water Needs/ Tolerance

Lowest Water Needs/ Tolerance

Fig. 3.30. Planting according to water needs and tolerance. Note how dry-stacked rock can be used to stabilize banks of water-harvesting basins.

Fig. 3.31. A tree needing better drainage (often a fruit tree) planted beside a basin where its roots can easily tap the harvested water, while leaving the base of the tree (root crown) well drained and dry to reduce the chance of crown rot

OVERVIEW OF CISTERN SYSTEMS

Cistern systems should follow the rainwater-harvesting principles described in chapter 1, but because they store a readily accessible body of water, you should also follow an additional set of principles specific to cisterns. The principles and components of cistern systems outlined below are discussed in detail in volume 3, as are various tank options (pre-manufactured or made on site). Search "Water Tanks/Cisterns" on www.HarvestingRainwater.com for links to some available pre-manufactured tanks and current costs.

TEN BASIC COMPONENTS OF A SIMPLE ABOVE-GROUND RESIDENTIAL CISTERN SYSTEM (FIG. 3.32)

1. Catchment surface
2. Gutters and downspouts
3. Screening of cistern and downspout openings
4. First-flush systems (optional)
5. Cistern
6. Vent
7. Overflow
8. Faucet and valve
9. Filters and pumps (optional and not pictured)
10. The maintenance team: You

TEN CISTERN SYSTEM PRINCIPLES

See figures 3.34 through 3.42 for examples of cisterns/tanks. See volume 3 of *Rainwater Harvesting for Drylands and Beyond* for more details on how to apply the cistern principles.

1. **Ensure adequate inflow**. Don't lose water. Size your gutters, downspout, and inflow pipe to handle the maximum rainfall intensity likely to occur in your area. (Search "Downspout & Gutter Sizing" on www.HarvestingRainwater.com.)

2. **Ensure adequate outflow and use it as a resource**. The diameter of the cistern overflow pipe must be equal to the diameter of the cistern's inflow pipe so your system does not back up.

Direct that overflow resource to another tank or mulched and vegetated infiltration basins.

3. **Design your system to collect high quality water**. The higher the quality of harvested water the more options you have for its potential use, so don't contaminate your water with any toxic or contaminated materials making up your system. Materials rated for contact with potable water yield the highest quality water.

4. **Design a closed system that passively filters itself**. Design or install "closed" cisterns screened off from sunlight, insects, and critters so algae and bacteria will not grow, mosquitoes will not propagate, and critters' waste or drowned bodies will not contaminate your water. Additionally, tank covers will reduce water loss to evaporation. Construct the outflow pipe from the cistern (the "supply" pipe) a minimum of 4 inches (10 cm) above the bottom of the cistern to keep the sludge of sediments (leaf litter, dust, etc.) from being pulled into the supply pipe. See figure 3.33 for an example.

5. **Maintain access to your tank and its interior**. You need access to check water levels, clean out the tank, and make repairs. Place above-ground cisterns so there is enough space to walk completely around them to check for (and repair) leaks and conduct inspections, especially if they are close to a building. All tanks should have access holes to get inside the tank for inspection, cleaning, or repair. Include lockable/secure lids to prevent accidental entry by children.

6. **Vent your tank**. All covered tanks with tight-fitting lids or tops must be vented to prevent a vacuum from forming within the tank when large quantities of water are quickly drawn from the tank.

7. **Use gravity to your advantage**. Place your tank at a location where you can utilize the elevation of the catchment surface and the free power of gravity to collect rainwater and distribute it around your site. Below-ground tanks may not be able to use gravity to distribute the stored

continues on page 86

Fig. 3.32. Basic components of a cistern system

KEY:

1. Catchment surface
2. Gutters and downspouts
3. Screening of cistern and downspout openings
4. First-flush systems (optional) – outletting where flush water is a convenient resource
5. Cistern
6. Vent – that is light-proof, critter-proof, and insect-proof
7. Overflow – to another tank, vegetated earthwork, or natural drainage
8. Faucet and valve
9. Filters and pumps (optional and not pictured)
10. The maintenance team: You

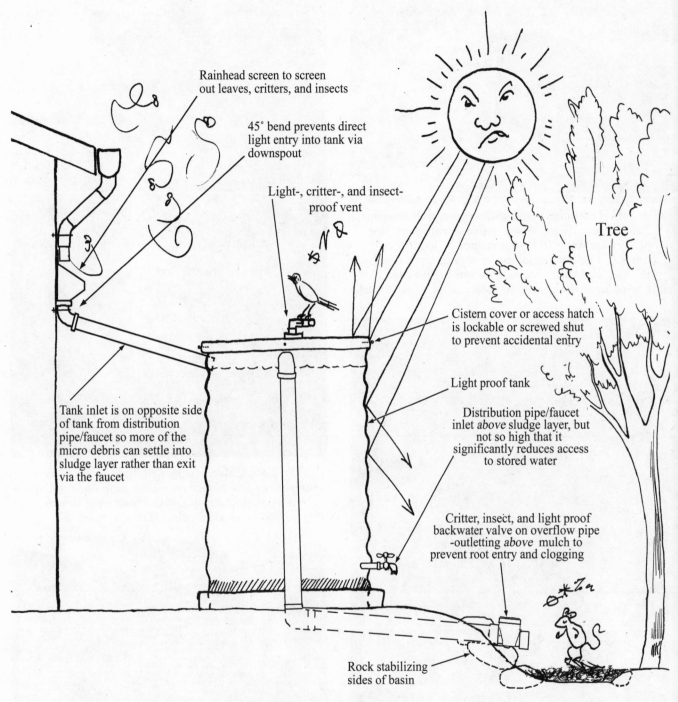

Rainhead screen to screen
out leaves, critters, and insects

45° bend prevents direct
light entry into tank via
downspout

Light-, critter-, and insect-
proof vent

Tree

Cistern cover or access hatch
is lockable or screwed shut
to prevent accidental entry

Light proof tank

Tank inlet is on opposite side
of tank from distribution
pipe/faucet so more of the
micro debris can settle into
sludge layer rather than exit
via the faucet

Distribution pipe/faucet
inlet *above* sludge layer, but
not so high that it
significantly reduces access
to stored water

Critter, insect, and light proof
backwater valve on overflow pipe
-outletting *above* mulch to
prevent root entry and clogging

Rock stabilizing
sides of basin

Fig. 3.33. A closed cistern system.
For video of this set up and more see the video page at www.HarvestingRainwater.com.

Fig. 3.34. Galvanized tank and *wet system* downspout, Tank Town, Texas. A wet system downspout pipe always has water in it so it cannot be used in cold climates. The horizontal run of pipe is supported by the soil, and then rises up (but not higher than its inflow point at the gutter) to enter a tank or earthwork.

Fig. 3.36. Culvert tanks with light-, insect-, and critter-proof lids. Overflow moves from one tank to the next, then out to vegetated basins not shown. Milagro Cohousing, Tucson, Arizona

Fig. 3.35. A 10,000-gallon (37,850 liter) steel tank/ mural and *dry system* downspout collecting roof runoff for irrigation, Children's Museum, Santa Fe, New Mexico. A dry system downspout drains all water out of the pipe, leaving it dry between rains. It does not collect sludge nor is it prone to freezing damage. For color photo see inside back cover.

Fig. 3.37. A 300-gallon (1,135-liter) corrugated steel WATERHARVEST SLIMLINE® tank. Reproduced by permission of BlueScope Water. Slim tanks work great for small spaces and narrow yards, and are available up to a 1,500-gallon (6,000-liter) capacity in metal, plastic, or fiberglass. They can double as garden walls or privacy screens.

Fig. 3.40. A 6,000-gallon (22,700-liter) ferrocement tank storing roof runoff for domestic use and irrigation, Meuli residence, Tijeras, New Mexico. The square piece of corrugated metal atop the tank is the lid to the access hole.

Fig. 3.38. A 300-gallon (1,135-liter) polyethylene tank with rainhead downspout screen collecting roof runoff from porch roof for courtyard landscape irrigation, Nature Conservancy offices, Tucson, Arizona. Two 45° bends in downspout pipe keep direct light out of tank. Access hole lid includes subtle vent. At bottom of tank on left side, a ball valve accesses tank water; beside the valve is a spigot accessing municipal water. Beside the spigot is the backwater valve directing overflow water to a mulched basin.

Fig. 3.41. A 7,000-gallon (26,460-liter) poured-in-place concrete tank faced with stone doubles as a patio and collects roof runoff for gravity-fed irrigation of citrus trees and other plants below. Note how the tank's mass creates a warmer microclimate for the frost-sensitive citrus trees on the south-facing side of this northern hemisphere tank. Tucson, Arizona. For color photo see inside back cover.

Fig. 3.39. A below-ground 1,200-gallon (4,542-liter) concrete cistern engineered to take weight of vehicles above. Only access lids are visible. Cistern collects roof runoff from home. Santa Fe, New Mexico

water, but they must be designed so overflow can occur with gravity. You can always add a pump to increase water pressure and performance if needed, but don't turn a pump into a crutch your system must depend on.

8. **Make rainwater use convenient**. Where feasible, select your tank location so it is near both the water source (roof) and the destination (garden, sink, etc.). This will minimize the length of downspouts, pipes, and hoses, which will save money and materials and help maintain water pressure. At the very least, place or plumb the cistern's faucet conveniently close to your point of use, even if the tank is inconveniently distant. In low-pressure gravity-fed systems, the distribution pipe from the tank should have an interior diameter of 1-inch (25 mm) to reduce the water pressure loss that occurs with the increased surface friction of smaller diameter pipe. Furthermore, use full-port valves that do not constrict to smaller interior diameters. (Search "Full Port Valves" on www.HarvestingRainwater.com for more.)

9. **Select and place your cistern so it does more than store water**. The more your cistern does, the more cost-effective it is. By designing a cistern to also act as part of a privacy screen, fence, retaining wall, or support pillar for a covered porch you eliminate the cost of buying other materials to make that section of screen, fence, wall, and/or pillar.

10. **Maintain your system**. Inspect and clean out your gutters before the rainy season, and in times of leaf drop from surrounding trees. (Pruning branches growing over the roof and gutter can reduce the cleaning job.) If you have a first flush diverter, drain it after rainstorms. (However, if rains in your area are typically very light, first flush draining can be more infrequent to reduce water loss from the first flush diversion.) Use the water you collect in your tank. A rainwater tank full of water you never use is typically a waste of resources. (An exception is a tank of water reserved for emergency use.) Periodically check your system to ensure there are no leaks, no broken inlet screens, no excessive build-up of sediment in the tank (up to the level

of the outlet pipe), and to make sure all lids are securely closed.

REAL LIFE EXAMPLE

THE RUNNING RAIN SOCIETY

I end this chapter with the story of a homesteading couple who harvest rainwater in both soil and tanks. Although they live on 40 acres (16.2 ha) of land, the principles they learned from living many years with an evolving system can apply equally well to a 1/10th-acre urban lot, an entire community, and everywhere in-between.

LIVING OFF RAINWATER HARVESTED IN TANKS AND SOIL

In 1978, Dan Howell found his ticket from the Southern California rat race to the country life. There on the laundromat bulletin board a 3 x 5"card read, "40 acres in beautiful, rural New Mexico—cheap."

Dan bought it sight unseen and set out to find his land in west-central New Mexico. At 7,000 feet elevation (2,134 m) the air was crisp beneath an expansive blue sky. The land was starkly beautiful but eroding. Hardy pinyon pine, juniper, and thin grasses sparsely peppered the land speaking of climate extremes. Summer day temperatures would soar over 100° F (38° C), and winter nights would drop well below freezing. Above all else, it was dry. Annual rainfall averaged just 14 inches (356 mm). Ground-water was so deep almost everyone in the area brought their water in on trucks. Soils were denuded from poorly managed cattle grazing and excessive timber cutting. Without the anchoring effect of vegetation, wind and stormwater ran over the land, taking the topsoil with them. Dan was determined to move to this land and live in a better way. He went back to Los Angeles to prepare. Five years later, in 1977, he and his new wife Karen headed to the land with a nest egg of money and determination.

Dan and Karen parked their trailer and built a small shed covered with 200 square feet (18.6 m²) of corrugated metal roofing. A gutter was attached, and drained rainwater to ten sealed 55-gallon (208-liter) plastic drums. When the July rains came, the water barrels filled instantly, allowing Dan and Karen a welcome break from hauling water. Their neighbors hadn't collected a drop of the rain and were back on the long road to town hauling water the next day.

Dan and Karen knew rain and snow would be erratic, ranging from 7 to 30 inches (178 to 762 mm) of precipitation a year, with an average annual rainfall of 14 inches (356 mm). Over their 40 acres (16.2 ha) of land this range of rainfall would yield 7,623,000 to 32,670,000 gallons (28,856,000 to 123,669,000 liters) of rainfall per year!

The Howells realized that rainfall was far more abundant than the volume of local municipal water they could haul. It was also free—and came to them! So the Howells decided to make rainwater the main water source for all their water needs, and named their homestead the "Running Rain Society."

Once they hooked up a 500-gallon (1,900-liter) galvanized steel tank to the gutter of their newly built home's metal roof, the water truck got a rest. That one tank, and the new 400-square-foot (37.2-m²) roof, were enough to provide all the Howells' domestic water needs for drinking, washing, bathing, and cooking for an entire year! (See figure 3.42.)

LIVING ON RAINWATER: TANK STORAGE

The Howells have been living on rainwater for over 20 years. It tastes great and has never needed to be filtered, though they did install a drip irrigation "Y-filter screen" three years ago for the sake of their guests, because tiny rust particles from the tank had started showing up in the water. There is very little air pollution in their area, and more importantly Dan and Karen's metal roof and whole water catchment system is kept clean and toxin-free from start to finish. They have been vigilant in avoiding potential sources of toxins such as asphalt roofing shingles or lead flashing. And they ensure that their tank is closed off to any sunlight, insects, or critters that could introduce or breed unwanted bacteria or diseases.

The Howells have made dramatic lifestyle changes tied to their water use. Dan and Karen use only 5

Fig. 3.42. The Howells' home and 500-gallon (1,895-l) galvanized steel cistern

gallons (18.9 liters) of water each per day within the home. Any water going down the sink or bath, drains outside to water plants. They built a waterless composting toilet to eliminate water consumption for sewage treatment, and use the composted humanure to enhance their land's fertility. By living within the limits of their on-site water resources the Howells don't impoverish their area from over-consumption, nor do they need to work full time to meet their basic needs.

As a guest cottage and workshop were built, the roof catchment surface grew from 400 to 2,000 square feet (37.2–185.8 m²). Two additional tanks raised the Running Rain Society's domestic water storage capacity to 4,500 gallons (17,000 liters). Dan and Karen started small and worked the kinks out of their system before expanding, and that's just how they recommend others proceed.

HOW THE HOWELLS DEAL WITH FROZEN WATER IN THEIR CISTERNS

In winter, the Howells drain enough water from their 500-gallon (1,900-liter) cisterns so the top 20% of the tanks are empty. This ensures that there is enough room for the water to expand if it freezes. One to two days is the longest Dan and Karen have gone without access to water that was frozen inside their outdoor tanks. They always have a back-up water supply stored inside the home for such occasions. That is the extent of the Howell's preparation for winter water storage, and it works well for them.

(Under colder conditions different strategies may be needed. To prepare for freezing weather, ask water-harvesting locals what they do, and look into local codes for water tank and plumbing installation in your area.)

HARVESTING RAINWATER IN THE SOIL

While the Howells were setting up their rooftop rainwater collection system they were simultaneously working in the landscape to harvest runoff water to serve their plants.

Dan and Karen were using all their roof runoff for their domestic water needs so they decided to set up a large food garden where the landscape naturally concentrated runoff water within a small arroyo (drainage) about 150 feet (45.7 m) from the trailer and house. To stabilize the arroyo and ensure the gardens would be secure from erosive stormwater, Dan and Karen constructed a series of check dams and gabions in the drainage. Check dams are permeable barriers placed *within* a drainage perpendicular to the flow of water. Rock gabions are further stabilized with a wire fence wrapping, but are prone to failure when the wire inevitably rusts or breaks.

The gabions and check dams were situated within the drainage so that once detritus and sediment backed up behind the small structure, they would create a series of level terraces stepping down from the top to the bottom of the arroyo. The flow of water was spread over a wide surface, slowing it down and allowing it to gently sink into the soil. Other earthworks slowed water flow further upslope within the local watershed draining to the arroyo. With erosion checked and water harvested in the soil, new vegetation started to appear in the once-deteriorating arroyo bed. As Dan points out, "If you have scouring in an arroyo and no vegetation, you know the situation is out of control. If vegetation is established on its banks, or sometimes on the bottom, you know the situation is more stabilized."

Within these stabilized level terraces Dan and Karen planted 600 square feet (55.7 m²) of gardens meeting 15–25% of their food needs. They had the most success with asparagus, garlic, Egyptian walking onions, and Jerusalem artichokes. The gardens are

Box 3.2. The System Is Low-Maintenance, But Definitely Not Maintenance-Free.

In 2002 a broken pipe resulted in the irrigation system losing 7,000 gallons (26,550 liters) of harvested rainwater. Unlike a municipal system where there is a backup water supply, Dan and Karen's system is limited and leaks must be dealt with right away. They made it through the season, but they had to be extra conservative until the rains renewed their water storage. Frequent inspection of your system is a very good idea.

watered primarily from rainwater stored in the soil, but for less drought-tolerant plants the Howells wanted a source of water for surface irrigation in dry times. To support this, they hand dug two 20,000-gallon (75,700-liter) dirt reservoirs high in their landscape. These reservoirs completely fill in one good rain, and provide all the irrigation water needed for an entire year! Once the reservoirs are full, water is pumped from the reservoirs to a 10,000-gallon (37,850-liter) fiberglass tank and a used 5,000-gallon (18,950-liter) steel tank also sitting high in the landscape. Water is then distributed from the tanks to the gardens using gravity flow.

To support water-needy vegetables, Dan and Karen arranged soaker hoses several inches below the land's surface and mulched the soil above to retain moisture. Ten pounds per square inch (psi) of water pressure is needed for the irrigation lines to function, and this is easily achieved by the tanks' placement above the elevation of the garden: *Each foot a water source (tank) is raised above its destination (garden), gravity provides 0.43 psi of pressure.* As Dan says, "No pumps, no utilities, and no pollution." (See the chapter on principles for cistern systems in volume 3 for a gravity-fed drip irrigation system that can work on less than 10 psi of pressure.)

After each rainy season surplus water in the reservoirs is siphoned out over the landscape where it's needed. Accumulated silts are removed from the bottom of the reservoir and wheel-barrowed downslope to make water-harvesting contour berms. Sometimes a basin is dug along the upslope side of these berms, making a berm 'n basin. These structures are laid out in lines of equal elevation along the land, where they intercept and quickly infiltrate rainwater into the soil where it is available to vegetation.

Dan and Karen are always experimenting to try to increase the productivity of their site and the efficiency of their water harvesting. A rain-fed orchard within a net and pan system of berms had limited success. From this they discovered that the microwatersheds needed to be enlarged to harvest more water, and that climate extremes limited the diversity of productive crops they could grow. More successful was the creation of different styles of berm 'n basins that sped up their revegetation work alongside their dirt roads where access, inspection, and maintenance is easy (for more on the earthworks techniques mentioned in this and the above paragraph, see the chapter on berm 'n basins in volume 2).

The Howells' work isn't over yet. They have a life far from the 9-to-5 office day, but not a life of leisure. The system needs to be maintained, and they have plans to dig out more reservoirs and encourage the growth of more native grasses that can harvest sediment in runoff flow before it enters the reservoirs. They plan to use the additional harvested water to cultivate native and medicinal plants. When the storm clouds break over their land, earthworks and tanks fill with water, check dams catch fertile silts, and vegetation bursts into new growth. More resources are now being generated on their site than drained away. It's been over twenty years since that fateful day in the California laundromat when Dan decided to run with a wild idea: Who would've guessed it would have them both running with the rain?

Integrated Design

This chapter shows you how to maximize the potential of your site's water resources by integrating harvested water with sun, shade, and vegetation at your site to help passively cool buildings in summer, heat them in winter, allow for on-site solar power and solar hot water, and enhance plants and gardens. By doing so, you realize the seventh principle of water harvesting: Maximize beneficial relationships and efficiency by "stacking functions."

Use the strategy of "integrated design" to provide on-site needs (e.g., water, shelter, food, aesthetics) with on-site elements (e.g., stormwater runoff, greywater, cooling shade, warming sun, vegetation, compost) by creating an efficient design that saves resources (e.g., energy, water, transportation, money) while improving the function and sustainability of the site.[1] Integrated design helps turn "problems" into solutions. For example, erosive floodwater runoff harvested into basins can provide water to grow shade trees, helping control flooding and erosion. The key is to *see, understand,* and *combine* on-site elements—such as stormwater runoff, vegetation, and solar exposure—to maximize their beneficial use.

To help you do just that, this chapter helps you develop an integrated design for your site. Once again we start with observation, this time focusing on the sun's path across your site, and mapping it along with the other observations you've plotted on the site plan you created in chapter 2. Why look at the sun in a water-harvesting book? That is so you can orient buildings, plantings, and more to maximize the degree to which they can *produce* resources, rather than consume them—by passively and freely heating, cooling, powering, growing, and maintaining themselves in a way that will make all your water resources (and time and money) go further. In that vein, *seven basic patterns* of integrated design are presented in this chapter to help you create a conceptual layout of water-harvesting earthworks, tanks, gardens, trees, and buildings that work off your observations and build on your site's existing resources while helping mitigate its challenges. The more patterns you incorporate into your design, the more integrated it becomes. (Appendix 5 provides worksheets prompting you for information to go along with the following seven patterns.) The chapter ends with tips on how to further refine your site's integrated conceptual design, and gives you an integrated plan example.

Many of you, especially story lovers, may want to skip ahead to chapter 5—the story of how my brother and I created and implemented an integrated plan for our urban lot using design patterns described in this chapter, then worked with others to expand this approach into our neighborhood. Chapter 5 shows you the power and fun of integrating the information and strategies in this book and will feed your hunger for, and understanding of, the important information in this chapter.

NOTE: All times used in this book and chapter are *solar time* based on the sun's actual location in the sky, which changes throughout the year. In contrast, clock time is based on an imaginary sun location that results in dividing time into equal 24-hour days throughout the year.

Since *solar* or *true* noon—the time of day when the sun is highest in the sky and which divides the daylight hours for that day exactly in half—changes throughout the year, it can be more than an hour different from *clock* noon. Other solar hours of the day relate directly to solar noon. For example, solar 11 A.M. is one hour before solar noon. The exact date, your location in a time zone, and consideration of Daylight Savings Time will all affect the clock time of your local solar noon and other solar times.

Websites where you can look up solar noon are:
• www.sunrisesunset.com (United States & Canada)
• www.timeanddate.com (World locations)
• www.esrl.noaa.gov/gmd/grad/solcalc/
• or the Winter Solstice Shadow Ratio Calculator in the Sun & Shade Harvesting section of www. HarvestingRainwater.com

THE PATH OF THE SUN

We live on a planet with about a 23.5° tilt—(actually closer to 23.44°) that travels completely around the sun each year. These characteristics result in a gradual shift throughout the year in the direction and time of day that the sun rises and sets, and the angle of the sun above the horizon. The degree of shift depends upon your site's latitude on earth. (See figure 4.1.)

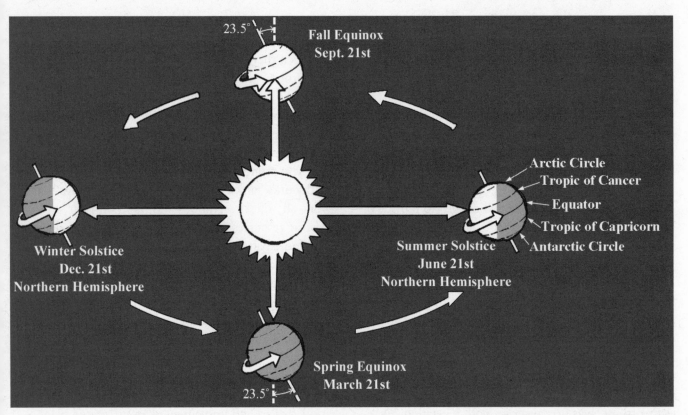

Fig. 4.1. The tilt of the earth's axis results in gradually changing angles of the sun above the horizon throughout the year. These changing angles of the sun's rays result in the changing seasons. The more directly overhead the sun is, the hotter it is. The more diffuse the angle of the sun, the cooler it is.

On the winter solstice in the northern hemisphere, the sun is low in the sky at noon and it does not even rise above the Arctic Circle. At the same time in the southern hemisphere, the noonday sun is directly overhead at the Tropic of Capricorn.

On the summer solstice in the northern hemisphere the sun is high in the sky at noon, the sun never sets within the Arctic Circle, and the noonday sun is directly overhead at the Tropic of Cancer.

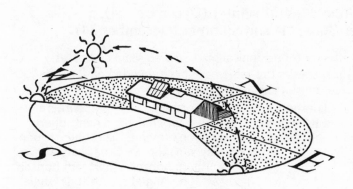

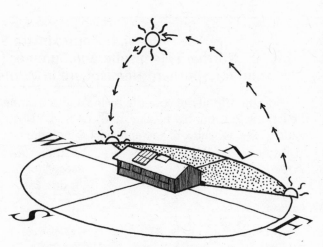

Fig. 4.2A. The *winter* solstice path of the sun at 32° N latitude. House shadow cast at noon. Dotted shading represents section of the sky the sun does NOT traverse. See appendix 7 for images of the sun's path at other latitudes.

Fig. 4.2B. The *summer* solstice path of the sun at 32° N latitude. House shadow cast at noon. See figures 4.5A and B for shadows cast throughout the day and year. For videos of these changing sun and shadow paths see the Sun & Shade Harvesting section at www.HarvestingRainwater.com.

In the northern hemisphere, the sun rises *south* of due east and sets *south* of due west in the winter, while it rises and sets *north* of due east and west, respectively, in the summer. (See figures 4.2A and 4.2B.)

In the southern hemisphere it is the opposite: The sun rises and sets north of due east and west in their winter (which occurs while the northern hemisphere is having summer), and it rises and sets south of due east and west in their summer. On the spring and fall equinox (March 21 and September 21 in the northern hemisphere; September 21 and March 21 in the southern hemisphere) the sun rises and sets due east and west.

ORIENT YOURSELF TO THE SUN'S "FLOW" THROUGHOUT THE YEAR

For those living in the northern hemisphere, the *south*-facing or equator-facing side of buildings, walls, and trees is the "winter-sun side" and the north-facing side is the "winter-shade side." This is because the winter sun stays in the *southern* sky all day. Midday the sun's angle off the horizon remains low (the angle gets lower the further north you are in latitude). (See figure 4.3 and figures A7.1A to A7.7B in appendix 7.)

It is the opposite in the southern hemisphere. The *north*-facing or equator-facing side of the buildings, walls and trees is the "winter-sun side" and the south-

Box 4.2. Approximate Sun Angles by Latitude for Northern Hemisphere in Summer (June 21) and Winter (December 21), and for Southern Hemisphere in Winter (June 21) and Summer (December 21)

Find the latitude of your site by looking at a globe, atlas, or topographical map, or web search "What is the latitude of (your town, state, and country)?" Note: The angles in this table have been rounded off for easier reading. See appendix 6, and appendix 7 for more resources on sun angles and passive solar design.

Latitude, N or S	Date	Location sun rises, N or S of due East	Location sun sets, N or S of due West	Northern hemisphere: Location of noon-day sun, above southern horizon (Altitude Angle)	Southern hemisphere: Location of noon-day sun, above northern horizon (Altitude Angle)
0°	June 21	23° N	23° N	113°	67°
	Dec. 21	23° S	23° S	67°	113°
4°	June 21	23° N	23° N	109°	63°
	Dec. 21	23° S	23° S	63°	109°
8°	June 21	24° N	24° N	105°	59°
	Dec. 21	24° S	24° S	59°	105°
12°	June 21	24° N	24° N	101°	55°
	Dec. 21	24° S	24° S	55°	101°
16°	June 21	24° N	24° N	97°	51°
	Dec. 21	24° S	24° S	51°	97°
20°	June 21	25° N	25° N	93°	47°
	Dec. 21	25° S	25° S	47°	93°
24°	June 21	26° N	26° N	89°	43°
	Dec. 21	26° S	26° S	43°	89°
28°	June 21	26° N	26° N	85°	39°
	Dec. 21	26° S	26° S	39°	85°
32°	June 21	28° N	28° N	81°	35°
	Dec. 21	28° S	28° S	35°	81°
36°	June 21	29° N	29° N	77°	31°
	Dec. 21	29° S	29° S	31°	77°
40°	June 21	31° N	31° N	73°	27°
	Dec. 21	31° S	31° S	27°	73°
44°	June 21	33° N	33° N	69°	23°
	Dec. 21	33° S	33° S	23°	69°
48°	June 21	36° N	36° N	65°	19°
	Dec. 21	36° S	36° S	19°	65°
52°	June 21	39° N	39° N	61°	15°
	Dec. 21	39° S	39° S	15°	61°
56°	June 21	44° N	44° N	57°	11°
	Dec. 21	44° S	44° S	11°	57°
60°	June 21	51° N	51° N	53°	7°
	Dec. 21	51° S	51° S	7°	53°
64°	June 21	63° N	63° N	49°	3°
	Dec. 21	63° S	63° S	3°	49°
68°	At 68° the sun never rises on the winter solstice, and never sets on the summer solstice. (This is true for all latitudes above 66.5°, which defines the Arctic/Antarctic circles.)				

Box 4.2. (continued)

Note 1: Within the tropics (less than 23.5° latitude), in the northern hemisphere, on the summer solstice, the sun rises *north* of east and stays in the *northern* half of the sky all day long. In the tropics of the *southern* hemisphere, on the summer solstice, the sun rises *south* of east and stays in the *southern* half of the sky all day long. This is shown as a summer noonday sun location of greater than 90° in the chart on the previous page. (Note that this does not occur the whole summer, only for a period of time around the solstice, which varies with latitude.)

Note 2: The difference between the location of the noonday sun on the summer and winter solstice is shown as 46° for every latitude. This is about twice the Earth's tilt of 23.44°, upon which all figures in this chart are based.

Note 3: Noon is *solar* noon (the point at which the sun is highest in the sky on that given day), not clock noon.

NOAA Solar Calculator: Find sunrise, sunset, solar noon, and solar position for any place on earth at http://www.esrl.noaa.gov/gmd/grad/solcalc/

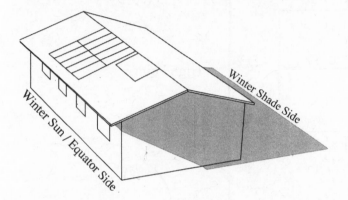

Fig. 4.3A. *Winter* sun exposure and shade cast at noon on the winter solstice at 32° latitude. Solar hot water heater and solar panels on roof

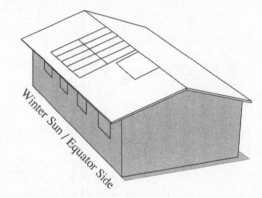

Fig. 4.3B. *Summer* sun exposure and shade cast at noon on the summer solstice at 32° latitude. Solar hot water heater and solar panels on roof

facing side is the "winter-shade side." The winter sun remains in the *northern* sky.

Identify the equator-facing "winter-sun side," and the "winter-shade side" of your home now! The rest of this chapter continually refers to this orientation, so get ready and get oriented now!

In summer north of 23.5° N latitude (Tropic of Cancer), the sun rises and sets north of due east and west, but at midday the sun is in the southern sky (and higher off the horizon than in winter). So the "summer-shade side" of buildings, walls and trees is the south-facing side in the early morning and late afternoon, but midday it is the north-facing side (with a much shorter shadow than is cast in winter). The converse is true in the southern hemisphere. (See

figure 4.2, box 4.2, and appendix 7 for sun angles by latitude and season, for the northern and southern hemispheres.)

Residents of the tropics (from 23.5° N and S latitude to the equator) have a different situation. In *summer* in the northern tropical latitudes, the sun also rises north of due east and west, but unlike the non-tropic latitudes, the sun stays in the northern part of the sky all day. So the "summer-shade side" of objects in the tropics is the south/equator-side all day long. The converse is true in the southern tropical latitudes (see figures A7.1A and A7.2A in appendix 7).

Identify the "summer-shade side(s)" of your home now!

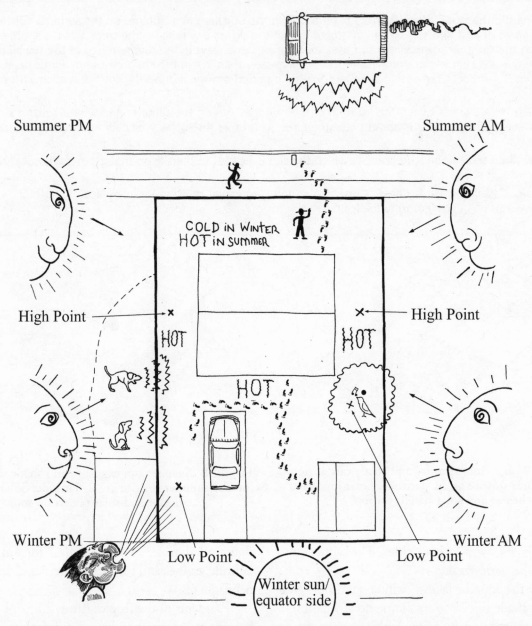

Summer PM

Summer AM

COLD IN WINTER
HOT IN SUMMER

High Point

High Point

HOT

HOT

HOT

Winter PM

Winter AM

Low Point

Low Point

Winter sun/
equator side

Fig. 4.4. Map the resources to harvest, and challenges to divert or diffuse on your site, sample map. See appendix 5 for worksheets. Overlay these observations with those of water flow (fig. 2.4) then create an integrated plan that harvests multiple resources while diffusing multiple challenges (fig. 4.27).

ADD THE LOCATION OR FLOW DIRECTION OF THE SUN AND OTHER OBSERVATIONS TO YOUR SITE PLAN

Map the location of the rising and setting sun on the summer and winter solstice; where you would like more shade or exposure to sun; the direction or location where prevailing winds, noise, or light come from; the foot traffic patterns of people, pets, or wildlife; and any other resources or challenges you may want to design for (fig. 4.4). By recording your site's existing resources and challenges you can improve the layout and design of water-harvesting earthworks, tanks, gardens, shade trees, paths, and buildings so they harvest more resources and diffuse or divert the challenges.

THE SEVEN INTEGRATED DESIGN PATTERNS

Now use the following integrated design patterns to get ideas on how you can efficiently arrange elements of your design to the unique conditions and needs of your site. You will want to have your site map (and worksheets from appendix 5) handy on which to write any additional information. All "Action Steps" discoveries and calculations should be written on your worksheets for further reference. These patterns have a sequence, as you will see.

INTEGRATED DESIGN PATTERN ONE

ORIENTING BUILDINGS AND LANDSCAPES TO THE SUN

Integrate the orientation of buildings, living spaces, and water-harvesting earthworks/planting areas with the sun to maximize passive heating and cooling while reducing water and power needs. A year-long study in Davis, California, monitored temperatures in two identical apartment buildings with different orientations to the sun. No heating or cooling systems were operated during this year. The study found that apartment units in the building with an *east-west orientation* (long walls facing south and north) and with small roof overhangs were *17°F (9.4°C)* **warmer in winter** *and 24°F (13.3°C)* **cooler in summer** than apartments in a similar building with a north-south orientation (long walls facing east and west).[2] (See figure 4.5.) That is a huge difference! Building, buying, or renting a home with correct solar orientation costs nothing extra, yet it can drastically reduce utility costs and increase comfort by maximizing winter sun warmth and minimizing summer heat.

Upon hearing about this friends in Boulder, Utah changed the orientation of their mobile home from north-south (figure 4.5B) to east-west (figure 4.5A), resulting in more wall and window surface area exposed

to the winter sun, with a resulting drop in winter heating bills from over $300 a month to under $50 a month. Inside temperatures also became cooler in summer due to less wall and window surface area exposed to the hot morning (east) and afternoon (west) sun, with year round comfort greatly increased.[3]

The buildings and trees in your landscape cast cooling shade during the day and the vegetation's canopy reduces radiant heat loss at night creating a diverse array of microclimates. When planting within water-harvesting earthworks identify these microclimates, and select and place vegetation appropriate to these microclimates. Cold-sensitive plants go on the warm winter-sun/equator-facing side of a tree or building. Hardy drought- and heat-tolerant plants go on the west side where afternoon sun is hottest and evapotranspiration is greatest. Cold-tolerant plants go on the cool winter-shade side. Vegetation needing more water goes on the east side where plants will get sun on cool mornings and shade on hot afternoons.

Action Steps

- How is your site and/or your home oriented? If you don't know, get a compass or ask the sun or the stars. The sun will orient you to the cardinal directions if you use the information in box 4.2 and appendix 7, and if you pay attention to where the sun is throughout the day. See the end of appendix 7 to find north using the sun and other stars. Put this information on your site map.

- When adding onto an existing building, try to orient the addition appropriately to the sun, even if the existing building is not ideally oriented. For example, a building with a north-south axis (fig. 4.5B) could have a room added to its east- or west-facing side (instead of its south- or north-facing sides) to lengthen/increase the building's winter-sun-facing exposure without lengthening its east- or west-facing exposure.

- When you know your building's orientation, and have a good grasp of where north, south, east, and west are in relation to your site, move on to the

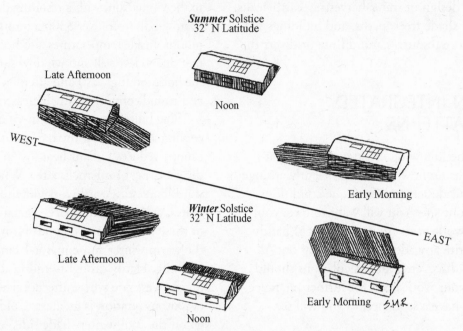

Fig. 4.5A. This orientation is *warmer* in winter (look *below* the west-east line) and *cooler* in summer (look *above* the west-east line). Sun exposure on, and shade cast by, an east-west oriented building at 32° N latitude. View of building's winter-sun/equator side. Note how the winter-sun side is shaded by the roof overhang at the summer solstice, but not the winter solstice. This ideal building orientation and roof slope is perfect for the inexpensive clustered roof top installation of solar PV panels and a solar hot water heater without special mounting racks.

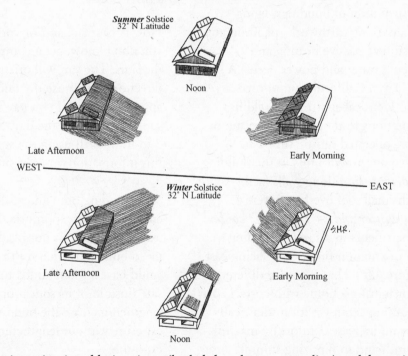

Fig. 4.5B. This orientation is *colder* in winter (look *below* the west-east line) and *hotter* in summer (look *above* the west-east line). Sun exposure on, and shade cast by, a north-south oriented building at 32° N latitude. View of building's winter-sun/equator side. This poor building/roof orientation necessitates a more expensive spread out installation of solar PV panels and water heater so they do not shade one another in winter. In addition, special mounting racks are needed to angle solar panels and water heater correctly toward the sun. More PV panels are also necessary to meet increased power needs of the house's mechanical heating/cooling system.

following patterns. They will help improve the performance of your home and landscape, even if their orientation is less than ideal.

INTEGRATED DESIGN PATTERN TWO

DESIGNING ROOF OVERHANGS AND AWNINGS TO OPTIMIZE WINTER SUN AND SUMMER SHADE

Like a broad-brimmed sun hat, roof overhangs and awnings on a building can improve your comfort. Roof overhangs can improve a building's passive cooling and heating performance regardless of orientation and increase roof area resulting in more roof runoff. Properly sized overhangs on a building's "winter-sun" equator-facing side let in low angle winter sun while blocking overhead summer sun. Overhangs on the east, west, and winter-shade side of buildings boost summer shading. Combine overhangs with correct solar orientation and you'll be even more comfortable and pay far less utilities. (See figure 4.6.)

Given your latitude, how far should window overhangs extend out from a building's *winter-sun/equator-facing side*? Here are two ways to figure it out.

One is with an equation. The other is with a to-scale drawing.

The equation way

The following equation is from *The Passive Solar Energy Book* by Edward Mazria.[4] (Note that the equation only applies to equator-facing windows.)

window height ÷ F = Overhang Projection (OP)

Window height is determined by measuring the vertical distance from the windowsill to the bottom height of the overhang's extension (see figure 4.6), while F is a factor selected from the table in box 4.3 according to your latitude and climate. (Note: If you will be installing gutters, be sure to include their width as part of the overhang's extension, since they will extend the overhang and the shadow it casts.)

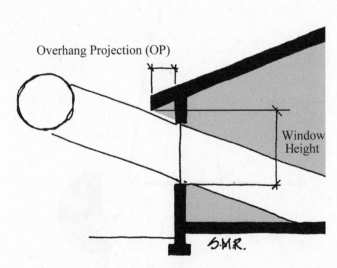

Overhang Projection (OP)

Window Height

S.M.R.

Noon at Winter Solstice

Fig. 4.6A. A winter-sun-side roof overhang allowing *winter* sun exposure for a window at 32° latitude

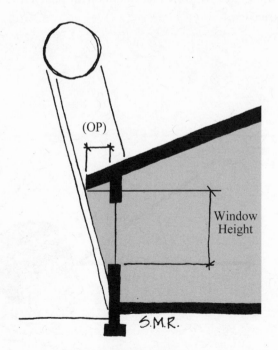

(OP)

Window Height

S.M.R.

Noon at Summer Solstice

Fig. 4.6B. A winter-sun-side roof overhang providing *summer* shade for a window at 32° latitude

Box 4.3. Latitude and F factor

North or South Latitude	F factor
28°	5.6–11.1
32°	4.0–6.3
36°	3.0–4.5
40°	2.5–3.4
44°	2.0–2.7
48°	1.7–2.2
52°	1.5–1.8
56°	1.3–1.5

Using lower F factors provides more shade for more of the summer. Those living in a climate of hot summers and mild winters will likely want to use the lower number in the F factor range when calculating overhang length; those living in climates of mild summers and cold winters will likely want to use the higher number in the F factor range.

The to-scale drawing way

This method is my favorite because I can see more exactly where the sun and shadows will be any time of the year, and I can quickly adjust proposed overhang/awning and window placements with pencil and eraser as needed.

Draw your building to scale.

Then look up your sun's location above the horizon (known as its altitude angle or elevation angle) at solar noon on both the winter solstice and the summer solstice (see figures 4.7A and 4.7B). You can look up the sun's altitude angle for your site based on the closest latitude to your site in box 4.2, the *sun path diagrams* in appendix 7, or you can calculate it using the following two equations (in which 23.44° is the earth's declination or tilt):

- summer solstice noon altitude angle =
  90° – (latitude + 23.44°)
- winter solstice noon altitude angle =
  90° – (latitude – 23.44°)

Conduct a web search for "latitude and longitude for your city and state" to find your site's latitude in decimal format. For example, 32.22° is Tucson, Arizona's, latitude.

Use a protractor to draw those sun angles as they relate to key design elements (overhangs/awnings, window frames) that might block the direct sun (figures 4.8A and 4.8B). Aim for overhang/awning lengths and window sizes and placements that allow the maximum amount of direct sun through the

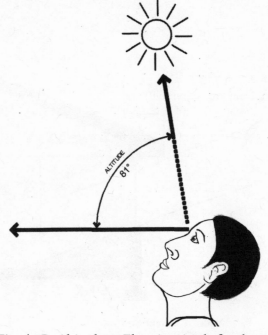

Fig. 4.7A. Altitude or Elevation Angle for the noonday sun's location above the equator-side horizon on the *winter* solstice at 32° latitude is 35° (see box 4.2).

Fig. 4.7B. Altitude or Elevation Angle for the noonday sun's location above the equator-side horizon on the *summer* solstice at 32° latitude is 81°.

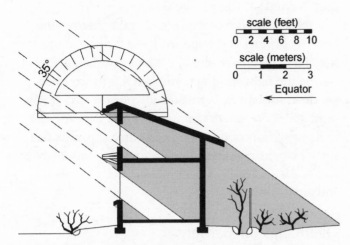

Fig. 4.8A. *Winter* solstice sun, heat, and light access illustrated with a to-scale drawing for 32° latitude. The lower awning is easier to access, is adjustable, and is retracted during the cold season to let in more sun and warmth.

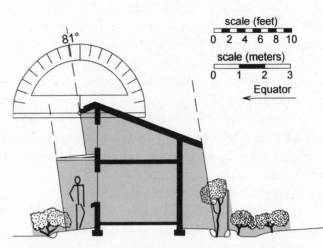

Fig. 4.8B. *Summer* solstice shade and cooling illustrated with a to-scale drawing for 32° latitude. The lower, adjustable awning has been extended to maximize shade for the hot season.
For video of this awning see Sun & Shade Harvesting page at www.HarvestingRainwater.com

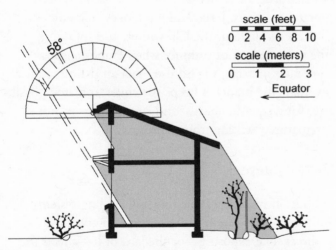

Fig. 4.8C. To-scale drawing for *spring* equinox. Temperatures are quickly warming up at this 32° latitude site, so less direct solar exposure is needed than in winter, but some direct passive heating is still provided by keeping the lower overhang retracted. Those in colder climates would want to design for more direct solar access.

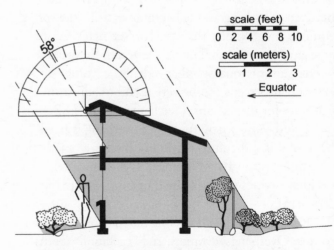

Fig. 4.8D. To-scale drawing for *fall* equinox. Temperatures are typically 20° to 30°F (11° to 16°C) warmer than at the spring equinox (see fig. 2.7 for equinox temperature fluctuations in Tucson, Arizona). This 32° latitude site experiences hot summers that linger into fall, so the lower awning is kept extended to provide needed shade and cooling.

equator-facing windows in winter (unless you're in the hot, wet tropics where you don't need winter heat), while shading out the sun in summer. Try moving different elements by extending, shortening, raising, or lowering the overhangs/awnings, and/or raising or lowering windows or making them taller or shorter to get the desired effect.

For solar design software that does this, see the Sun Angles and Passive Solar Design section of appendix 6.

Refine your drawing/design further by adding the sun's noonday altitude angle for the spring and fall equinoxes (March 21 and September 21).

You can use the following equation in which 0° is the earth's equinox declination or lack of tilt:

- spring/fall equinox noonday altitude angle = 90° – your latitude

Aim to shade most or all of the equator-facing windows on spring and fall equinoxes in climates where ambient temperatures are hot on those days. Let most of the sun through the windows in climates where ambient temperatures are cold on the equinoxes.

Note that in many areas there can be up to a 30° F (16° C) difference in average temperatures between the spring and fall equinoxes, with the spring equinox much cooler than the fall (see figure 2.7 for an example of this in Tucson, Arizona). Retractable winter sun- or equator-side overhangs, awnings, or exterior sunscreens can be very useful to adjust sun exposure for the seasons. (See figures 4.8A, 4.8B, 4.8C, and 4.8D where the upper overhang is fixed because it is difficult to reach, but the lower overhang/awning is seasonally adjustable.) Trellised winter-deciduous vines can be trained to approximate the same effect.

Passive heating and cooling ranges from low-cost to free. Overhangs, awnings, and vegetation screens consume no power or off-site water, create no pollution, and make no noise. They can be enhanced further with curtains or blinds opened when you want free daylight and heat, or closed if you want to deflect heat on hot days or to hold in warmth on cold nights. Likewise, fully operable windows can be opened or closed for ventilation and to function as inlets and outlets for indoor or outdoor warmth or cooling.

These practices enable you to sail your home like a ship in winds of free on-site resources.

See the real life example in chapter 5 describing how my brother and I applied these features at our site. To maximize potential further, read box 4.4 to choose the right windows, and box 4.5 to size your windows for best performance in your climate. If you want still more, see the Sun Angles and Passive Solar Designs section of this book's resource appendix 6 at www.HarvestingRainwater.com for information about additional strategies at the building, neighborhood, and city scale.

Overhangs for Gardens

Here in the low desert of southern Arizona, living overhangs can allow entry of low-angle winter sun while creating diffuse summer shade, conditions that vegetable gardens crave. To accomplish this, dig sunken garden beds under winter-deciduous, nitrogen-fixing, native mesquite trees (*Prosposis velutina*) that allow winter sun into the garden and place your more sun-loving plants in the south and southeast beds. Using this layout, I can harvest salad greens, artichokes, herbs, snow peas, garlic, onions, potatoes, carrots, Jerusalem artichokes, and edible flowers throughout the fall, winter, and spring. In the extreme heat of summer when evapotranspiration rates increase, the diffuse shade of the mesquite's overhanging branches keeps the chiles, tomatoes, basil, eggplant, squash, gourds, and summer greens from prematurely wilting from exposure.

Action Steps

- Shut off mechanical heating and cooling systems once in each season of the year to observe how direct solar exposure—or the lack of it—affects the comfort of your home. Don't do this, however, when indoor pipes could freeze (perhaps due in part to poor winter sun exposure).

- Use the overhang projection calculation or to-scale drawing method to determine the appropriate overhang sizes for your area. Compare the existing overhangs to what the calculation or drawing recommends. Observe how the overhangs affect your comfort.

- When installing or extending an overhang, consider the following options:

  - An overhang extending across the whole *summer-sun-side* (equator-facing side in latitudes *above* 23.5° latitude, or the side opposite the equator in latitudes *below* 23.5° latitude) of your building to benefit your wall as well as your windows.

  - Awnings for your summer-sun-side windows and/or wall (fig. 4.9).

  - A seasonally retractable/extendable summer-sun-side awning (see figures 4.8A through 4.8D)

- Consider a very sparse trellis to support dense winter-deciduous vines which will provide leafy shade in summer. (See figure 4.10.) In the winter, remove and mulch the uppermost sections of the leafless vines so your windows get direct winter sun (see "Integrated Design Pattern Five: Maintaining Winter Sun Access"). Use rooftop runoff to irrigate the vines. (Note: While water-harvesting earthworks

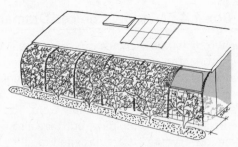

Fig. 4.10A. Winter-sun-side trellis in *summer*

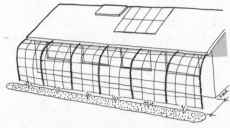

Fig. 4.10B. Sparse winter-sun-side trellis in *winter*

should be placed a minimum 10 feet (3 m) from the house, vegetation can be planted closer to a building. Just train the roots of the plants to find the water harvested in the more distant earthworks. Do this by applying irrigation water on the side of the plant closest to the earthworks. Then every month or two in the growing season move your irrigation emitter or hose a foot further from the plant and a foot closer to the earthworks, until you eventually end up watering the plant via its extended roots within the earthworks.)

- Position *exterior* blinds or shutters on east- and west-facing windows to help block direct summer sun before it enters and heats the building (fig. 4.11).

- Install covered porches and/or shade trees on the east, west, and summer-sun sides of your building as presented in the next pattern.

- Aim to design, purchase, retrofit, rent, or use buildings in which equator-facing rooms have appropriately sized equator-facing window(s) (box 4.5) based on the building's floor area and winter heating

Fig. 4.9. Window awning shading/cooling windows and wall in summer

Box 4.4. Window Choices Dramatically Affect Passive/Free Heating and Cooling

The wrong window choice can dramatically diminish passive heating and cooling potential even if your building is oriented perfectly and has ideal overhangs. I've stood in the full winter sun on the inside of equator-facing windows and felt no heat. The problem is the use of inappropriate low-emissivity (low-E) windows.

Low-E glazing/windows have a transparent film that reduces heat flow through the glass by reflecting radiant heat back in the direction it came from. The enhanced insulation qualities of these windows can hold onto heat generated inside a building, but they also keep out radiant heat from the winter sun.

Some low-E windows are designed to allow higher solar heat gain for passive heating. However, these are hard to find because most manufacturers and builders focus on conserving energy by using higher reflection-value windows. This approach ignores the potential to integrate energy conservation and energy production through lower reflection-value, equator-facing windows that harvest winter heat.

Before installing any equator-facing window, check its National Fenestration Rating Council (NFRC) sticker, which lists its transmittance characteristics. Beware of salvaged windows that may no longer have NFRC stickers. If a window is being installed facing the equator with the goal of achieving passive solar heating, the window must meet the following criteria:

- A Solar Heat Gain Coefficient (SHGC) greater than 0.60. The higher the number (1 is the highest), the greater the solar gain.

- A U-factor rating of potential heat loss of 0.35 or less. The lower the number, the better the window keeps heat inside.

- A Visible Light Transmittance (VT) greater than 0.65.[5] Below 0.40 is considered a tinted window.

To achieve maximum passive heating potential for my projects in the southwest U.S. that have well-designed overhangs, I've chosen operable, non-low-E, untreated, clear double-glazed or double-paned windows for all *equator-facing* windows. To achieve maximum insulation value for windows facing all other directions, I've chosen operable, double-glazed low-E windows. To reduce loss of heat on winter nights through the non-low-E windows I do not install an excessive amount of equator-facing glass (see box 4.5) and I close curtains over these windows nightly in winter then open the curtains in the morning.

Whatever the coating (or lack of it) on your glass, don't overlook your window's ventilation potential. Choose operable windows you can open and close to let in warmer or cooler outside air as needed. The window section of appendix 5 expands on this further, while appendix 8 shows you how you can direct or divert more air flow in and out through your windows.

Box 4.5. Size Your Equator-Facing Windows to Maximize Free Heating and Cooling

For buildings in various climates using direct-gain/distributed mass systems, the table below gives you guidelines on optimal equator-facing window surface areas and optimal interior mass surface areas to reduce daytime overheating and nighttime overcooling.

Direct-gain buildings use the direct entry of winter sun coming through equator-facing windows to heat the interior space and the mass within. Thermal mass absorbs and holds heat, reducing interior temperature fluctuations.

Distributed mass is thermal mass that is distributed throughout a large interior area in the form of exposed concrete, masonry, earth, or stone floors (which work best if insulated from the earth below), walls, and plasters (which work best if insulated from the exterior). Low-mass gypsum plaster or wallboard can even be used as distributed mass when installed in two 5/8-inch (15-mm) thick layers, but due to its lower ability to store heat, divide that wallboard area by 4 when determining area of mass per area of equator-facing window (aperture) in the table below.[6]

Examples following the table illustrate how to use it.

Table reprinted with permission from *Passive Solar Architecture: Heating Cooling, Ventilation, Daylighting, and More Using Natural Flows* by David A. Bainbridge and Ken Haggard, Chelsea Green Publishing, 2011.

Box 4.5. (continued)

Climate	Very Cold	Cold	Temperate			Tropical Dry	Tropical Wet
Thermal Load / Heating and cooling needs	Heating only	Heating only	Heating and some cooling	Balanced heating and cooling through-out the year	Cooling and some heating	Cooling with small heating	Cooling only
Percent of floor area needed to equal the total equator-facing window area	10–20%	10–25%	14–20%	9–15%	8–13%	6–11%	0%
Area of exposed interior mass (in ft² or m²) needed per area of equator-facing glass (in ft² or m²)	5–10 ft² (0.46–0.92 m²)	6–11 ft² (0.5–1.02 m²)	8–12 ft² (0.74–1.11 m²)	9–14 ft² (0.83–1.3 m²)	9–14 ft² (0.83–1.3 m²)	9–14 ft² (0.83–1.3 m²)	9–14 ft² (0.83–1.3 m²)

I redesigned an old 200 ft² (18.5 m²) cinderblock one-car garage into a direct-gain/distributed mass "cottage" that I now call a "garottage." Using the table above, I calculated its direct-gain/distributed mass window area as follows:

• I live in a climate where homes typically need cooling and some heating. According to the table, this requires a total equator-facing window area 8–13% of the floor area.

• My garottage floor area equals 200 ft² (18.5 m²), and I used 12% of my floor area as my window area goal. So, 200 x 0.12 = 24 ft² (2.2 m²) of needed equator-facing window area is needed.
    I attained this using
        - One 2.5' x 3.5' window = 8.75 ft² (0.81 m²)
        - One 1.5' x 3' door window = 4.5 ft² (0.41 m²)
        - Two 2.3' x 2.3' windows = 10.58 ft² (0.98 m²)
        - Total equator-facing glass = 23.83 ft² (2.2 m²)

Note: I only measured the actual area of glass, and did not include any part of the windows' frame or sash.

According to the table, the recommended area of interior mass for my site is 9–14 ft² (0.8 –1.3 m²) of mass / ft² (m²) of equator-facing window area. Within the range, I used 14 ft² (1.3 m²) so my goal was 14 x 23.83 = 333.62 ft² (in metric, 1.3 x 2.2 = 2.86 m²) of needed exposed interior mass.

I exceeded that with the interior exposed cinder block wall area minus the area of windows, doors, cabinets, etc. = 343 ft² (3.8 m²) of mass. The cinder block wall was *exsulated*—insulated on the exterior.

A thermally exposed concrete floor slab (uncovered by carpet, closets, etc.) could count as more mass, but it would ideally be insulated from the ground below in my climate, and in colder climates must be insulated in order to count as thermal mass.

If I had installed 5/8-inch (15-mm) gypsum plaster or wallboard two-layers thick over framed insulated walls, I could have counted that 359 ft² (33.3 m²) of exposed wallboard area as mass. However, due to its lower ability to store heat, I would have to divide the wallboard's area by 4 (359 ÷ 4 = 89.75 ft² [or 8.3 m²] of mass). That's not enough mass by itself, but if I added 200 ft² (18.5 m²) of thermally exposed concrete slab floor, I would attain 12 ft² (1.1 m²) of exposed interior mass per 12 ft² (1.1 m²) of equator-facing glass—well within the recommended range.

See more on the garottage in the Real Life Example of chapter 5.

Fig. 4.11. East-facing exterior blinds doubling as pollinator habitat. Beneficial carpenter bees live in the hesperaloe flower stalks of which the blinds are made. Note dappled shade from trees to east provides additional cooling!

needs. In addition, make sure the rooms that are used most during the day are located on the winter-*sun*- or equator-facing side of the building for passive winter heating and lighting. Night-use bedrooms should be placed on the winter-*shade*-facing side of buildings, since blankets and others can warm you.

INTEGRATED DESIGN PATTERN THREE

SOLAR ARCS

A solar arc is created using a number of shading elements such as trees, cisterns, trellises, covered porches, and overhangs laid out in a *C-shape*, like that of a big open-armed hug that welcomes the full potential of the winter sun (fig. 4.12). At the same time,

it deflects much of the summer sun using the arc's "back, shoulders, arms, and hands." Water-harvesting earthworks are the foundation of solar arcs when vegetation is used as the sheltering element (figure 4.13 shows the growth of the solar arc from basin to young trees to mature trees). Situate the earthworks and trees close enough to buildings to use roof runoff as the primary source of irrigation water and household greywater as the secondary source. As the shade trees grow they beautify your yard, clean the air, and dramatically cool summer temperatures (see boxes 4.6 and 4.7).

Solar arcs shade buildings, gardens, and gathering spots in the yard from the summer's northeastern morning and southwestern afternoon sun. Put trees that drop their leaves in winter on the east and west arms of the arc to allow filtered sun to penetrate the arc early in spring and late into fall. Evergreen trees work well on the northern band of the arc to block summer sun and deflect cold northerly winter wind. The same principle works in the southern hemisphere, only the directions change.

Action Steps

• See if you have any elements of a solar arc in place around your home or garden, such as an existing shade tree or building.

Box 4.6. "Cool and Clean" or "Cool and Polluted"

The generation of electricity used to mechanically air condition an average household *causes* about 3,500 lbs (1,587 kg) of carbon dioxide and 31 lbs (14 kg) of sulfur dioxide to be released from powerplant smoke stacks each year.[7] Cooling solar arcs of trees *consume* carbon dioxide and produce oxygen—up to 5 lbs (2.2 kg) of oxygen per day per tree.[8] According to the National Arbor Day Foundation, over a 50-year period, a well-placed shade tree can generate $31,250 worth of oxygen and provide $62,000 worth of air pollution control.[8]

See The Value of Trees for additional benefits at www.dvrpc.org/Green/pdf/ValueofTreesStatsSheet.pdf

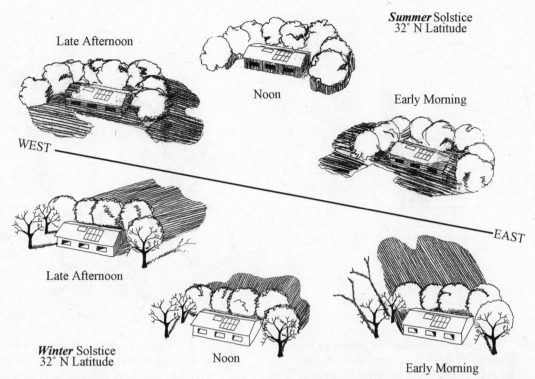

Fig. 4.12A. Solar arc of trees with an *east-west* oriented building at 32° N latitude. View of building's winter-sun side. Roof-mounted solar panels and solar water heater receive full sun all year round. Winter-deciduous trees on the east and west sides of building expose it to more winter morning and afternoon heat and light than evergreens would. Compare to figure 4.5A.

For video of this see the Sun & Shade Harvesting section at www.HarvestingRainwater.com.

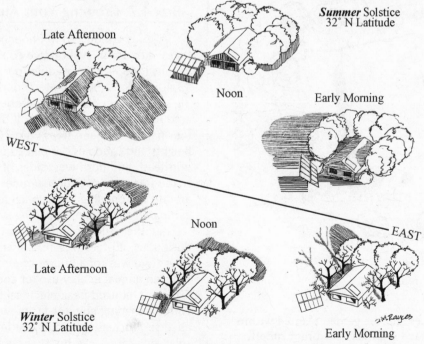

Fig. 4.12B. Solar arc of trees with a *north-south* oriented building at 32° N latitude. View of building's winter-sun side. Solar panels are mounted on a special sun-tracking rack on south-side of trees to receive full sun all year round. Winter-deciduous trees on the east and west sides of building expose it to more winter morning and afternoon heat and light than evergreens would. Compare to figure 4.5B.

Fig. 4.13A. Water-harvesting basins placed to help grow a solar arc of trees, and one basin on the winter-sun side of the house to grow a sunken winter garden (32°N latitude)

Fig. 4.13C. Trees at full size forming a living solar arc, and a thriving winter garden. Solar panels and solar water heater have full access to sun all year.

Fig. 4.13B. Young trees and the garden planted within the basins and irrigated with harvested roof runoff. Dotted lines represent supplementary greywater lines. Greywater is directed to fruit trees where people gather for convenient irrigation and fruit harvest.

Box 4.7. Growing Your Air Conditioner

A study conducted in Phoenix, Arizona found that water use in evaporative coolers averages 65 gallons (246 liters) per cooler per day, or about 13,400 gallons (50,725 liters) during the cooling season from March to October.[9] That same amount of water could fulfill all the water needs of four native mesquite trees with 20-foot (6-meter) heights and canopies.[10] If placed on the east, west, northeast, and northwest sides of a home, these shade trees could reduce summer temperatures around the building by as much as 20°F (11°C) compared to the same building without shade.[11]

Another study found that just three winter-deciduous, 25-foot (7.6-m) tall and 25-foot wide trees (two west of a house, one on the east) reduced annual energy use for cooling by 10 to 50% and reduced peak electrical use up to 23%, without shading equator-facing windows and adversely increasing winter heating needs.[12] A solar arc of trees can boost these benefits further (see figure 4.14).

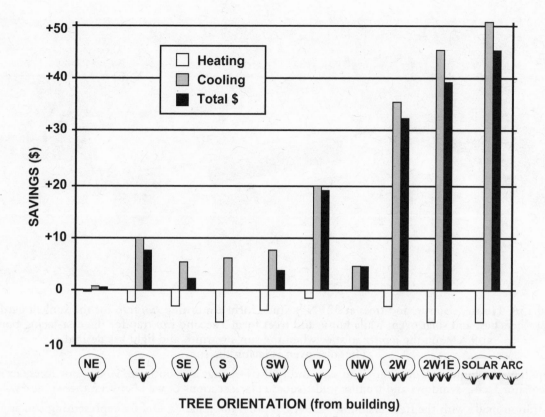

Fig. 4.14. Place trees wisely. Graph illustrates annual heating, cooling, and total energy use differences (energy savings are positive, energy use increases are negative) for an east-west oriented building in Sacramento, California, based on placement of adjoining winter-deciduous trees. Savings are expressed as a percentage of energy (kWh) saved for a shaded building compared to an unshaded building, with single shade trees located at cardinal (E, S, W) and inter-cardinal (NE, SE, SW, NW) points around the building. Results are also given for two west shade trees combined with one east shade tree (2W1E), and for a solar arc of trees—one east, one northeast, one northwest, and two west, but NO trees to south (SOLAR ARC). Note how a misplaced shade tree on the south side can result in heating increases that offset cooling savings (this increases in cooler/colder climates). Adapted from "Potential of Tree Shade for Reducing Residential Energy Use in California" by James R. Simpson and E. Gregory McPherson, *Journal of Arboriculture* 22(1), 1996

• Map where missing pieces of a solar arc should be located to complete it and benefit your home or garden. Create water-harvesting earthworks and/or install cisterns to sustain shade trees.

• Consider a design for *layers* of solar arcs. One layer can be close to the house to shelter the house. The next layer can be along the fence line of the yard to shelter the yard, you outside, tender veggies, or fruit trees planted within the arc. The further you are from the house and its resources (roof runoff, greywater, caregivers living in the home) the hardier (more drought-, heat-, and cold-tolerant) the arc's plant species should be. Typically hardy native perennials are best for the peripheral arc.

INTEGRATED DESIGN PATTERN FOUR

SUN & SHADE TRAPS

A sun & shade trap creates a nice place to plant a garden or take an *outdoor* nap. Elements making up this trap can include cisterns, tall low-water-use vegetation, a house, trellis, shed, or other shading elements. A sun & shade trap is more open to the morning sun than a solar arc. In the northern hemisphere the *L-shaped* trap is open to the east and south, and closed to the north and west. In the southern hemisphere it is open to the east and north, while closed to the south and west. (See figure 4.15.) The sun & shade trap creates microclimates ideal for gardens, sensitive plants, cozy outdoor gathering areas, and hammock roosts.

Fig. 4.15A. House, cisterns, and trees at 32° N latitude form a morning *sun* trap for the sunken garden, cistern benches, and solar oven; while house and trees form a second *sun* trap for the east-facing porch at 9 A.M. on the *winter* solstice when the sun's warmth and light are desired.
The solar oven is facing the sun.

Note how the building's orientation to sun, and roof overhangs also optimize *winter sun* access to its equator-facing windows and trombe walls (square black sections of wall between the windows).

Illustration made with the free version of Sketch Up program, a handy tool for representing sun angles and shadow lengths for any inputted date and time.

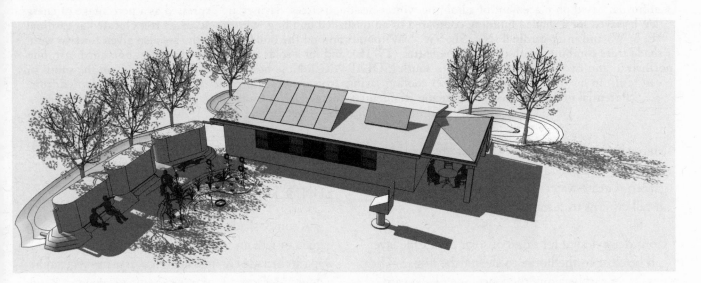

Fig. 4.15B. Same house, cisterns, and trees form an afternoon *shade* trap for the sunken garden and cistern benches; while house and trees form a second *shade* trap for the porch at 4 P.M. on the *summer* solstice when cooling shadows are desired on hot afternoons.
The solar oven is facing the sun.

Note how the building's orientation to sun, and roof overhangs also optimize *summer shade* access to its equator-facing windows and trombe walls.

The sun & shade trap effect illustrated in these two figures works every day of the year, with sun and shadow angles changing seasonally. In climates or situations where more shade is desired, particularly in the morning, the solar arc pattern could be used instead of the sun & shade trap pattern.

Fig. 4.16A. 9 *A.M.*, garden sun trap, Lancaster residence, Tucson, Arizona, February 22. In cold morning hours, the winter garden and the home's equator-facing windows and solar panels are in full sun to quickly warm up, grow crops, and produce energy.

Fig. 4.16B. *Noon*, garden sun trap, Tucson, Arizona, February 22. As temperatures have warmed up, winter garden is still largely in direct sun, though the leafless winter-deciduous mesquite tree casts some diffuse shade over a third of the garden. The sun has set on the east face of the two 1,300-gallon (4,920-liter) ferrocement cisterns behind the chairs west of the garden.

Box 4.8. Sun & Shade Trap Cisterns Acting as a Windbreak, Firebreak, Sun Screen, Privacy Screen, Property Fence, and Economic Stimulus

Creating myriad beneficial relationships (the seventh water-harvesting principle) make static cisterns dynamic. Thus the cisterns in figures 4.16A, 4.16B, and 4.16C were selected and placed to transform them into much more than water tanks. We selected and purchased locally made tanks to strengthen our local water-harvesting economy. Then we placed the tanks on our western property line to act as a section of property fence. In this location the tanks deflect hot, dry prevailing winds from the southwest; shade the garden and a seating nook from the hot afternoon sun; give our neighbors and ourselves a privacy screen; and serve as a sound-deflecting, concrete, water-filled firebreak.

Fig. 4.16C. 3 *P.M.*, garden shade trap, Tucson, Arizona, February 22. During the hottest hours of the day, the winter garden is in the cool shade cast by the sun & shade trap created by the cisterns, orange tree, and ramada/shade structure (the northeast corner of which is showing in the upper left corner of the photo) west of the garden. This shading reduces the plants' heat stress and evapotranspiration, making the sole source of water—harvested rainwater—go further.

My sun & shade trap *captures winter sun in the mornings* to burn off frosts and warm things up, but *shades out sun on hot afternoons*. This extends the cool growing season of my low desert garden by two months. I can plant a month earlier and get an extra month of growth by keeping my vegetables from bolting (going to seed). The afternoon shade reduces plant evapotranspiration, heat stress and drought stress, and reduces pest problems. I enjoy hanging out in my sun & shade trap garden, because, like the plants, I get warmth in the morning when I want it, and shade in the afternoon when I need a break from the sun.

Action Steps

• Before you plant, identify and map the areas of your site where a sun & shade trap might make sense, and map any existing elements already in place.

• Harvest rainwater to support vegetative elements in the sun & shade trap. As Mr. Phiri says, "Plant the water before you plant the trees."

• Determine if a new cistern could be placed to help create a sun & shade trap for a garden or patio.

INTEGRATED DESIGN PATTERN FIVE

MAINTAINING WINTER SUN ACCESS

Just as you harness the full potential of the rain, do the same for the winter sun in order to reduce your energy consumption and simultaneously reduce water consumed to generate that energy (as explained in appendix 9). Keep solar exposure open for winter sun-facing windows, gardens, solar water heaters, solar panels, and solar ovens. The Village Homes development in Davis, California, found that with winter solar access retained, simple solar homes can achieve 40 to 50% of their winter heating needs from the sun, while more sophisticated designs can meet 85% of their heating needs.[13] I rely on south-/equator-facing windows for most of my home's heating needs. A solar water heater provides all my hot water and ten solar panels provide all power needs. But misplaced

trees could seriously cripple this performance since even the *shadow cast by a leafless tree could block out over 50% of the potential heat and light* (see figures 4.17A and 4.17B).[14]

When planning where to place above-ground cisterns or trees, and the water-harvesting earthworks that will sustain them, be conscious of the shadows they will cast. Place them an appropriate distance from the house so they won't block the winter-sun/equator-facing side of your home, solar water heaters, and solar PV panels (figures 4.17C and 4.17D). Placed even farther from the house, they will allow access to all of the above plus to lower-elevation trombe walls, solar hot air collectors, solar ovens (figures 4.17E and 4.17F), winter gardens, and solar greenhouses/cold frames.

Winter solstice shadow ratio: determining shadow lengths

You can determine the longest shadow an object will cast on the winter solstice (December 21 in the northern hemisphere, June 21 in the southern) by looking up your latitude in box 4.9, and multiplying the object's height by the associated factor of the winter solstice shadow ratio. For example, at 32° N latitude the solar noon ratio is 1:1.45, so for every foot (or meter) of an object's height, the shadow cast at noon on December 21 will be 1.45 feet (or meters) long. Multiply the height of a mature 20-foot (6-m) tall tree by 1.45 to get 29 feet (8.7 m)—the length of the noonday shadow cast to the north at this latitude. (See figures 4.18A and 4.18B.)

Shadow ratios for the benchmark times of solar 9 A.M. and 3 P.M. are given. Since the sun is much lower in the sky at these times, the shadows will be much longer than those at noon. The hours between 9 A.M. and 3 P.M. are ideal for winter sun access for passive heating, solar cooking, producing solar power, and solar heating of water. Before 9 A.M. and after 3 P.M. the potential use and efficiency of the sun is greatly reduced since the sun is lower in the sky, resulting in its rays traveling through more heat- and light-diffusing atmosphere.

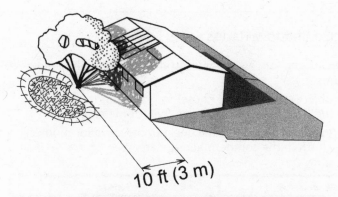

Fig. 4.17A. Majority of winter solar access lost. At 32° latitude from 9 A.M. to 3 P.M. on the winter solstice, a misplaced mature *evergreen* tree blocks winter sun exposure for winter-sun/equator-facing windows and rooftop solar water heater and solar panels. Shadows are shown as gray for 9 A.M. and 3 P.M. Solar noon shadows are black.

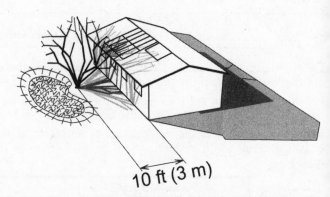

Fig. 4.17B. Over half of winter solar access lost. At 32° from 9 A.M. to 3 P.M. on the winter solstice, the bare branches of a misplaced mature *deciduous* tree shade out over 50% of winter sun exposure, increasing heating costs and severely hampering solar power production.

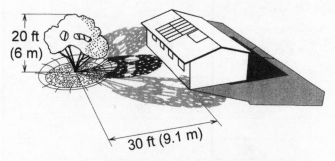

Fig. 4.17C. Winter solar access retained. At 32° latitude from 9 A.M. to 3 P.M. on the winter solstice, an evergreen tree planted sufficiently far from the house, so solar access is retained from the bottom of the winter-sun/equator-facing windows up to the equator-facing slope of the roof. See box 4.9 to estimate winter-shadow lengths.

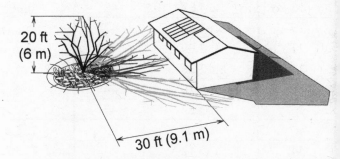

Fig. 4.17D. Winter solar access retained. At 32° latitude from 9 A.M. to 3 P.M. on winter solstice, a winter-deciduous tree lets more warming winter light onto the ground than an evergreen tree, while creating better growing conditions for a winter garden beneath the tree.

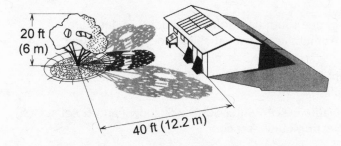

Fig. 4.17E. Winter solar access retained for entire winter-sun/equator-facing wall, solar oven, solar hot air collectors beneath windows, and roof at 32° latitude

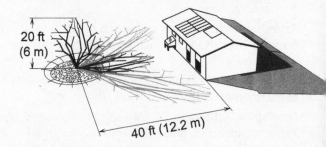

Fig. 4.17F. Winter solar access retained for entire winter-sun/equator-facing wall, solar oven, trombe walls beneath windows, and roof at 32° latitude. If you miscalculate the distance of the tree from the building, a winter-deciduous tree will not be as bad as an evergreen.

Box 4.9. Winter-Solstice Shadow Ratios

Determine objects' shadow lengths and direction on the day with the longest shadows of the year using the following shadow ratios and the sun's azimuth angles for your latitude.

Use the ratio 1 : x to get the *length* of the shadow cast by an object (tree, cistern, house) for each foot (or meter) of height. Multiplying the height of the object times x tells you the length of the shadow in feet (or meters).

To get the *direction* of the shadow cast by an object, look at the Azimuth angle—the angle of the sun's location east or west of True South in the northern hemisphere or of True North in the southern hemisphere. (See figures 4.19 and 4.20.)

Latitude °N or °S	9 A.M. & 3 P.M.		10 A.M. & 2 P.M.		Noon	
	Winter-Solstice Shadow Ratio[1]	Azimuth Angle[2]	Winter-Solstice Shadow Ratio[1]	Azimuth Angle[2]	Winter-Solstice Shadow Ratio[1]	Azimuth Angle[2]
0°	1: 1.17	58°	Angles & ratios for 10 A.M. & 2 P.M. are typically less important at lower latitudes where shadows are shorter than those cast at higher latitudes.		1: 0.43	0°
4°	1: 1.27	56°			1: 0.52	0°
8°	1: 1.38	53°			1: 0.61	0°
12°	1: 1.51	51°			1: 0.71	0°
16°	1: 1.67	49°			1: 0.82	0°
20°	1: 1.86	47°			1: 0.95	0°
24°	1: 2.09	46°			1: 1.09	0°
28°	1: 2.38	45°			1: 1.25	0°
32°	1: 2.76	44°			1: 1.45	0°
36°	1: 3.27	43°			1: 1.69	0°
40°	1: 4.00	42°	1: 2.64	29°	1: 2.00	0°
44°	1: 5.12	41°	1: 3.22	29°	1: 2.41	0°
48°	1: 7.04	41°	1: 4.09	28°	1: 2.98	0°
52°	1: 11.2	41°	1: 5.55	28°	1: 3.85	0°
56°	1: 26.15	40°	1: 8.50	27°	1: 5.36	0°
60°	Sun below horizon on winter soltice[1]		1: 17.56	27°	1: 8.70	0°
64°			Sun at horizon on winter solstice[1]		1: 22.37	0°
68°	At & beyond Arctic/Antarctic Circles (66.56°N/S), sun never shines on winter solstice.[3]					

Noon is solar noon, **9 A.M. & 3 P.M.** are 3 hours before & after solar noon, **10 A.M. & 2 P.M.** are 2 hours before & after solar noon.

Azimuth angle: Degrees sun is east (A.M.) or west (P.M.) of due south (in northern latitudes) or due north (in southern latitudes) on winter solstice at given times. Azimuth angle is 0° at all latitudes at **solar noon** *on all dates*.

Winter-solstice shadow ratio: Ratio of an object's height to the length of its shadow cast on winter solstice at given times

For shadow ratios and azimuth angles for *every degree of latitude* see Winter-Solstice Shadow Ratio & Azimuth Table on the Sun & Shade Harvesting page at **www.HarvestingRainwater.com**.

1. Shadow ratio = 1 ÷ tangent of altitude angle.
    At noon on the winter solstice, the altitude angle = 90° − (latitude + 23.44°).

2. http://www.esrl.noaa.gov/gmd/grad/solcalc/, accessed 7/5/2012

3. http://en.wikipedia.org/wiki/Arctic_Circle, accessed 7/16/2012

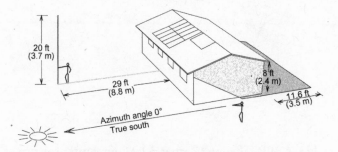

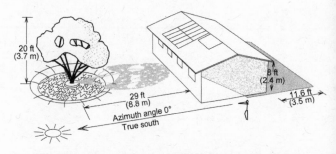

Fig. 4.18A. Using the winter-solstice noon shadow ratio of 1:1.45 for 32° N latitude from box 4.9 to determine length of solar noon shadows cast, multiply mature tree height of 20 ft (6 m) by 1.45 = solar noon shadow length of 29 feet (8.7 m). Stand at proposed tree location (with or without a pole as tall as proposed tree height) and walk off its estimated shadow length to determine where the tree will or will not adversely shade the building.

Fig. 4.18B. Shadows cast at noon on winter solstice at 32° N latitude once tree reaches mature size. For video of this see the Sun & Shade Harvesting page at www.HarvestingRainwater.com.

Shadow ratios are also given for the hours of 10 A.M. and 2 P.M. in box 4.9 for higher latitudes with longer winter shadows where the challenge of maintaining solar access increases, particularly in dense urban settings. Due to this increased challenge, some solar rights requirements—such as the Living Building Challenge Imperative 18: Rights to Nature—only require direct solar access to be maintained between these winter solar hours.[15] Interestingly, research has found that solar zoning maintaining this type of winter solar access does not reduce desired high urban densities.[16]

Azimuth: determining sun and shadow angles

The sun's Azimuth angle—the angle of the sun's location east or west of True South in the northern hemisphere or of True North in the southern hemisphere—is also given in Box 4.9, since shadows are cast in the opposite direction of the Azimuth. For example, at 32° N latitude at noon, the Azimuth is 0°, so the sun is due True South and the shadow will be cast due True North. While at 3 P.M., the sun is 44° west of True South, so the shadow will be cast 44° east of True North at that time. (See figures 4.19 and 4.20.)

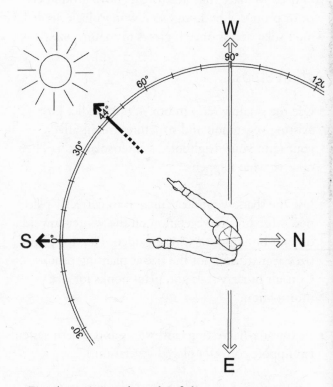

Fig. 4.19. Azimuth angle of the sun at 3 P.M. on the winter solstice at 32° N latitude. The Azimuth is the sun's location in degrees east or west of True South in the northern hemisphere or of True North in the southern hemisphere.

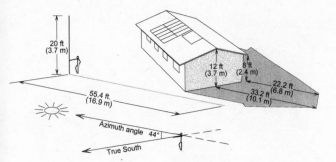

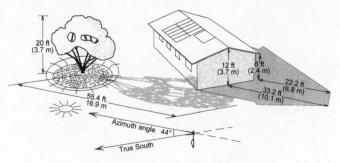

Fig. 4.20A. Using winter-solstice 3 P.M. shadow ratio of 1:2.77 from box 4.9 to determine length of shadow cast at 32° N latitude, and using Azimuth angle of sun's location from True South to determine direction of shadow

Fig. 4.20B. Shadows cast at 3 P.M. on winter solstice at 32° N latitude once tree reaches mature size

Using the shadow ratios and azimuth angles in box 4.9 you can determine how far away from your home in the direction of the equator and winter sun you need to place that tree (or an above-ground cistern or proposed building) so it won't block needed winter solar access once it grows to mature size.

Action Steps

• Use the shadow ratio in box 4.9 to predict how existing vegetation and structures might affect your (and your neighbor's) winter solar gain. Then observe what happens.

• Use the shadow ratio calculation to correctly place new structures or vegetation on the winter-sun side of your buildings or winter garden. Plan for plants' sizes at maturity, not the size at planting time. Consult nursery staff and plant books for size information.

• Plan water-harvesting earthworks and cistern systems to support correctly placed vegetation.

Box 4.11. Pruning for Winter Sun

If you currently have mature trees blocking your winter heat and light you can prune them to regain winter solar access. Use box 4.2 to determine what part of the sun's path you could reopen with pruning, then prune if appropriate; see also figure 4.21.

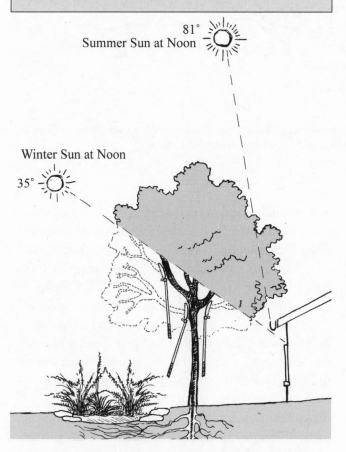

81°
Summer Sun at Noon

Winter Sun at Noon

35°

Fig. 4.21. A tree pruned to allow direct winter sun access through winter-sun/equator-facing windows, while shading out the summer sun at 32° latitude

INTEGRATED DESIGN PATTERN SIX

RAISED PATHS, SUNKEN BASINS

Keep access ways "high and dry" and planted areas "sunken and moist." Always pair a raised path with a sunken basin to capture runoff and grow shelter and beauty for the path (fig. 4.22). Trees planted in these water-harvesting basins shade and beautify the adjacent roadways, paths, and patios. This reduces excessive solar exposure and in turn the risk of skin cancer, the fastest growing form of cancer in the United States,[20, 21] while creating a comfortable place to drive, walk, ride a bike, or converse. In cold climates raised paths and roads also help keep them ice free.

Shade trees absorb much of the rain falling within the diameter of the tree canopy and runoff flowing around their bases. This creates a living flood control system and filters runoff-borne contaminants like nitrates, phosphorus, and potassium, which trees consider food.[22]

A parking lot can be planted with a "living carport" of flood-controlling, pollutant-filtering native shade trees irrigated solely by the parking lot's runoff. If low-water-use food-producing trees such as the velvet mesquite are planted, it becomes a "parking orchard." (See figure 4.23.)

Action Steps

• Observe the relative height of paths, sidewalks, driveways and streets compared to adjacent planting areas in your home and community. Do you see the "raised path, sunken basin" pattern or a

Fig. 4.22. Raise pathways, and sink mulched and vegetated basins.

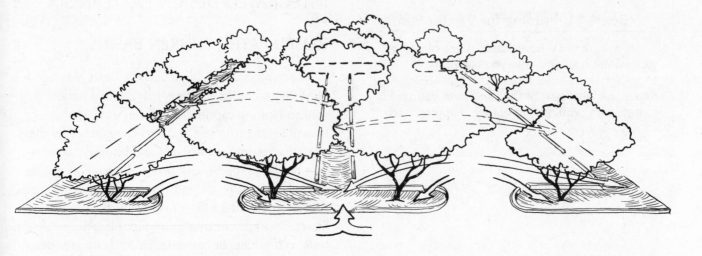

Fig. 4.23. A parking orchard of low-water-use, food-producing shade trees passively irrigated by runoff from a raised parking area harvested within sunken, mulched basins

sunken path, raised planting area pattern? Is stormwater being directed to vegetation, asphalt, or storm drains?

• Observe undisturbed natural areas. You will most likely find the largest, densest vegetation in depressed areas and along drainages where water concentrates.

• Identify and map areas where you can develop the raised path, sunken basin pattern at home. Create water-harvesting earthworks by digging sunken basins, then use the newly available dirt to create the raised paths.

INTEGRATED DESIGN PATTERN SEVEN

REDUCE PAVING AND MAKE IT PERMEABLE

Rainwater is like a naked person—it won't stick around if you put it on hot pavement in summer. So, we must reduce the amount of impervious paving on our sites and neighborhoods, while increasing shady vegetation (compare figures 4.24A and 4.24B).

Much of the heat stored in urban concrete and asphalt during the day is released in the late afternoon and evening, keeping temperatures high. Excessively wide, unshaded streets and dense unshaded development directly contribute to the heat island effect, and were found to raise maximum daytime temperatures by 10°F (5.5°C) in Davis, California.[23] (See figure 4.25A for an example of a wide, tract-home street.)

Typical residential streets in the western U.S. are up to 40 feet (12.2 m) wide, but there are alternatives. At the innovative Village Homes housing development in Davis, California, streets are 20 feet (6.1 m) wide.[24] By narrowing streets, using cul-de-sacs, and limiting driveways to the length of a vehicle, Village Homes made 15% more land available for community gardens, orchards, tree-lined walkways, and bike routes.[25] Adjoining trees create a canopy over the narrow streets, shading 80% of the streets' area, reducing summertime temperatures, reducing the need for buzzing air conditioners, and creating a much quieter and more pleasant neighborhood (fig. 4.25B).

Direct street runoff to adjoining water-harvesting earthworks, try to reduce the amount of paving, and see if some of remaining pavement should/could be made permeable. In volume 2, the chapter "Reducing Hardscape and Creating Permeable Paving" features a number of permeable pavements that can reduce runoff by up to 90% and yield significantly cleaner stormwater compared to impervious pavements.[26]

Fig. 4.24A. A dehydrated and exposed residential lot dominated by impervious pavement, sparse vegetation, and bare compacted earth graded to drain all runoff to the street

Fig. 4.24B. A hydrated and sheltered residential lot dominated by water-harvesting earthworks, native low-water-use vegetation, and permeable paving graded to minimize runoff and utilize it on-site. See figures 3.16 to 3.18 for more strategies.

Fig. 4.25A. Creating the summer *heat*-island effect, which can *increase* temperatures by up to 10°F (5.5°C). A wide, exposed, solar-oven-like street in Tucson, Arizona. There are no trees in the public right-of-way.

Fig. 4.25B. Creating the summer *cool*-island effect, which can *reduce* temperatures by up to 10°F (5.5°C). A narrow, mature tree-lined, and shaded street in Village Homes, Davis, California. Trees are planted in the public right-of-way just a couple of feet from street curb to shade more of the street. In winter these trees will drop their leaves to let in more heat and light when needed. See Fig. 3.17 for a narrow, tree-lined street in Tucson, Arizona.

Fig. 4.26. Grow a classroom instead of building one, and grow an air conditioner instead of buying one. Admiring the birds in the canopy of oak trees forming a living, self-cooling outdoor classroom at the Wildflower Center in Austin, Texas

Action Steps

• Look for examples of pervious and impervious paving around your home and community.

• Determine how you can reduce the paving on your site, and make remaining pavement more permeable. Implement these changes.

• Consider turning your driveway into a park-way by limiting its size to the length and width of your vehicle.

• Consider using porous brick, cobbles, or stabilized gravel in place of impervious paving materials.

• Plan water-harvesting earthworks and appropriate plantings in areas where paving is removed, and adjacent to areas where paving remains.

• Consider growing outdoor "rooms" of shade trees to replace or reduce the need for more buildings (fig. 4.26).

• Daylight or uncover buried waterways. Show the flow.

• In volume 2, read the chapter "Reducing Hardscape and Creating Permeable Paving."

INTEGRATING THE ELEMENTS AND PATTERNS OF YOUR SITE TO CREATE A REGENERATIVE LANDSCAPE DESIGN

The preceding integrated design patterns should have you thinking about, mapping, and planning *where* you can lay out water-, sun-, and shade-harvesting strategies to create a more integrated, efficient, and productive design. These patterns are meant to give you a conceptual framework for creating "regenerative designs." Regenerative designs arise from thoughtful integration of elements with natural local resource flows so they can enhance and recreate themselves without dependence on imported inputs such as piped-in fuel or pumped-in water. Once established, these regenerative designs do the bulk of the work, so you don't have to. They maximize the return on your on-site resource-harvesting investments.

In order to continually progress your own understanding of what is possible through design, the following question and its three criteria are useful for assessing design success:

Are the elements, systems, and patterns of design relationships that you have created **degenerative**, **generative**, or **regenerative**? Compare their characteristics below, and apply these concepts to your thinking and design.

A **degenerative** investment:
- Starts to degrade or break down as soon as it is made;
- Requires ongoing investments of energy and outside inputs to keep it functional;
- Consumes more resources than it produces;
- Degrades the health of its surroundings and/or the world;
- Typically serves only one function.
- Examples include: ornamental lawns and landscapes dependent on chemical pesticides, fertilizers, and energy-consumptive irrigation water imported from wells, surface waters, or utilities—contributing to water extraction rates that exceed natural water recharge rates; mechanically heated, cooled, ventilated, and lit buildings powered by polluting imported energy; and conventional single-use parking lots.

A **generative** investment:
- Starts to degrade as soon as it is made, but can be used to make or repair other investments (as is the case with tools);
- Requires ongoing investments of energy and outside inputs to keep it functional;
- Produces more resources than it consumes;
- Conserves other resources;
- Typically serves multiple functions.
- Examples include: multi-use landscapes (producing multiple resources such as food, beauty, and wildlife habitat); passively heated, cooled, ventilated, and lit buildings; durable, renewable,

non-polluting energy products such as solar, micro-hydro, and wind power systems (turning buildings and communities into clean energy producers, instead of dirty energy consumers); parking lots that grow a carport orchard of food-producing shade trees using harvested stormwater; and water-harvesting earthworks, cisterns, and gravity-fed greywater plumbing that increase the multiple cycling and accessibility of on-site water resources, while conserving overall regional waters and other resources.

A **regenerative** investment:
- Can repair, reproduce, and/or regenerate itself;
- Starts to grow or improve once it is made;
- Does not require ongoing investments of imported energy and outside inputs to keep it functional;
- Produces more resources than it consumes;
- Improves the health of its surroundings and the world;
- Typically serves multiple functions;
- Is alive or part of a living culture, that continually evolves/regenerates the investment.
- Examples include: multi-use landscapes living solely off natural rainfall and stormwater runon, and requiring no additional outside resources after establishment; self-regenerating natural mulches, soils, forests and ecosystems; infrastructure, laws, and cultures maintaining equal access and rights for (and stewardship by) all—to such natural resources as the sun, prevailing winds, rainfall, and surface waters; revolving community loan funds building the capacities of individuals, their community, and its watershed; saving seed from each year's hardiest and most productive crops (then planting, and again saving seed year after year) to evolve a steadily hardier and more productive seed stock specifically adapted to the unique conditions of your land and climate; vegetative rainwater-harvesting structures that grow and repair themselves after establishment, and the songs and stories that help ensure the continuation of these progressing practices and investments.

Strive to make all your water-harvesting endeavors regenerative, and as the water farmer Mr. Phiri would say, "You'll be rhyming with nature." You may not get there right away, but just by passively harvesting water, sun, wind, and shade the way this book suggests, you'll rise from the degenerative to the generative level.

TYING IT ALL TOGETHER: CREATING AN INTEGRATED DESIGN

Use your site plan, with its mapped resources and challenges, as the foundation for selecting and placing water-harvesting structures to create an integrated design that increases site efficiency and maximizes site potential (see figure 4.27 on the next page for an idealized site plan).
To do so:

1. Make multiple copies of your site plan (drawn to scale) to use as base maps for draft observations and ideas.

2. Play with different conceptual water-harvesting plan layouts. I recommend two options:

   • Make cut-outs (in the same scale as your site plan) of the trees, cisterns, patios, gardens, and other elements you want to add to your site. Move these cut-outs around your site plan imagining how they will interact with the flows of your on-site resources (rainwater, greywater, sun, wind, etc.).

   • Put tracing paper over your site plan and sketch where you could place various elements (trees, cisterns, patios, gardens, and other elements) you want to introduce to your site, and then see how they interact with on-site resource flows.

As you keep playing with various arrangements, ask yourself, "Where do I need water, where do I have it, how much do I have, and how/where can I best utilize it?" Remember, your goal is to increase site efficiency and maximize site potential. Refer to earlier sections of this chapter for conceptual ideas. See the resource appendix at www.HarvestingRainwater.com for more on integrated design patterns.

3. *Refine your design by planning the water-harvesting details.* After figuring out where you want to harvest water, it's time to figure out how to do so. Re-read chapter 3 to determine what specific water-harvesting strategies are most appropriate for your needs, whether to harvest water in soil, tanks, or both. (Volume 2 provides a more thorough overview of water-harvesting earthwork strategies and their appropriate use, detailed implementation, and additional information incorporating greywater harvesting and use.)

Now, walk your land again, imagining how various strategies could work within the unique context of your site. Play more with ideas and layouts on paper—it's much easier to make changes with a pencil and eraser than with a shovel. When you feel you've got your plan set, scratch out, stake, or otherwise mark locations of paths, trees, water-harvesting strategies, and other elements in the dirt at your site. Walk around your site, feeling what it would be like to inhabit this system. Are their adjustments that could make it more generative or even regenerative? Could you integrate any more sun and shade harvesting? Make any needed changes and if all feels good—go for it!

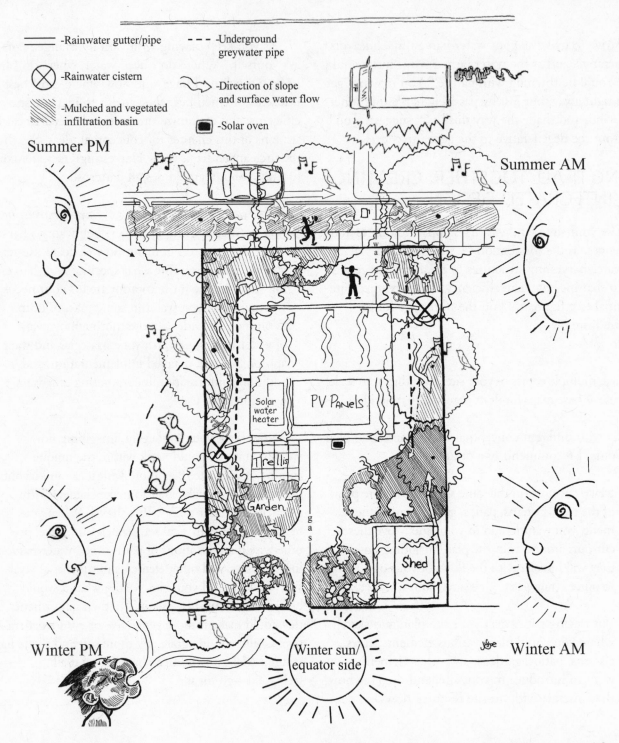

Legend:
—— -Rainwater gutter/pipe
- - - -Underground greywater pipe
⊗ -Rainwater cistern
▨ -Mulched and vegetated infiltration basin
〰 -Direction of slope and surface water flow
▬ -Solar oven

Summer PM

Summer AM

Solar water heater

PV Panels

Trellis

Garden

Shed

Winter PM

Winter sun/ equator side

Winter AM

Fig. 4.27. An integrated rainwater- and greywater-harvesting landscape plan. Rain- and greywater-irrigated food-producing shade trees placed in a cooling solar arc around the home. Winter solar access retained for heating, cooking, power, and light. Impermeable hardscape reduced by removing driveway, planting a garden, and parking car on street. Garden placed in a sheltered sun & shade trap and irrigated by a cistern forming part of the trap. A wind- and noise-break of hardy, low-water-use, song-bird-attracting vegetation placed within basins on the perimeter to diffuse and cool the wind. The vegetation and a cistern also screen out the barking dog's view so they stop barking. Pathways are raised to direct runoff to adjoining vegetation in basins. Greywater pipes and underground utilities are buried in raised pathways to prevent conflicts with vegetation. Greywater outlets into sunken basins. Roofs are guttered and sections of the land are regraded or bermed to utilize runoff flowing from off-site, and to retain all on-site rainwater on site. Street- and sidewalk-runoff irrigated shade trees buffer the home from the street. All of this invites friends to walk over and visit.

An Integrated Urban Home and Neighborhood Retrofit

This place is so cool! I want to live like this!

—Sam, a 4th grade student from Davis Elementary
visiting our homestead on a school field trip

This chapter tells the story (in three parts) of the experiences my brother's family and I have had while striving to incorporate all the strategies and concepts laid out in prior chapters of this book. As you'll see, you don't need any prior experience or knowledge to begin—we sure didn't have any. And yet the overall results have been amazing.

REAL-LIFE EXAMPLE

PART ONE – OUR BEGINNING: REALIZING THE *WATER-HARVESTING PRINCIPLES*

Right after my brother Rodd and I bought our east-west-oriented fixer-upper of a house in 1994, the summer rains poured from the sky. I opened the front door and it came off the hinges and fell onto a wet floor. I worriedly walked around, noting all the spots where the roof leaked, then went to sit down on the toilet (the only chair in the empty house) to gather my thoughts. Upon taking a seat, I promptly fell through the rotted wood floor along with the toilet. Startled, but not hurt, I looked around from my new vantage point to see all the rusted water pipes below the floor. It was then I realized I was going to be reading a lot more "how-to" books than I had originally estimated. Thinking back, I sure wish I had back then the book you are reading now.

Rodd and I bought this house together because neither of us could afford a home of our own. Only severely run-down properties were a financial option, but this was perfect because there was plenty of potential to make things better and gain new skills. We had both recently attended a permaculture course in which we learned all kinds of sustainable-design and -living concepts that we were anxious to put into practice. While we lacked practical experience and knowledge, the house and yard soon became our laboratory, playground, and vocational school, where we did 90% of the work ourselves, learning and saving money the entire way.

Our first step was to inventory what was happening on the site to inform what might need to be done—*long and thoughtful observation*. Rodd and I noted where runoff pooled against the outside of the house and where termite damage was severe. We also saw how the bulk of rain falling on our site picked up sparse topsoil and ran off into the street with it. At the same time, stormwater from the street flowed onto our driveway, through the one-car garage, and out the side door to flood the neighbors' properties. Yet most of the year our property and the adjoining streetscape were very hot, dry, dusty, and exposed. On a plan we made of the property, we mapped these observations and others—the noise, headlights, and pollution coming from the speeding traffic on the street; where we wanted privacy; where we needed shade; and where

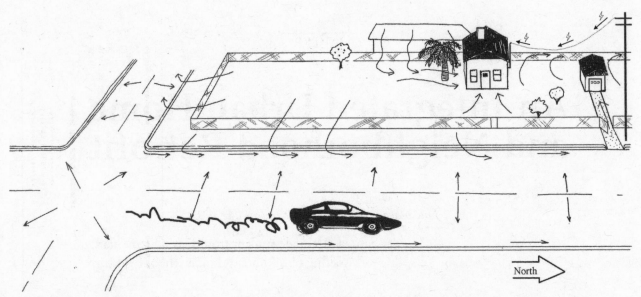

Fig. 5.1A. Our site at time of purchase in 1994. Most rainfall drains off site. Runoff flows up against house, to the street, or through the garage. All greywater goes down the sewer. Palm tree blocks home's winter solar access.

Exposed and unwelcoming street and walkway heat up like a solar oven most of the year.

Fig. 5.1B. Our site in 2013. No rainfall leaves site. On-site runoff is infiltrated into the soil before it can get to house or garage, and water drains away from buildings. Street runoff is directed to basins and trees along the curb. All greywater is directed to—and recycled within—the landscape.

Winter solar access regained after removing palm tree. Solar panels installed on roof, solar clothes dryer (clothes line) installed along fence, and solar oven and solar hot water heater installed on the ground south of south-facing trellis. (Solar hot water heater installed on the ground because old roof could not bear the water heater's weight.)

Along with our neighbors we took to the streets. More native food-bearing trees were planted in the public right-of-way. Street curbs were cut to direct street runoff to tree basins. Traffic circles and chicanes or pullouts were installed to calm traffic, harvest water, and grow shade. All strategies build community as we enhance and enjoy our community together.

we needed to enhance the solar exposure to equator-/south-facing windows, etc.

More observations were added as we spent more time at the site, and we continue to add observations today. We spent many hours imagining how we could improve the site by converting its challenges—such as flooding, overexposure to the sun, and its location on a busy intersection—into resources by growing shade and food with the floodwaters, harnessing the power of the sun, and working with our neighbors to create a high-visibility demonstration project to improve our community. (See figures 5.1A and B, "before" and "after.")

Rodd and I *began at the top of our watershed*—the (unfortunately) porous roof. We removed the deteriorated asphalt roofing and hauled it to the dump; not realizing it could be recycled. We then expanded the roof area by installing an extension of our roof overhang, guesstimating it would be long enough to create more summer shade, but short enough to allow the free heat and light of the winter sun to enter south-facing windows. The results were pretty good, but hardly perfect, since we lacked the overhang-sizing tools in chapter 4. When determining our roof-overhang length, we also forgot to include the width of the gutter that we intended to install. With the gutter now installed, a bit more of the winter sun is shaded than is ideal. But our choice of roofing material was excellent. We installed 26-gauge galvanized-steel metal roofing that should last for the rest of our lives. It is very durable, easy to install, structurally strong, and above all, the water it sheds is non-toxic, allowing for more uses of the high-quality rainfall runoff.

With the roof fixed, we now started to crave—rather than fear—storms. And we wanted to know how much rain we could harvest. Our calculations astonished us. We calculated that, with an average annual rainfall rate of 11.12 inches (282 mm), the rain falling on our site's watershed in an average year totaled over 95,800 gallons (362,500 liters)! That annual figure is the sum of the following:

- over 6,000 gallons (22,700 liters) of rain running off our 990-ft² (92-m²) roof,

- an additional 35,000 gallons (132,400 liters) falling on our 132- x 46-foot yard,

- 27,000 gallons (102,200 liters) falling on the 20-foot (6.1-meter) wide public right-of-way on the south and east sides of our corner lot just outside our fence.

- 26,000 gallons (98,400 liters) harvested off the adjoining streets,

- 1,800 gallons (6,800 liters) from a 270-ft² (25.1-m²) section of our neighbor's roof draining runoff into our yard.

"This is awesome!" we thought. We went right to work trying to capture and utilize rainwater as close as possible to locations where we needed plants and the water to support them (see box 5.1). However, it would be another year before I came upon the simple calculations (shown in appendix 4) that enabled me to figure out how many of our plants' water "expenses" could be met by our rainwater "income" of 95,000 gallons (see box 5.2).

We started using our rainfall by hand-digging simple earthworks in the high points of our yard and at those points where concentrated stormwater runoff from roof and street edges entered the landscape. This combined the principles of starting *at the top* and *starting small and simple* and solved the majority of our flooding problems. A berm redirected street runoff away from the garage and focused it into a large linear basin located along our property line. Here we planted shelterbelt trees that naturally pumped floodwater up to their food-bearing canopies while shading the public footpath. Another berm prevented the runoff from the higher, southern half of our lot from draining downslope into our house on the lower, northern half. This provided more water for an existing orange tree on the higher half of our lot and for our future garden site. We regraded the area around the house and garage to ensure positive drainage of roof runoff away from the buildings. We placed water-harvesting basins 10 feet (3 m) or farther from the structures to avoid the risk of saturating the soil under and near our walls, then planted and mulched these basins. Overflow water was directed from the upper

Box 5.1. Prioritizing Local Water Harvesting Over Costly Utility Imports

Rodd and I committed to a simple but powerful shift that would be the main driver of the design of our landscape and gardens—radically enlarging its sustainable potential:

Contrary to the trend in the U.S. whereby 30 to 70% of the drinking water consumed by a typical single-family household is the *primary* water source for landscape irrigation,[1] we would not use drinking water from our water utility (or well) as our sole, primary, or even secondary source of irrigation water.

Instead we would use:

- rainwater and stormwater as our *primary* irrigation sources,
- greywater as our *secondary* irrigation source, and
- drinking water only as a *supplementary* back-up irrigation source reserved for use strictly when the primary and secondary sources went dry.

Making this commitment meant all our plantings would be designed around the availability of on-site rainwater and greywater instead of piped-in drinking water. Therefore, all our plants were placed in association with water- and mulch-harvesting earthworks, instead of water- and soil-draining raised mounds or flat bare earth. Native low-water-use plants were placed in areas receiving only rainfall. Higher-water-use plants were located in or beside water harvesting basins receiving rainfall, stormwater runon, and/or greywater. Drought-susceptible exotic fruit trees were placed exclusively next to the house where they could tap the year-round greywater flow generated by us household "fruit eaters."

Box 5.2. Balancing Water Expenses (Irrigation) with On-Site Water Income (Rainfall)

Furthering our long and thoughtful observation, we calculated that our site's average of 95,000 gallons (362,500 liters) of annual rainfall and available stormwater flow would be enough to meet the average annual irrigation demand of around 30 low-water-use native trees, 10 high-water-use exotic fruit trees, or a vegetable garden about 1,500-ft^2 (138-m^2) in size (see appendix 4 for tree lists and exact calculations of water needs). We opted for a mix of plantings that fit within the budget of our on-site water income, while reducing our overall water expenses by using mulch, afternoon shade, windbreaks, and on-site greywater sources.

wider to expand their water-harvesting capacity. And my advice to you is this: If you are wondering whether your earthworks are deep and wide enough, always err on the side of greater water-harvesting capacity (see volume 2 for sizing rules of thumb and calculations).

The harvesting became even more fun and dynamic as we challenged ourselves to view all our on-site materials as potential resources to harvest and utilize. For example, the soil we dug from the basins was used to form our berms, raised water-harvesting paths placed on contour, and raised gathering areas that drained water toward adjoining plantings. We focused the earthworks and the water in places where we wanted to *stack functions* and *maximize living groundcover* with multi-use vegetation. These areas included earthworks on the east, north, and west sides of our home to support a solar arc of cooling shade trees; along the property line where they formed a beautiful living fence of native plants and wildlife habitat; and around an existing citrus tree that would form part of a sun & shade trap to shade the future organic garden from the afternoon sun.

We placed a 200-ft^2 (18.6-m^2) garden constructed of sunken, mulched basins just south of our home where the low-growing veggies and greens would allow winter sun to enter the equator-/south-facing windows. But first we had to remove a palm tree that was blocking the windows' winter sun. We cut the

basins to the lower basins in long zigzagging patterns to *utilize overflow as a resource* and to spread and *infiltrate* the water still more.

Our initial earthwork creations were soon tested in a big storm. No more runoff pooled against the house, all basins rapidly filled up and safely overflowed into one another, and the last of the overflow went into the street. It worked! But we hated losing the overflow to the street. So, Rodd and I dug our basins deeper and

Fig. 5.2. The winter-sun/equator-facing side of our home at noon on the winter solstice 2012. The solar panels (on roof), solar hot-water heater (on ground to right of house), homemade solar oven (in front of water heater), and winter garden are all placed within the same solar envelope that maintains open access to the winter sun for the south-/equator-facing windows. Where winter solar exposure does not need to be maintained, we shade the site with trees.

Note how the gutter extends the overhang and casts more winter shadow on the windows than we intended.

tree down, chipped it, and placed the resulting mulch within our basins. With the palm removed we had ideal winter solar exposure for winter and cool-season gardening (the most productive and least water-consuming gardening season in Tucson), passive solar heating of our home and water, and the potential to put solar panels on our roof in the future to generate our own electricity.

Between the garden and house we built a sparse trellis structure. We grow winter-deciduous vines on the trellis to create heavy shade in summer. These same vines die back in winter, then we remove and mulch their above-ground biomass so low-angle winter sun can provide heat and light to our house. More summer shade is cast by trees and a ramada—placed far enough south of the home and garden to maintain the winter's beneficial solar gain at and near the house

(thanks to using our winter-solstice shadow ratio from chapter 4's box 4.9).

Having expanded and retained our winter-sun access, our east-west oriented home's main source of winter heat is the sun coming through the south-/equator-facing windows. The sun also helps grow abundant winter greens, heats our solar oven, heats the solar hot-water heater (our sole source of hot water), and—with the installation of solar panels—is our primary source of electricity. Our main sources of cooling are the roof's extended overhang, the solar arc of shading vegetation, and passive strategies such as nighttime ventilation through operable windows and screen doors. Our electric bills no longer exist, and our gas and water bills are little more than the service charges, since our consumption is negligible. (See figure 5.2.)

Fig. 5.3. Late-winter garden grown entirely from rainfall and roof runoff harvested in 1,300-gallon (4,920-liter) tank in 2005. For color photo see inside back cover.

Along with sun and water, we also harvest the free power of gravity. In the past, water would pour off our roof and our neighbor's roof, collecting in a low spot between our homes (spaced just 3 feet [1 meter] apart) where the accumulated moisture was deteriorating our walls. We solved this with our new earthworks and by using the elevation of each roof to our advantage. With our neighbor's permission, we guttered his roof, sloping the gutter to redirect his runoff from the low spot between the buildings to a citrus tree on the high point of our yard. Water that used to land far away from the tree now passively drains toward it.

We guttered our roof too, directing the runoff to an above-ground cistern positioned beside that same citrus tree. Since we put the cistern on the high point of the yard, we can use gravity to move water from the roof's gutter to the tank, and from the tank to the garden—or anywhere else on the property. No costly pumps are needed! To further boost tank performance we built an earthen platform 1-foot (30-cm) high on which to place the tank. The higher the tank, the greater the gravity-fed water pressure from the tank will be. (For every foot of elevation that the water level in the tank is above the water's destination we gain 0.43 pounds per square inch of gravity-fed water pressure [for every meter we gain 9.7 kilopascals of pressure].)

Needless to say, gravity pressure gets lower when the water level in the tank is low. So we maintain optimal water flow in our system by avoiding flow-reducing internal friction from excessively narrow-diameter hose or pipe, long hose, or highly constrictive valves. Having the cistern (or the extended valve/faucet location that taps the cistern) right next to the garden keeps our hose length to a short 25 feet (7.6 m) and we use only non-restrictive full-port valves. The result is fast, consistent gravity-fed flow from the cistern—always free, convenient, and enjoyable to use. (See figure 5.3, and the "Garden Hose Dangers and Recommendations" blog on my website for hose, pipe, and valve recommendations and sources.)

Fig. 5.4. Three-way diverter valve under sink gives the convenient option of sending greywater to landscape or sewer depending on what goes down the drain.

Fig. 5.5. Greywater drains (marked with destinations: olive + pomegranate, orange, fig, or white sapote) conveniently placed beside washing machine. Drain hose from washer is placed in a different pipe with every load of laundry.

Note: include an additional marked drainpipe going to the sewer (capped when not in use) as an option for households that occasionally use non-biocompatible detergents or that have seasonally saturated or frozen soils.

Our original cistern has a 1,300-gallon (4,920-liter) capacity. We selected this size after calculating average annual roof runoff, assessing our water needs, and determining the resources we wanted to commit to the system. We knew that we did not have enough runoff to meet both domestic water needs and those of the garden and landscape, so we reduced water demand and increased supply by implementing conservation strategies such as installing a composting toilet, installing a greywater system to reuse drain water in the landscape (see figures 5.4 through 5.7 and the greywater chapter in *Rainwater Harvesting for Drylands and Beyond, Volume 2*), and replacing some water-hungry exotic plants with drought-tolerant natives. This helped, but we still didn't have enough roof runoff to meet all our water needs. So we decided to use the roof water just for irrigation as part of an experimental start-up cistern system. Wanting to keep the tank affordable—and under 1,500-gallon (5,700-

liter) capacity—we decided to size the tank to capture the runoff only from the 650-square-foot (60.4-m²) section of roof that sloped toward the garden, rather than the whole roof. From that section of roof we then wanted to be able to store at least all the runoff from a 3-inch (76-mm) rainfall event (or two 1.5-inch [38-mm] summer storms hitting a week-or-so apart). Using this information, we calculated we needed a 1,300-gallon (4,920-liter) cistern. (See Equation 6 in appendix 3 for this calculation.)

Knowing the tank size we needed, the next question was what type of tank we wanted to use. Locally available pre-manufactured tanks included metal, plastic, fiberglass, or precast septic tanks. We also had the option of making our own ferrocement tank or culvert cistern. (For more on tank options search "Water Tanks/Cisterns" at www.HarvestingRainwater.com.) We opted for a precast septic tank for the following reasons:

1. In terms of locally available ready-made tanks, this was our cheapest option for the volume, dimensions, and material we needed.

Fig. 5.6. Each greywater drain outlet empties into a mulched and vegetated infiltration basin that also harvests rainwater and runoff. The pipe outlet discharges 3 inches (7.6 cm) above the surface of the mulch to keep roots from growing into and clogging the discharge pipe. Greywater immediately infiltrates through the mulch to the soil below. Greywater is dispersed to multiple basins to reduce flow to any single basin, enhancing infiltration further. Basins are sized to contain and infiltrate the peak surge of greywater that could potentially be discharged in a short time period. (For more information and additional variations see the Greywater chapter of volume 2.)

2. The 5-foot (1.5-m) tall, 10-foot (3-m) long, 4-foot (1.2-m) wide septic tank had the right dimensions for our space. To make sure the tank inlet was lower than the roof edge, the tank could be no more than 6.5-feet (2-m) tall to fit on the elevated pad we built for it. This septic tank's dimensions fit within that height limitation, and enabled us to maintain direct, free, gravity-fed inflow to the tank.

3. We wanted to use the tank as part of our western-property-line fence, privacy screen, and western sunscreen. The 10-foot (3-m) length of the tank worked well for these uses. (See figures 4.16A, B, C and 5.3.)

4. Concrete is relatively fireproof so the tank could act as a firebreak in case the neighbor to the west sparked a fire—and the water in the tank could help fight a fire. (Many insurance companies offer fire-insurance discounts to property owners who have

a dedicated fire-protection cistern with a valve sized to match the local fire department's standard fire hose.)

5. We wanted to play with the idea of using a locally made septic tank as a cistern in order to get to know its assets and faults, as well as to expand the local manufacturer's awareness and promotion of water harvesting. We figured this could be a viable ready-made system that we could potentially install for future clients.

The septic tank was custom-made for use as a cistern. We wanted only non-toxic construction materials, so we avoided concrete with fly ash, which could contain toxic heavy metals. The concrete was reinforced with steel to allow above-ground installation. (See volume 3 for details.) The cost in 1996 was $600, which included delivery and placement.

In 2009 we put in a second, identical septic-tank cistern (with the same dimensions as the first) along

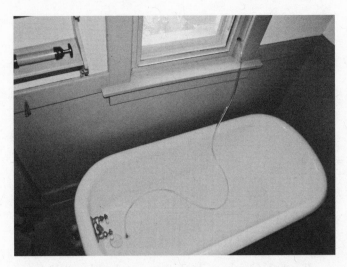

Fig. 5.7A. Vinyl tube is suction-cupped to bottom of bathtub and runs through window frame to convey greywater to landscape, bypassing inaccessible tub-drain plumbing. Note this type of system is not very convenient—you must go outside and pump to get the siphon going (see figure 5.7B)—though it is useful for accessing otherwise-inaccessible greywater.

Fig. 5.7B. Siphon hand pump (from auto parts store) pulls tub greywater through vinyl tube. Once water is flowing the pump is pulled off, the tube is placed beside the fig tree, and the siphon effect continues to draw the water. Purchase correct tube diameter to fit the pump. (See volume 2 for other, more convenient variations.)

the property line next to the original cistern, and connected the two tanks together with piping. We then directed more runoff into the tanks by adding the runoff from the guttered roof of an adjoining 120-ft^2 (11-m^2) ramada. The tank cost was $1,200 this time and well worth it. The added water storage has extended our growing season in droughts, and adds flood-control capacity in wet years. Lesson learned: design tank-system overflow points, elevations, and locations so the tank systems can be expanded later if needed.

The expanded system works great, with over 95% of our garden's water needs now provided by harvested rainwater (figure 5.8). In addition, the garden's delicious results often crown us "Kings of the Potluck" when we bring our rain-irrigated mixed heirloom salads topped with edible flowers to neighborhood gatherings.

I no longer feel that Rodd's family and I are living entirely out of balance with the water resources of our dryland environment. Instead of degeneratively getting all our water from overdrafted groundwater and Colorado River water imported from distant watersheds, we are living more and more within our

rainwater budget and the natural limits of our local environment.

Within our generative landscape, rainwater has become our primary water source, greywater has become our secondary water source, and utility water is strictly and infrequently used as a supplemental source for multi-use vegetation. Most of our established native-plant landscape is now regenerative, thriving and spreading on rainwater alone. The further we go, the easier and more fun the game becomes, as we creatively get ever more out of our site while giving back more than we get. In that spirit, we set up an outdoor shower that gives the bather the option of using cistern water distributed from a nozzled shower bucket hanging from a hook. All shower greywater

Fig. 5.8. Food grown and processed on-site and irrigated with rainwater. For color photo see inside back cover.

Fig. 5.9. Outdoor shower screened with salvaged palm fronds. Bather can wash with water from the water-utility, which we solar heat then send through the showerhead, or a bather can use cistern water placed within the nozzled shower bucket. Three drains at the bottom of the shower divert greywater to three different plantings in the landscape.

goes directly to trees surrounding, and screening, the shower. (See figure 5.9.)

In cooperation with our neighbors, we have established a solar-powered, one-washing-machine "laundromat" in our backyard, used by up to seven neighboring households. We harvest both greywater from the laundromat and fruit from the trees it irrigates. We have reduced impermeable hardscape by replacing our asphalt driveway with lush plantings placed within infiltration basins. We worked with the City to remove 26% of the pavement in the center of the intersection adjacent to our site and replace it with a water-harvesting, traffic-calming circle planted with native vegetation. We cut openings in the street curbs so we could direct street runoff into mulched street-side basins growing a greenbelt of street-shading trees irrigated with this street runoff. (See figures

5.10A and B and the "Reducing Hardscape and Creating Permeable Paving" chapter in volume 2.) The street-watered landscape is growing quickly and is abundantly productive. Now, similar water-harvesting landscapes are appearing in our neighbors' yards and along their sections of the street. Summer heating, utility-energy use, and the cost of living are all steadily dropping for our community, instead of the water table. The value of this was recognized when we won "First Place - Homeowner Landscape under $10,000," "Best Water Harvesting," and the "J.D. DiMeglio Artistry in Landscaping" awards in the 2005 Arizona Department of Water Resources/Tohono Chul Park Xeriscape Contest.

We are always pushing to improve the efficiency, convenience, and enjoyment of our system. We continually *reassess what we have done* as we strive to implement all the principles, ethics, and integrated-design patterns to make our work more effective. As part of this process, we have expanded or modified basins where needed, replaced or relocated poorly selected or planted vegetation, and have made our cistern and greywater systems simpler and more accessible to use. We are also moving toward making rainwater the primary water source *inside* our home.

To that end, we have our cistern water tested periodically (as we recommend everyone do if they are going to drink the water from their tanks), and thanks to following all of the "Ten Cistern-System Principles" described in chapter 3, the results have always shown the water to be safe to drink right out of the cistern tap—which I do. We test our cistern water right after the first rainfall occurs after a long dry spell. We expect cistern water to be at its lowest quality at that time because the maximum number of potential contaminants would have built up on the roof between rains and potentially would have flowed into the tank with the first rain. To be safer still, we use a Potters for Peace gravity-fed ceramic water filter to treat our primary drinking rainwater and the rainwater we offer to all guests (see box 5.3, and figure 5.11). Our guests love it because it tastes great—which is why in most of the world naturally salt- and chemical-free rainwater is known as "sweet water."

Creating, living with, and evolving these systems has given us the real-life experience and knowledge we lacked when we began, and has led us to our vocations. Rodd is now Restoration Field Supervisor for Tucson Audubon Society's watershed and wildlife-habitat restoration projects. Building on lessons

Fig. 5.10A. Before the 1996 planting of rain and trees. Public right-of-way adjoining property, with asphalt driveway freshly removed, 1994.
For color photo see inside front cover.

Fig. 5.10B. After planting of rain and trees. Tree-lined footpath reviving the once sterile right-of-way, 2006.
For color photo see inside front cover.

Fig. 5.11A. Chi Lancaster holding up our Potters for Peace / Filtron ceramic water filter. The filter fits inside the top of the bucket and the filtered water collects in the bottom.

Fig. 5.11B. Chi Lancaster pouring harvested rainwater into the filter, as her son Vaughan drinks some of the filtered rainwater. Water moves through the filter by the free force of gravity at a rate of 0.26 to 0.79 gallons (1 to 3 liters) per hour.

Box 5.3. A Simple Water Filter Using Local Materials to Improve Local Conditions

Potters for Peace (www.PottersForPeace.com) is a non-profit organization that shows local potters around the world how to use local clays to make simple, low-fired, low-cost, gravity-fed ceramic water filters painted with colloidal silver to filter out 99.98% of bacterial contamination such as *fecal coliform*. These filters can last a lifetime—unlike waste-generating, money-consuming single-use disposable filters, disposable filter cartridges, or bottled water. And they enhance local conditions—and local economies when made by local manufacturers—by improving the health of people using them. This is particularly true in areas with poor water quality, and where water infrastructure is of inadequate construction or where it has been temporarily damaged by a storm, malfunctions, pipe breaks, etc.

Because I have not found a manufacturer in the U.S., I purchased my Nicaraguan-made filter through Peace by Piece Fair Trade at http://peacefairtrade.com/product-catalog/water-purifier-potters-for-peace.html/.

Note: Unlike activated-charcoal filters, these ceramic filters are not designed to filter out human-made solvents, pesticides, or chemical contaminants found in areas that are affected by industrial and/or agricultural emissions and pollution. In such areas, use filters designed to remove these contaminants from the water, and work to eliminate and remediate the pollution sources that are affecting the water.

learned at home, I started a business designing and implementing these strategies, then wrote this book, and now travel the globe teaching the book's contents and learning from others who are harvesting and enhancing local resources in ways that are well-adapted to their local environment and culture.

More and more people throughout the world are developing their systems and gaining experiences in much the same way Rodd and I did. We help, teach, and encourage one another, often visiting during joy-inducing rainstorms to see how things are working and to expand earthworks in the downpour—when this work is the most fun! The movement grows as the practice, examples, and knowledge we gain in each of our backyard microwatersheds begins to overflow and nourish the greater communities and watersheds around us. Do likewise. Start at the top. Start small. Start now.

PART TWO – OUR EVOLUTION: TAKING IT FURTHER WITH THE *INTEGRATED DESIGN PATTERNS*

Things were going great after publishing the first edition of this book. Interest in water harvesting was growing more than ever. Yet my unrest was also growing. Rodd and I had begun transforming our property hoping to test, learn from, and help create the kind of abundant and regenerative environment—for ourselves, our home, our neighborhood, and our local environment—that occurs when everything and everyone works *together* for the betterment of the whole. We came a long way by harvesting the rain, but not nearly far enough in more fully understanding and partnering with the greater, more integrated potential of this Place. What else could we engage with that, also creates the unique character and quality of life of Tucson? At the very least, I recognized that I needed to deepen my relationship with the sun and more consciously integrate this element with our harvesting of water. In drylands, the seemingly abundant sun is the natural partner to the seemingly scarce water, yet both—the world over—can be equally abundant or scarce depending on how we interact with them. So I—a fair-skinned, blue-eyed, sunburn- and skin cancer-prone

guy—committed to the sun, though I wasn't sure how this relationship was going to play out.

In this same time period our family was growing. Rodd married Chi, and Chi gave birth to Vaughan, my giggly nephew, making us a very happy extended family. Vaughan's birth—and thoughts of my own yet-unborn offspring—motivated us ever more to live the life that would help generate the world we wanted for him and other children to follow. However, with the arrival of our new family members, space was getting very tight in our shared 740-ft² (68-m²) house, within which my assistant Megan and I also ran my business.

Thankfully I had sold my truck in 1996 and gotten an Xtracycle cargo bicycle and have been carless ever since. Rodd and Chi park their small truck on the street and bicycle regularly with Vaughan. As a result, our car-less 200-ft² (18.5-m²) one-car garage, built in 1962, was potential workspace/livespace waiting to materialize. However, I didn't initially see this potential and was depressed by the idea of moving my workspace into the dark, stuffy, junk-cluttered garage. Yet as I began to play with ideas about the space and how its transformation could stretch my understanding and ability to live more in balance with my family and neighbors in this Place, I realized this was a fantastic opportunity to apply all the integrative site design I had absorbed since beginning this work. This would be my chance to redo some things and correct errors I made the first time around (*continually reassess your system: the "feedback loop"*), but what really set me on fire was realizing this was my chance to more fully understand and build on this planet's—and my own—relationship with the sun.

Converting the garage to a working/living space would lead to many unexpected challenges, such as how to increase the building's interior height, space, and winter solar gain without shading my neighbors (figures 5.12A and B), and realizing that a wrong choice of windows could wipe out most passive-heating benefits otherwise gained by the building's ideal solar orientation and overhangs [box 4.4]). But these challenges sparked many new discoveries and developments, including the resources and design tools I've incorporated into this book's revision throughout chapter 4 and appendices 5, 6, 7, 8, and 9.

I threw myself into more *long and thoughtful observation* and more research on topics ranging from intensive passive-solar design, to the integration of water harvesting with sun and shade harvesting, the design and construction of small-to-tiny houses, the Living Building Challenge, and how our daily choices and actions converge in the cumulative impact of the water-energy-carbon nexus (appendix 9). The farther I strove, the more exciting everything got.

Months later I had a design to retrofit the garage into a 90% passively/freely heated and cooled, day-lighted, harvested-rainwater-watered workshop/live-shop (see box 5.4). The vast majority of the labor and materials came from our community, as skilled friends helped with the work and over 80% of the materials used were salvaged from around town or our own site—much of it once stored in the building. And that is how a garage-turned-"cottage" became our *garottage*.

Starting with the sun to maximize integrated passive heating and cooling

We designed and implemented the garage's conversion to preserve and generate resources for ourselves, our neighbors, and future generations—including Vaughan's. The roof was raised to increase space without increasing the building's footprint, thereby preserving land for growing food and shade, which we irrigate with waters harvested on-site. The roof angle and height were designed to increase our solar access in winter, while maintaining solar rights/access for our neighbor to the north. This ensured they too could freely and passively heat and cool their home without using polluting, water- and money-consuming fossil fuels. This in turn allows all of us to produce on-site renewable, zero-water-use solar power for ourselves and the community, with surplus power feeding to the community electrical grid via a grid-tied, rooftop solar PV system. (See box 4.10, and figures 5.12A and B.)

City zoning would have allowed us to raise the garottage roof from the original 10 feet 7 inches (3.2 m) to 25 feet (7.6 m), but this would have wiped out nearly all winter solar access for our northern neighbor's south-/equator-facing windows in the morning and at noonday. Instead we preserved their abundance of winter sunlight, heat, and power by raising the roof to just 15 feet 9 inches (4.8 m), with its peak height placed as far from the neighbor's as we could get it. (See how to do this in box 4.9, *Integrated Design Pattern Five—Maintaining Winter Sun Access*, and figures 4.17 to 4.20 in chapter 4.)

Box 5.4. Prioritizing Local Sun, Shade, and Wind Harvests Over Costly Utility Imports

To maximize the sustainable potential of our buildings and make their operating costs plummet, Rodd, Chi, and I committed to:

Not use mechanical systems such as a furnace or air conditioner—or even a woodstove—as *sole* or *primary* heating or cooling strategies of our buildings. (90% of a typical U.S. home's energy is consumed to run the house, not to build it.[2])

We committed instead to:

• Use passively harvested sun (done in multiple, beneficially redundant and integrated ways) as the *primary* heating source, and use shade and breezes (harvested in multiple, beneficially redundant and integrated ways) as the primary cooling sources.

• Use conservation strategies—such as insulation, wearing sandals and short sleeves (in summer) or slippers and sweaters (in winter), and cooking outside (in summer) or inside (in winter)—as the *secondary* sources.

• Use small active backup systems—such as a woodstove, solar-powered fan, or a one-room evaporative cooler—only for *supplemental* heating and cooling when such needs could not be entirely met by the primary and secondary sources.

Making this commitment ensured we designed and retrofitted our buildings to make the most of free, on-site potential (such as with sun, shade, and breezes) in order to create naturally comfortable buildings. This is in sharp contrast to the frequent practice of ignoring this free potential, and making buildings artificially comfortable with nearly complete dependence on costly purchased appliances and imported energy.

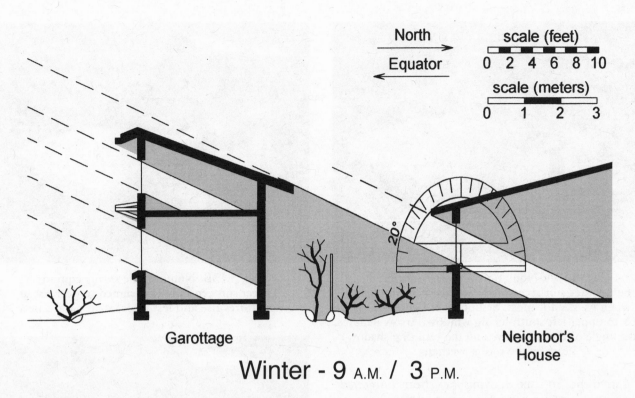

North →

Equator ←

scale (feet)
0 2 4 6 8 10

scale (meters)
0 1 2 3

20°

Garottage

Neighbor's House

Winter - 9 A.M. / 3 P.M.

Fig. 5.12A. Solar rights/access for all. Shadow cast by garottage on neighbor's house at 9 A.M. and 3 P.M. on the winter solstice does not extend above neighbor's windowsill, retaining neighbor's key winter-sun access.

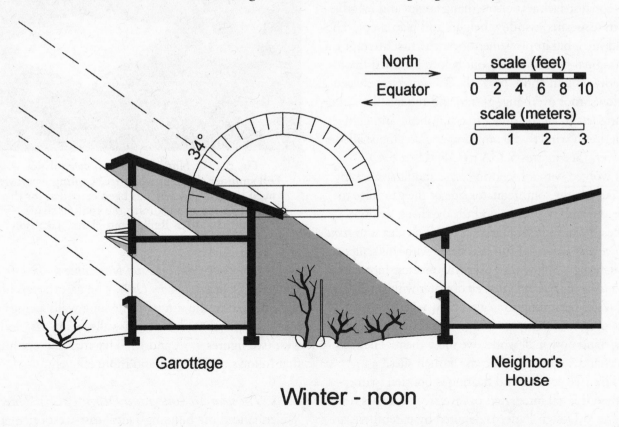

North →

Equator ←

scale (feet)
0 2 4 6 8 10

scale (meters)
0 1 2 3

34°

Garottage

Neighbor's House

Winter - noon

Fig. 5.12B. Shadow cast by garottage on neighbor's house at noon on the winter solstice. Note how sun is higher and shadows are shorter at noon than in morning and afternoon (figure 5.12A).

Fig. 5.13A. Noon on the *winter* solstice.
Full winter-sun access provides free heat and light
when we need it most. Sculpture in figure 5.14 is
next to upper left south-facing window. Arrow denotes
the angle of the sun's rays and the resulting shadow
cast by the roof's overhang.

Fig. 5.13B. Noon on the *spring* equinox.
Direct sun exposure is designed to decrease as
temperatures rise and less direct solar gain is needed.

I used the "to-scale drawing way" from *Integrated Design Pattern Two—Overhangs on Buildings* (see chapter 4) to determine the garottage's ideal winter-sun-/equator-facing roof-overhang angles and lengths, and its associated window heights and placement. The results are a big improvement over my first attempt on our original house, carried out before I learned the to-scale drawing method. Now the relationship between windows, roof overhang, gutter, and retractable awning lets in all the direct free sun, heat, and light we desire in winter while maximizing cooling shade in summer. (See figures 5.13A to C and box 5.4.)

I worked with a friend to create and hang a sculpture outside the south/equator side of the garottage to make the path of the sun and the overhang shadows visible all year long (see figure 5.14). The idea is to *show the flow and harvest* of sun and shade with this sculpture highlighting window-roof overhang relationships, just as we *show the flow and harvest* of water with surface-level water-harvesting strategies. Now when guests come over, love the space and its comfort, and want to understand how it all works, we show them. They see, understand, and can do likewise at their sites.

This free cooling and heating is boosted with seven additional integrated passive/free strategies (the Integrated Design Patterns denoted by parenthesis are described in chapter 4):

Fig. 5.13C. Noon on the *summer* solstice.
Full summer shade provides free cooling when we
need it most. Awning has been extended for the
hot season to maximize the cooling shade.
Awning by Allen Reilley and Henry Jacobson

1. *Orienting buildings and landscapes to the sun (Integrated Design Pattern One).* The garottage is ideally oriented to the sun with the long walls facing south and north and the short walls facing east and west (see figures 4.5A and B). This makes everything that follows much easier and more efficient.

2. *The solar arc (Integrated Design Pattern Three).* We enhanced the building's ideal east-west orientation by building a deep covered porch on the east side of

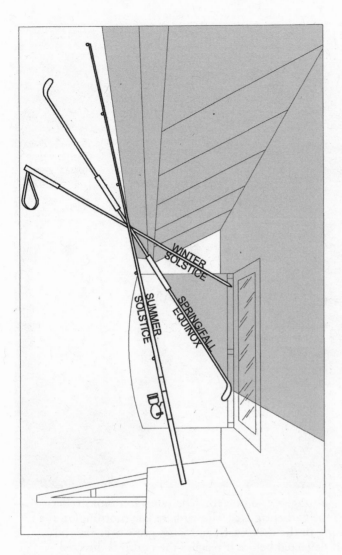

Fig. 5.14. Sculpture *showing the flow and harvest* of sun and shade. It also creates a privacy screen just west of the garottage's upper equator-facing window, blocking our view of the neighbor's yard to the west. Seasonal sports equipment denotes the sun angle and subsequent shadow line cast at noon by the roof overhang and gutter on the winter solstice (ski pole), spring/fall equinoxes (golf club), and summer solstice (fishing pole). We let in the sun when we need it, and shade it out when we don't. Shadow shown is cast on spring/fall equinox. Sculpture by Allen Reilley

the garottage. This serves as the east side of the solar arc, shading 90% of the east-facing windows from the summer's hot morning sun. The porch roof serves as a terrace and stargazing perch, a play deck for Vaughan and friends, and a summer sleeping pad (figure 5.15 and box 5.5). A closet-like shed on the *exterior* of the western wall serves as the west side of the solar arc,

Fig. 5.15. Garottage porch-roof bed. Summer night. Due to dramatic nighttime radiant-heat loss in our dry climate, summer-night temperatures on the roof are 10° F (5.5° C) cooler than below the roof, and 5° F (2.7° C) cooler than the open ground. This provides a comfortable, passively cooled night of sleep on the porch-roof bed. Note that all windows and doors (behind the security screens) are opened to encourage free ventilation by the cool night air.

Box 5.5. Unexpected Benefit of Outdoor Sleeping

We got our roof-bed idea (figure 5.15) from our elderly neighbor, Thomas Lewis, who told me stories of how great it was to grow up in our neighborhood before there was air conditioning, and how there was no crime. "Why no crime?" I asked. "Because everyone slept together," he replied. "What?!" "No, not like that," he explained. "In the summer we all slept outside in the cool night air. You heard everything. So no one could commit a crime without being noticed." Sure enough, later, while sleeping on the roof, I was awakened by, and I stopped, an attempted car break-in on the street below.

deflecting the summer's hot afternoon sun with the additional exterior insulation created by the shed and stuff inside.

Fig. 5.16. Placed along the northern property line, the garottage's two polyethylene rainwater tanks reflect light from the *winter* sun (rising in the *south*east) toward the gathering area, shown here late morning on the winter solstice. The tanks shade out the *summer* sun (rising in the *north*east), and vines grow over the trellis to shade the tanks. First-flush water (dirty roof runoff deflected before it reaches the tanks) is directed to the grape vine between the tanks.

Note: The pipe entering the porch-roof gutter is part of a "wet-system" downspout/conveyance pipe that drops down from the roof gutter of the main house, travels underground, then rises up to the garottage's gutter. This allows us to gravity-feed additional roof runoff from the main house into the garottage cisterns. This system works great because the downspout-pipe inlet point on the house gutter is higher than the pipe outlet on the garottage-porch gutter. Note that a wet system is appropriate for our hot climate; however, do not use it in cold pipe-freezing climates, where "dry-system" downspouts are best. (For video of this wet system see the "Image, Video & Audio" page at www.HarvestingRainwater.com.)

The solar arc is further enhanced with full-sized, winter-deciduous, native food-bearing shade trees planted east and north of the porch, and with shorter trees planted west and north of the shed—all positioned to *maintain winter-sun* exposure for neighbors to the north (*Integrated Design Pattern Five*). As a result, we get summer shade when we need it and our neighbors to the north get winter sun when *they* need it.

3. *Sun & shade traps* (*Integrated Design Pattern Four*). Two 1,000-gallon (3,785-liter) rainwater tanks are placed northeast of the garottage porch along the northern fence line. This cools the porch by shading out the early-morning summer sun that rises in the *north*east. These same tanks enhance the capture of winter warmth and light by forming a sun trap open to the winter-morning sun that rises in the *south*east. These tanks also provide a privacy screen in all seasons (figure 5.16).

Fig. 5.17. Indoor and outdoor thermometers on either side of the window by the main door let us know when to open windows to let in desired exterior warmth or cooling, or close windows to keep out unwanted exterior temperatures. Maximum/minimum thermometers monitor the extremes.

4. *Cool, non-toxic roof.* We used a reflective Galvalume 26-gauge metal roof on the garottage's main structure and Safecoat white elastomeric paint on the porch roof, both of which not only qualified for an Energy Star-compliant cool-roof tax credit (box 5.6), but also provide non-toxic surfaces from which to collect high-quality roof runoff. Passive cooling is taken further by also using light colors and reflective materials on the building's exterior walls. Furthermore, the metal roof and metal siding are fire resistant, qualifying for home fire-insurance rate reductions. The metal roof has the additional benefit of readily collecting dew on humid evenings and early mornings, and draining it right to the cisterns. (See the "Roofing" page under "Materials" on the "Rainwater Harvesting" page of my website for more.)

5. *Exsulation.* Chlorinated fluorocarbon (CFC)-free and hydro-chlorofluorocarbon (HCFC)-free foam insulation was installed on the exterior of the brick walls of the garottage and covered in stucco plaster. This buffers the mass of the brick—which readily absorbs and later re-radiates heat and cold—from unwanted extreme exterior temperatures, while enabling the brick mass to better hold moderate interior temperatures.

We were cautious in our selection of foam insulation. While ozone-depleting CFCs have been generally phased out, HCFCs are still used, and while their atmospheric life is shorter, they continue to release chlorine and contribute to ozone depletion.[5] A 12-inch (30-cm) layer of R-42 recycled-cotton insulation placed in the ceiling, along with an 8-inch (20-cm) layer of R-30 recycled-cotton insulation placed in the salvaged wood-stud walls above the original brick, further deflect extreme outdoor temperatures.

6. *Full-window ventilation.* All windows are the operable casement style, and all doors are operable to their full aperture for maximum free ventilation and fresh air. Insect screens control what comes in. In hot months, we open the doors and windows at night to release hot air and let in cool night air (figure 5.15). In winter, we close these at night but open them on any warm, sunny winter days. Indoor and outdoor thermometers signal when to open and close (see figure 5.17).

7. *Window-area limits.* Our equator-facing window area uses the correct type of glass (box 4.4) and casts sunlight on 12% of the floor area, thus falling within the rule-of-thumb specifications for temperate climates needing cooling and some heating (see box 4.5). The correct selection and sizing of windows—coupled with well-designed overhangs or awnings (*Integrated Design Pattern Two*)—allows us to avoid excessive nighttime cooling in winter and daytime heating in summer, while amplifying desired cooling, heating, and daylight. For example, the guidelines in box 4.5 enabled me to correctly size, enlarge (by three times), and replace one of the main house's undersized equator-facing windows, making the home's living room dramatically brighter and more comfortable. My clients have reaped similar benefits by using these guidelines to correctly size and install equator-facing windows in rooms and buildings where there were previously no such windows.

Committing to maximizing the garottage's beneficial relationships with the sun made the project buzz with life for me, and deepened my understanding of how to apply integrated-design patterns more effectively. Just as water harvesting heightens my awareness of the flows of water all around me, sun harvesting heightens my awareness of the flows of sun and shadow. Now when I'm outside I look for hints of rain, feel for wind direction, and check the position of the sun. It has been so much fun to stretch my observation, understanding, and learning throughout this building process—and has resulted in a comfortable structure I don't have to pay to mechanically heat and cool. While in the past I may have turned on a rattling air conditioner, now I simply open a window or extend an overhang in blissful silence.

Even better, I'm far more aware of how the sun moves through the sky and changes with the seasons, and how I can dance with that movement in the way I design, build, grow, and live. It roots me in Place to an incredible degree. I'm immediately reminded of the season, where I am on the planet, how my current position relates to the Earth's tilt and rotation around the sun, and how I—this little speck on this little planet in this massive solar system—can work and live in concert with the whole. And the more I live in concert with the whole, the more I benefit the whole—as it benefits me and all others...

Water-energy-carbon nexus—the bigger picture

All these integrated passive heating, cooling, lighting, and water-harvesting and water-conserving strategies throughout our site have saved resources and reduced our negative impacts on water, energy, and carbon, and overall costs.

• We have reduced our site's *electricity consumption* to about one-tenth (10%) of the typical Arizona household's electrical consumption of 1,095 kWh/month. (This later enabled us to install a much smaller and more affordable solar-power system that still meets all our needs.)

• We have reduced our *municipal-water consumption* to about one-sixth (17%) of a typical U.S. household's rate of 7,716 gallons (29,208 liters) per month.

• We have reduced our household's *carbon dioxide* (CO_2) *emissions* by about 2,100 pounds (952 kg) a month due to our reduced power and municipal-water consumption alone. (This number does not include the additional CO_2 reductions resulting from the installation and use of our rooftop solar PV system.) Without our energy and water conservation, this 2,100 pounds of CO_2 would've been emitted through our electricity and water utilities' coal- and gas-burning power plants. This reduction is the equivalent of parking/not using two-plus gasoline-fueled cars that would otherwise be driven 1,000 miles (1,600 km) per month, or planting 345 carbon-sequestering trees a month all of which grow to their full maturity.

Just as our household-scale changes save a *lot* of money, water, and power—while substantially reducing CO_2 emissions and other air, water, and soil contamination—we can collectively scale up to greatly improve our quality of life and that of the larger community. For example, if 100,000 households in Tucson reduced *their imported power consumption* by a similar volume, we would collectively save 39–57 *million* gallons

(149–218 *million* liters) of *water* per month that would otherwise be consumed at the power plant. Each month these power and associated water savings would reduce power-plant emissions equivalent to parking 207,000 gasoline-fueled cars otherwise driven 1,000 miles (1,600 km) per month, or planting and growing over 31 million trees.

In addition, if 100,000 Tucson households harvested and cycled on-site water to reduce their utility water consumption by a similar volume as our household, this would save about 670 *million* gallons (2.5 *billion* liters) of utility-provided water per month. In turn, this would reduce CO_2 emissions otherwise produced at the power plant to generate energy to pump and treat that water—by an amount equivalent to parking an additional 19,000 to 23,000 cars, or planting nearly 3 million additional trees every month.

See appendix 9 for simple charts that helped Rodd, Chi, and me figure this out, to see how you can do the same, and to find additional examples including how much water, energy, and carbon you can save by carefully choosing your water and power sources.

PART THREE—OUR NEIGHBORHOOD: COLLABORATING FOR LEVERAGED POTENTIAL IN THE WIDER COMMUNITY

What I find truly juicy is that our consciously integrated actions and harvests *increase* our *positive* impacts and renewable resource production, as they simultaneously reduce negative impacts and consumption.

For example, our home has become a non-polluting, neighborhood-based, renewable-power plant. Thanks to the reduction in our power consumption, our ten-PV-panel rooftop solar power system *produces* over three times the power our household consumes, which further reduces water consumption and carbon emissions at the utilities' power plants. Since we've hooked our solar panels to the utility grid (a grid-tied system), our surplus power feeds back into the grid and goes straight to our closest neighbors to help offset their portion of the coal-burning utilities' electricity and water consumption. We get credit for this power production so we don't get a power bill. Similar producers

of surplus solar power in communities with *feed-in tariffs*, such as Los Angeles, California; Gainesville, Florida; Oahu, Hawaii; and those throughout Germany even get paid for the surplus renewable power they provide to their local utilities.[6]

We reduce both water and energy consumption by using our zero-energy-consuming, integrated water-harvesting earthworks and tanks to transform dehydrating stormwater drainage into rehydrating stormwater *infiltration*. We do this by harvesting, filtering, reusing, and enhancing 100% of the rain that falls on our site, the adjoining earthen public right-of-way, plus the rain that runs onto our property from our neighbor's property. In addition, more than 90% of the stormwater flowing down our side of the street is harvested in our street-side water-harvesting basins (*Integrated Design Pattern Six*). In these basins the harvested stormwater grows food- and medicine-producing, wildlife-habitat-enriching, soil-building shade trees that cool the street, walkway, and neighborhood, thereby reducing the generated power and pumped water needed to cool neighboring homes. In this way our site has become a *net-benefit multiplier*. I write this while eating delicious mesquite-meal cookies, made with mesquite pods harvested from a native velvet-mesquite tree growing in the public right-of-way in front of my house, watered entirely with stormwater runoff from the street (see figures 5.18 and 5.19)!

Such integrated harvests are generating new community-based economies and augmenting existing ones such as tourism, education, and agriculture. In my neighborhood, a new mesquite economy has been

Fig. 5.18. Mesquite cookies.
Credit: Christian Timmerman

Box 5.7. Community Seeds Germinate, and Roots Deepen, When You Talk, Work, and Play with Your Neighbors

When Rodd and I moved to the Dunbar/Spring neighborhood, we wanted to get to know both the Place, and the People who made it what it was, in order to discover its essence, gain new perspectives, and find out how best to contribute to our new community. We'd greet all the people we saw, distribute home-baked goods around the holidays, and help neighbors with repairs and other needs. Chi later involved neighborhood kids by coordinating Halloween trick-or-treating and Easter-egg hunts. These efforts began to forge connections. Conversing one-on-one and at neighborhood meetings and potlucks, we deepened these connections and heard wild stories that opened windows into the community's past.

For example, we learned about the time a neighbor working in the crawl space under his floor fell through a rotted coffin lid and onto the skeleton within the coffin. He threw his head back as he started to yell, knocked himself out against the underside of the floorboards, collapsed atop the skeleton again, and finally scrambled away after regaining consciousness. It turns out many of the houses here were built over an old cemetery from which all the bodies had supposedly been removed, but we still find some when digging trenches and tree holes.

Stories of the community's past led to questions and conversations about the community's present and future. What were our neighborhood's problems? How might we solve them? What did we want as a community? What were our dreams? How could we realize them? And is so doing, how could we be something better? How could we *produce* something better? How could we do it in better way? This led to brainstorming, and if a juicy idea emerged and resonated with the group, strategizing might then follow. How could design, implementation, and outcomes tap into and grow the passions of all those involved? How could the project maximize the community's potential? How could it be made more dynamic, fun, and effective? More regenerative?

If things clicked, we'd go for it. It doesn't take much to start. A community tree planting, for example, just takes a few tree planters and a tree seedling or two. But we have found that more folks show up (and sometimes grant money too!) when you build on what folks love. So we recruited amazing cooks to prepare baked goods for the tree planters; birders to help select native songbird-attracting trees; and bicy-

cle enthusiasts to transport the trees, tools, snacks, and mulch with bicycle trailers. This generated more buzz. More people talked it up. More people stepped up to participate. As a result the planting produced something better than trees alone—people became aware of other practices and skills in the neighborhood, new community bonds and friendships were made, and more resources were generated in the planting of rainwater and food.

Leveraging the potential further, we made sure everyone could contribute. Volunteer crews that included—and trained—local youths helped plant trees for less-physically-able residents. These residents in turn provided homemade lemonade and neighborhood tales for the planters. This activity and the conversations it generated actively demonstrated how seemingly small efforts were reinvigorating the neighborhood and spurring our community's evolution as people with new (or renewed) caring, ideas, and skills stepped forward.

We found this was also true with our neighborhood mesquite milling. The first year we began with just one farm-scale hammermill (used to rapidly grind locally-grown mesquite seed pods into flour), a couple dozen passionate pioneers milling their hand-picked mesquite pods, and three great cooks preparing mesquite pancakes on three camp stoves to show everyone how tasty mesquite flour can be.

Eight years later the word and practice had spread, and our mesquite fiesta grew from a neighborhood event to a unique regional celebration rooted to the Sonoran Desert's rain-irrigated food plants and the cultures that developed these foods. It took three hammermills (two from other groups doing similar work) running throughout the day to mill all the harvested pods, and we served over 1,500 mesquite pancakes (see figure 5.20). By then we were offering bilingual information on how to plant the rain and food trees, and harvest and prepare the bounty. Many volunteers, non-profits, and emerging businesses were invited to showcase their unique uses of mesquite and other native wild foods and medicinals. Live music and puppet shows injected plenty of fun into our community garden/nature park/event space, which the neighborhood and broader community partners had created twelve years before from what had been a dirt parking lot of a closed-down school!

One year later the event was even better, as the

Box 5.7 continued

pancake breakfast organized by a few had evolved into a community-supported bake sale. Many people brought and shared their favorite wild-food creations, and we were overjoyed by the volume, diversity, and deliciousness of contributions that ranged from mesquite dog biscuits to acorn cookies to ironwood-tree edamame to mesquite naan bread topped with prickly-pear chutney! Everyone was getting exposed to new ways of utilizing neighborhood-grown wild

foods, along with the information and inspiration to realize still more. Ardent instructors led hourly workshops on growing and processing wild, dry-farmed foods. Many of the day's favorite recipes went home with people in the *Eat Mesquite!* cookbook, generated entirely from recipes developed in home kitchens and those of regionally famous chefs (which we had thoroughly sampled and vetted at community tasting parties). See www.DesertHarvesters.org for more.

Fig. 5.19. Harvesting mesquite pods

Fig. 5.20. Garth Mackzum enjoying mesquite pancakes with prickly pear syrup at a Desert Harvesters Mesquite Fiesta

created in which our neighbors and we can earn $25 an hour harvesting and processing native mesquite pods from community-planted, passively rain-irrigated trees in our streets and yards. These earnings are realized when we grind those pods into highly nutritious and naturally sweet flour, which we sell at our community's popular and ever-growing Mesquite Milling Fiesta and Bake Sale. No commute required for this job! (See box 5.7, www.DesertHarvesters.org, and the Vegetation chapter of volume 2 for more on this.)

While we do this only seasonally, some friends sell local dry-farmed mesquite flour year-round at farmers' markets, to local businesses and restaurants, and through their websites. The demand for this and other local foods draws an increasing number of people from across the U.S. and Mexico to join Tucsonans in attending the Desert Harvesters' Fiesta and similar events sponsored by an expanding number of groups in Tucson and around southern Arizona. At the fiestas, participants can attend demonstrations by local businesses and non-profits on water harvesting, wild foods, seed saving, solar cooking, and composting; and see examples of integrated water-, sun-, wind-, and shade-harvesting strategies at nearby sites within the fiestas' neighborhoods. An ever-increasing number of related workshops and conferences also promote these strategies, luring still more locals and travelers into the community and the movement, enabling them to generate similar transformations, skills, and jobs back home. It's all about connecting People with Place in a way that enhances the capacity of both.

Taking it to the public streets

Emboldened by such successes, our neighborhoods are taking integrated resource harvesting and enrichment to the streets—the primary public land, or "commons," of the urban environment. Since 1996 when we started the annual Dunbar/Spring neighborhood rain- and tree-planting project, our neighbors have planted over 1,275 food-bearing native trees and annually infiltrated hundreds of thousands of gallons (millions of liters) of stormwater within street-side water-harvesting earthworks. These improvements—mostly along formerly barren sun-baked streets—have dropped summer temperatures by as much as 10° F (5° C) in the now-shaded areas, and have significantly reduced flooding. (See figures 5.10A and B, and the "Reducing Hardscape" and "Vegetation" chapters of volume 2.) Best of all, ever-more neighbors are meeting, working, and recreating together. We talk to one another. We get to know one another. Our neighborhood and our neighboring neighborhoods now form a stronger community.

Making connections is easy these days. When I need help moving a heavy object, want to share an idea, or yearn to reconnect with neighbors after a long trip, I do a little tree pruning and soil mulching on a neighborhood street or traffic circle. Within minutes I'm talking with neighbors walking or riding by (see figure 5.21).

Taking that public pruning further, our neighborhood partnered with local non-profit Watershed Management Group to turn thrown-away community *brush and bulky prunings* into *chipped and mulchy gleanings*—helping harvest our share of the 12% to 14% of a typical U.S. municipality's solid-waste stream made up of nutrient-rich yard trimmings.[7,8] In the past people would pile all their brush and bulky yard trimmings on the street where they'd get picked up by garbage trucks and hauled to the dump.

Fig. 5.21. Public amenities and activities encourage public interaction. Vaughan Lancaster harvests prickly pear fruit in street-side rain garden as mama Chi and neighbor Jon converse.

Now rather than export this organic resource, we harvest and enhance it with twice-a-year neighborhood pruning workshops, during which community members prune their own trees with help from volunteers, clear obstructions to public pedestrian foot paths, and place prunings on the street curbs. A chipping service then chips the prunings and returns them to the curb as mulch, which neighbors can then apply to the surface of their water-harvesting earthworks. There, this organic matter naturally breaks down to become water-absorbing, mycelium-feeding mulch and humus—the foundations of soil, fertility, hydrology, and food systems (see box 5.8 and figure 5.22).

Mulched water-harvesting traffic circles and chicanes/pull-outs located *within the curbs* of our neighborhood streets help build soil, calm traffic, replace impervious pavement with living sponges of vegetation (*Integrated Design Pattern Seven*), and encourage bicycling and walking rather than driving (see figures 5.1B and 5.23). Walking and biking our calmer, shadier streets we meet, greet, and get to know still more neighbors. Everyone's sense of community grows, crime rates drop, we have fewer traffic accidents, and we are more active and healthy.

Fig. 5.22. Harvesting organic matter and more. Prunings are cut up and dropped below the plant from which they came to cycle the nutrients back into the soil and its beneficial microorganisms, which in turn feed and water the plant.

Those in surrounding neighborhoods and the larger community enjoy this too, since our improvements have enabled two of our neighborhood streets to be designated by the city as bicycle boulevards—corridors of bicycle- and pedestrian-friendly, low/slow-traffic residential streets that interconnect many city neighborhoods with safe bike/pedestrian crossings at major intersections. Add the native food-bearing vegetation supported by harvested street runoff, and these become *living* corridors of people, water, wildlife, shade, and food.

These living bicycle corridors are so alive, they are quickly becoming the preferred transportation modes for an increasing number of residents and visitors who are experiencing daily journeys every bit as interesting and inspiring as their destinations.

It takes off

This is taking off! In 1993 when I started *talking* about harvesting and enhancing on-site water, sun, shade, mulch/soil, and wind using integrated-design principles and techniques, I often felt like a lone freak—there were so few examples. So I focused on *doing*: creating examples, fixing mistakes, improving and further integrating my systems, teaching, working with colleagues, and writing this book to show the way. Others soon did the same, among them the non-profit groups Watershed Management Group and Sonoran Permaculture Guild. Now all kinds of people—along with businesses and institutions including our city, agricultural extension service, and university—have begun harvesting water in a way that *slows, spreads, and sinks the water* as it *shows the flow.* Fifteen years ago, our home was the only one on our block harvesting rainwater. Now over 80% of the neighbors on our block do it—not yet to their fullest or most-integrated potential—but they're on their way.

Within my neighborhood all this positive activity is due to many individual and community efforts along with a huge boost from a Pima County Neighborhood Reinvestment grant. By the end of 2012 this grant had funded the installation of three more water-harvesting traffic-calming circles (for a total of 10), 33 water-harvesting chicanes/pull-outs (figure 5.23), 50 water-har-

Fig. 5.23. A newly constructed water-harvesting chicane (and a water-harvesting traffic circle behind) along a bicycle boulevard just after a summer storm in the Dunbar/Spring neighborhood, Tucson, Arizona, 2012

Fig. 5.24. Vaughan Lancaster plays near a rain garden filled with street runoff flowing over the low point of a curb where there was once a driveway. A second rain garden is just outside the right hand side of the photo, 2011.

5.25. A 16-inch (40-cm) curb cut made in street curb to direct street runoff to street-side tree

5.26. A 4-inch (10-cm) core cut through the street curb to direct street runoff to a street-side tree in a newly constructed rain garden. Native flowering shrubs and grasses are yet to be planted in the basin and on the terrace, 2012.

vesting basins in the public right-of-way (for a current neighborhood total of over 85), 50 water-harvesting curb cuts and 35 curb cores directing stormwater to associated street-side basins (figures 5.25 to 5.26), and the planting of 150 additional native trees and shrubs (for a total of well over a 1,000) that turned those mulched basins into true rain gardens. These combined public right-of-way water-harvesting strategies annually harvest over *660,000 gallons (2,500,000 liters) of stormwater* that used to contribute to the flooding of our streets (figure I.9). Though this is just the beginning, for we could easily harvest another 20 times this volume within the neighborhood's rights-of-way with more rain-planting efforts.

Public art, created by local artists and funded by the grant, helps celebrate these achievements, connecting with and enhancing People and Place. This art also generates mini-playgrounds, which become destinations for families' daily walks or bike rides and conversation (see figures 5.27 and 5.28). For example, a sculpture by Joseph Lupiani within a water-harvesting traffic-calming chicane features the native Sonora Sucker fish above a Texas Horned Lizard, symbolizing our past and our potential (figure 5.28).

The sculptured fish remind us of their once-abundant presence (they were even fished!) in the Santa Cruz River—located less than a half mile (0.9 km) west of our neighborhood. But by 1937, the fish were extinct in this area[13] because the river's flows

Fig. 5.27. Gila monster bench beside a community book exchange box in the public right-of-way. The Gila monster (a native lizard) is reading a poem by Paul Laurence Dunbar in book held open by its foot. Our Dunbar/Spring neighborhood is named after the Dunbar school (named after the poet) built here in 1918. Sculpture/bench by artists Hiro Tashima and Jason Butler. Book box by neighbors Bill Moeller, Gavin Troy, Turtle, and Ian Fritz

had subsided due to the degenerative overpumping of groundwater and the mismanagement of the river and its watershed.

The Texas Horned Lizard demonstrates a regenerative strategy it evolved in response to dry conditions.

Fig. 5.28. Vaughan Lancaster inside a sculpture by Joseph Lupiani showing locally extinct Sonora Sucker Fish and a rainwater-harvesting Texas Horned Lizard (in front of Vaughan) within a water-harvesting traffic-calming chicane in the Dunbar/Spring neighborhood, Tucson, Arizona, 2012

This lizard can harvest rainwater off its wide back and direct it to its mouth via gutter-like channels between its scales.[14] As we, in our own ways, harvest the local rain, we decrease the "need" to pump groundwater, we lose less rainfall to runoff and evaporation, and we get closer to sustainable rates of water use and reuse. We start *giving water back* to our Place.

This leads to a path of abundance in action. When all of us who share this river and watershed live in a way that gives back more than we take, that stewards and regenerates our shared ecosystem—our shared home—then our river, its fish, and all its related bounty could flow again in whole new ways.

Planting local rain and magnifying its potential has begun in neighborhoods all over our city, country, and Earth. Seeing so many people caring gives me great hope. When we translate this caring into intelligent and appropriate action, everyone benefits. When you create, live in, and *evolve* thriving examples that people can see, walk through, experience, understand, and replicate, they can't help but notice, get inspired, do likewise, and/or launch new efforts. As you embark on this path, continually strive to make your efforts as integrated and regenerative as you can, and your path's dynamic potential will soar. It is the ever-growing and improving examples that keep sparking and germinating more positive evolution. Engage and partner with *your* natural surroundings and with your community—plant the rain, dance with the sun, grow the shade, feed the soil, sail with the wind, and be one of your community's sparks!

Appendix 1
Patterns of Water and Sediment Flow
with Their Potential Water-Harvesting Response

*A thing is right when it tends to preserve the integrity, stability, and beauty of the
biotic community. It is wrong when it tends otherwise.*

—Aldo Leopold, from "The Land Ethic" in *A Sand County Almanac*

PATTERNS OR TRACKS

The patterns or "tracks" left by the flow of water
and sediment are excellent guides to direct you in
selecting and placing appropriate water-harvesting
strategies. Both erosion and deposition are natural
occurrences that create characteristic patterns. Erosion
creates concave landforms when sediment is carried
away by runoff water, leaving depressions such as
rills and gullies behind. When sediment-laden runoff
water is slowed or spread out, it loses the energy
(some may think "force") needed to carry sediment.
The largest sediment drops out of slowing water
first. If the water continues to slow, smaller sediment
is deposited. Thus, in places where water flow has
slowed, deposited sediment creates convex landforms
such as alluvial fans and depositional bars.

Erosion and deposition are naturally kept in
balance in healthy watersheds by the presence of
vegetation, porous living soils, and the slowly occurring erosion that is a normal part of a dynamic
equilibrium, and which gradually moves sediment
downslope. As the components of sediment—organic
matter, soil, and inorganic material—migrate and
deposit downslope, they are replaced upslope by on-
site leaf drop from vegetation; and leaf drop, soil, and
inorganic material migrating downwards from higher
upslope. At the same time material "migrates upslope"
in the form of animal droppings containing organic
material and seeds that build soil and grow vegetation.

In unhealthy watersheds, erosion far surpasses
deposition and can carve deep cuts in the land, leading to a rapid loss of water, organic matter, soil, and
inorganic material. A deep erosion cut is a sign of an
unstable landscape that is losing valuable organic and
inorganic resources. Learning to recognize erosion
patterns, and understand their causes, is essential in
planning effective water-harvesting strategies that
break the erosion triangle shown in figure A1.1.

SPEED, DEPTH, and VOLUME refer to three
characteristics of water flowing on the broad surface of
the land and within channels. Reduce any of the three
by SLOWING, SPREADING, and/or SINKING the
water's flow, and you begin to break the erosion cycle.
The more places you break the cycle, the more you
reduce erosion. If you put a water-harvesting strategy—such as a berm 'n basin, one-rock check dam,

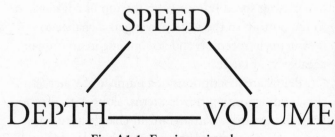

Fig. A1.1. Erosion triangle

or infiltration basin—in the path of flowing water, you will reduce erosion by *reducing* the SPEED of water flow, while increasing the time the water has to linger, cycle, and SINK into the earth below where it becomes available to support the growth of plants. The more plant life the harvested water supports, the more roots hold the soil, the more plant canopy shelters the soil, and the more biomass catches and builds the soil. Your efforts will help generate many living "speed bumps" on the erosion highway.

The greater the DEPTH of flowing water at any given point, the greater the water's power to erode sediment when compared to shallower water moving at the same speed. If numerous water-harvesting strategies are placed to SPREAD the water's flow from top-to-bottom throughout a watershed, then erosion will be decreased by *reducing* the DEPTH of flowing water. At the same time, SPREADING the water increases the *time* period and *spatial* area of contact between water and earth, *increasing* the total VOLUME of water infiltrating into the subsurface. Water that SINKS into the subsurface migrates laterally and downward through the force of capillary action and gravity. This leads to longer duration water storage and higher soil-moisture availability than occurs when water flows rapidly across the land surface or within channels and quickly evaporates.

The greater the VOLUME of water flowing over land or within a channel, the more sediment it can carry. If you place a number of water-harvesting and water-infiltrating strategies starting at the top of the watershed, you will SINK a larger VOLUME of water higher in the watershed, reducing erosion by reducing the overall VOLUME of unchecked surface water flow throughout the watershed. With more contact time and contact area between water and earth, more water will be absorbed into the subsurface. This is like placing speed bumps from the top of a watershed to the bottom so the water never gets a chance to destructively speed up and accumulate in an unmanageable VOLUME.

Below are descriptions of a number of water, erosion, and deposition flow patterns, along with descriptions of how these patterns might change in response to constructing water-harvesting earthworks. Many of the earthwork techniques are introduced and illustrated in chapter 3. These are covered in more detail in *Rainwater Harvesting for Drylands and Beyond, Volume 2: Water-Harvesting Earthworks*, along with the publications *Let the Water Do the Work: Induced Meandering, an Evolving Method for Restoring Incised Chanels* by Bill Zeedyk and Van Clothier (found at www.QuiviraCoalition.org and www.StreamDynamics.us) and *Erosion Control Field Guide* by Craig Sponholtz and Avery C. Anderson, (found at www.QuiviraCoalition.org and www.DrylandSolutions.com). Bill Zeedyk and Van Clothier greatly improved this appendix with their review, editing, images, and sharing of their evolving water- and sediment-harvesting strategies and insights, as did Craig Sponholtz, who also contributed much of the writing.

PATTERNS AND RESPONSE

SEDIMENT SIZE AND TRANSPORT

PATTERN: Sediment moves—or "flows"—along with water flows. Always consider both water flow and sediment flow in design. The lower the velocity or energy of the flow, the smaller the size of sediments the water can carry. Velocity is reduced when the steepness of the flow's path becomes more gradual, the depth of the flow is widened as its depth is lessened, and/or when the channel's bed and banks come to be rougher. Steep, fast-moving rivers have enough energy to easily transport boulders. Gently sloped, slow-moving streams may only be able to transport grains of sand and smaller particles. The velocity of water flowing across the broad landscape also affects the size of sediment it can transport, but the size range of transported materials is much smaller. Stationary, ponded water has no energy to hold particles except for the finest suspended clay particles.

RESPONSE: The size of sediment that you observe has been moved by flowing water indicates the velocity of water flows that have occurred there—primarily during the last flood. Determining typical water flow velocity—along with the quantity and likely sediment sizes the flow will transport—is key to selecting, placing, and sizing appropriate water-harvesting and erosion-control structures. It is almost always better to stay out of boulder-strewn drainages, which have potentially intense flows. Instead focus water-harvesting and erosion-control efforts on more gentle slopes in the broad landscape and in the smaller drainages. Rock structures placed within channels should be constructed using rocks that are larger than 90% of the typical naturally deposited rock. If you are building a rock structure in a channel full of sand with no naturally deposited rock, you may want to import and use rock having a median dimension (not the shortest width nor the longest width of the rock) that measures approximately 1/8th of the total channel width (for example, in a 6-foot wide channel appropriate for hand work, 8- to 10-inch rock would be used to construct a one-rock dam across the channel width). The rocks in these structures should be tightly and carefully placed so they do not easily wash away. New rock structures are structurally weakest in the time period before sediment carried by smaller water flows has been deposited in and around the structures. To avoid impacts from large storms occurring soon after construction, spaces between the rocks, and on the upstream side of rock structures, can be pre-filled with gravel and/or smaller sediment to mimic the deposition that would occur in smaller flows.

When we harvest or manage flowing water, we are also harvesting and managing the sediment transported with that water. This is especially true in dryland watersheds where the relative sparseness of vegetation results in a sediment-rich environment. An ample supply of sediment is the unavoidable companion of any water harvested in erosive landscapes. Harvesting sediment-rich water can be a challenge but can also yield assets.

Sediment is an asset in in-channel sediment-harvesting structures such as one-rock dams when it fills in the spaces between rocks and forms a sediment-deposit on the upstream side of the structure, which readily absorbs and holds water for plants. It is an asset on alluvial fans and the broad landscape where media luna- (crescent moon-) shaped rock berms/sheet flow spreaders, or dense bands of native grass, slow water and collect sediment to provide a well-hydrated growing medium for more vegetation (see figures in the *sheet flow spreader* section of chapter 3). Here, the vegetation's stabilizing roots and above-ground growth

slow, spread, and harvest still more water and sediment. See Floodplains, Incised Channels, Crossover Riffles, and Alluvial Fans below for more information.

Harvesting sediment-laden water can be a challenge when water-harvesting earthworks such as basins, swales, and other earthen structures that "pond" water, fill in with sediment. This can reduce water-holding capacity, create unexpected overflow, and reduce the effective lifespan of water-harvesting earthworks, while creating the on-going job of removing the sediment. To reduce unwanted sedimentation of earthworks, revegetate the watershed above the earthworks using the sediment as a soil-building resource, especially if it is composed of fine particles, organic matter, and topsoil. Wherever possible direct the flow path of sediment-laden water across well-vegetated, gently sloping landforms before the flow enters basins and swales (see Alluvial Fans pattern section further on). This will allow fertile sediment to be deposited and utilized rather than being collected in water-pooling earthworks.

Sediment traps can be incorporated into the design of water-harvesting structures if it is not possible to intercept sediment prior to runoff water entering a basin or other earthwork. Sediment traps are basins designed to pond runoff and collect sediment, while allowing relatively clear runoff water to overflow to other water-harvesting basins. Sediment traps should be easily accessible and should be cleaned out frequently between storms. Sediment traps should not be planted. Depending on sediment sizes, the collected sediment can meet a variety of needs ranging from fine-grained material for plastering to a useful supply of gravel. Mr. Phiri similarly uses the sand collected in his reservoirs (chapter 1).

SHEET FLOW

Sheet flow is the relatively even distribution of runoff water flowing down the slope of the land rather than being focused into distinct channels. If after a large rainfall, you don't see distinct channels in an area of sloping land, water has flowed across the landscape as sheet flow. Such flow is beneficial unless the landscape is so disturbed or degraded that sheet flow leads to sheet erosion. Microdetritus berms and pedestals are indicators of sheet flow and sheet erosion, respectively, and are described below.

Microdetritus Berms

PATTERN: Microdetritus berms are small, curved lines of leaf duff, animal scat, twigs and other organic matter that were carried by sheet flow then deposited perpendicular to the direction of flow. The outer bow of the curve usually points downslope. These are typically less than two inches (51 mm) high, and often remain only a few weeks or less after a rainfall. They may be found in the broad landscape and on gentle slopes in yards, but are not found in drainages. (See figure A1.2.)

RESPONSE: These tiny berms of organic matter indicate recent sheet flow. Their presence typically signals there is no pressing need for erosion control. They do help indicate the direction of land slope and water flow—information which is useful when laying out contour berm 'n basins, boomerang berms, mulch berms, and sheet flow spreaders, and when planting on-contour.

Pedestals

PATTERN: As stabilizing vegetation in an area is impacted by overgrazing, changes in land use, or other destabilizing factors, once-benign sheet flow can begin to cause sheet erosion. Pedestals are isolated mounds of soil that remain after the exposed ground outside the perimeter of the pedestals is eroded away (fig. A1.3). A close look reveals that a pebble cap or plant roots and a canopy of leaves and branches have kept the earthen pedestal from washing away from the direct blow of falling raindrops and the resulting runoff.

RESPONSE: The presence of pedestals indicates that year-round soil-sheltering and soil-anchoring groundcover is lacking, and sheet flow has become sheet erosion. To respond, utilize strategies that re-establish dispersed—rather than concentrated—sheet flow, slow the speed and depth of runoff, increase infiltration, and support more vegetation. You can also protect the soil surface from the impact of raindrops using mulch and you can change land management practices. Grazing practices may need to be changed, access by ATVs and other off-road vehicles may need to be blocked, and even human foot traffic may need to be curtailed or redirected, if these activities are preventing re-establishment of perennial groundcover.

To stabilize eroding sheet flow areas in home landscapes and smaller landholdings, it may be appropriate to construct infiltration basins and contour or boomerang berms using earth, brush, mulch, or one-course structures of readily available rock.

Fig. A1.2. Microdetritus berms after sheet flow has ceased. Sheet flow was from top of photo to bottom of photo. Credit: Craig Sponholtz

Fig. A1.3. Pedestal pattern on Empire Ranch in southeastern Arizona

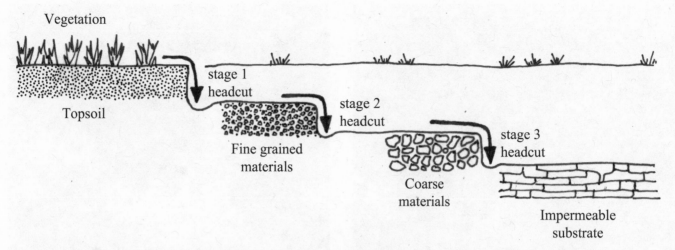

Vegetation

Topsoil

stage 1
headcut

Fine grained
materials

stage 2
headcut

Coarse
materials

stage 3
headcut

Impermeable
substrate

Fig. A1.4. Headcut progression. Uplands headcuts tend to advance headward in series or waves of stepped cuts and propagate headward/upslope. Adapted from *Let the Water Do the Work* by Bill Zeedyk and Van Clothier, Quivira Coalition, 2009

Imprinting and/or hydroseeding is better suited for badly disturbed broad-scale applications.

Monitor the sheet erosion by driving an array of large landscape nails into the ground up to the nails' heads. If pedestals form below the nail heads over time then sheet erosion is still occurring. The rate of soil loss can be measured in inches or millimeters per year.

CHANNEL FLOW

Channel flow consists of water concentrated within distinct channels or drainages. The source of the water may be localized rainfall, remote rainfall, springs, shallow groundwater, snowmelt from mountains, or other sources. The water may flow all the time (perennial flow), it may cease flow periodically (intermittent flow), or it may flow only in response to storm events (ephemeral flow). When trying to find erosional and depositional patterns in channels, look at Sediment Size and Transport (above), and look for Headcuts, Rills, Gullies, Floodplains, Incised Channels, Meandering Waterways, Bank Cutting and Pools, Crossover Riffles, and other patterns discussed below.

Headcuts

PATTERN: A headcut is an erosional feature that cuts a deep gouge into the earth, which quickly drains water away from the area resulting in a drying of the landscape. Headcuts can occur in uplands and within channels. Headcuts start small, but can grow to be severe. Those found within channels cut upstream or "headward" toward the headwaters of the stream, opposite the direction of water flow. *Uplands headcuts* and *in-channel headcuts* are described below.

An *uplands headcut* is found at the point—often within a gently sloping landform—where sheet flow has concentrated into a deep, erosive concentrated flow, such as at the leading eroding edges of rills and gullies. Upland headcuts often occur in a series of stepped cuts or "waves." They start in the topsoil, but once the cut extends down to the first layer of subsoil, another headcut frequently erodes into an even deeper layer of subsoil that is often composed of courser material than the layer above. This can lead to yet another headcut eventually digging down to a less permeable layer such as bedrock. In this way, a small headcut in the topsoil gets extended by larger headcuts in subsoil layers, and the gully gets deeper with every wave of headcuts. Uplands headcuts continue cutting upslope as long as there is soil to cut, or until they are corrected and stabilized (fig. A1.4).

In-channel headcuts can result from disturbances that change the stability and/or elevation of part of a channel, such as road crossings that drop the bed of a drainage, loss of stabilizing vegetation due to removal

Fig. A1.5A. Water flow pouring over a headcut. Credit: Craig Sponholtz, www.DrylandSolutions.com

Fig. A1.5B. Dry headcut long after water flow has ceased. Credit: Craig Sponholtz, www.DrylandSolutions.com

or die off, or a failed dam that creates an erosive scour hole, among other causes. Once initiated, an in-channel headcut is found within an existing channel at the headcut's pour-over lip—the location where the channel elevation changes abruptly at the point of disturbance, or even some distance downstream (possibly even out of sight) of the original disturbance. The result can be a lowering of the stream bed (see Incised Channels below) to the point where small floods no longer rise to the top of the channel's banks and spread out on the stream's floodplain (see Floodplains below). As the bed of the channel drops due to the initial headcut, it can spark headcuts that migrate up every tributary flowing into it. And as these tributaries drop they spark headcuts up every tributary flowing into them. During runoff events, the upland and in-channel headcuts are among the wettest parts of the landscape or channel, with water flowing over the pour-over lip or drop-off locations, scouring material away from the higher edge of the cut, while also falling as a waterfall causing turbulent scour at the base of the cut (see figure A1.5A). All of this erodes the headcut further headward. As the runoff diminishes, a headcut will continue to be one of the wettest places because this is where water will weep—gradually draining from adjacent sediments and subsoils and collecting in the low area. However, after the runoff event is over and the weeping of water has stopped, the headcut can become the dri-

est part of the landscape because the cut opens up, or exposes, the substrata of the channel bed, and the subsoil of the uplands, to more rapid moisture loss to evaporation. As this occurs, the lip of the pour-over may further collapse due to drying and cracking as surrounding plant roots die (fig. A1.5B). These are key points to take into account of when considering response strategies.

RESPONSE: Headcuts are an indicator of serious erosion problems. They need to be addressed quickly to stop channel incision. Within a gently sloping landform, one way to stop an upland headcut is to starve it of water by redirecting flow elsewhere, while taking great care not to direct the problem to a new location. Overland sheet flow should be slowed, spread out, and infiltrated into the soil to the greatest extent possible before it reaches the headcut. Water-harvesting strategies such as mulch, vegetation, infiltration basins, berm 'n basins, and sheet flow spreaders may work well in home landscapes and smaller landholdings. Imprinting, hydroseeding, and increased vegetation are better strategies for badly disturbed broad-acre areas located upslope of a headcut and outside of the channel. Perhaps the headcut is in a ditch or dirt road that has concentrated flow down a new erosive path. Look for an opportunity to redirect the water back into its more stable historic flow path.

Fig. A1.5C. Rock-lined plunge pool or Zuni bowl stabilizing headcut from fig. A1.5B, just after Zuni bowl was built. During water flows, water will pool in the bowl and infiltrate between the rocks to prolong the free irrigation of vegetation that will grow through the rock and help stabilize the structure. Arrows denote water flow.
Credit: Craig Sponholtz, www.DrylandSolutions.com

Within the eroding channel, stabilize the headcut itself, while also spreading and infiltrating the water flow with permeable barriers appropriate to the scale of the channel. During runoff events, a *rock-mulch rundown* (at an uppermost uplands headcut), *rock-lined plunge pool* (or *Zuni bowl* at in-channel head-cuts), or a *cross-vane* prevents headward erosion by preventing the drop-off or pour-over lip from migrating. As the runoff diminishes, a rock-lined plunge pool stays full of water, supersaturating the vicinity of the headcut with water, prolonging the time soil moisture is available for plants, and maximizing stabilizing plant growth. See figures A1.5C, A1.6, and A1.7.

Since headcuts often advance headward in waves of different heights, several different types of structures may be appropriate. A rock-mulch rundown is only appropriate on low-energy headcuts where sheet flow collects and *first* enters a channel. Zuni bowls can be used to control both upper headcuts at the point where sheet flow enters a channel and can control a headcut within a channel at the middle drop of a three-stage headcut. A one-rock dam could control a third, shallower drop at the lower end of the cutting sequence.

Fig. A1.6. Rock-mulch rundown stabilizing a previously erosive headcut where sheet flow transitioned to erosive channel flow at the cut. Note people are walking on a one-rock (high) dam below the rundown, and there is a second one-rock dam to the right of the lower person's waistline. Northern New Mexico. Installation by Craig Sponholtz

Fig. A1.7. A cross-vane (keyed into bedrock on left side of photo and native boulders on the right) controlling a headcut that used to erode the edge of the road, while creating a scour pool that clears sediment and collects water for wildlife. Designed by Bill Zeedyk. Hand-built by Pima County road crew in 2007. Altar Valley, Arizona

Rill Erosion or Runnels

PATTERN: Rills or runnels are tiny erosive drainages from which loose soil has washed away. They are the first stage of localized channel formation and are created when sheetflow collects into channelized flow due to loss of groundcover, and/or the steepness

A1.8. Rill erosion. Credit: Craig Sponholtz

of the slope increases. They are very common on eroding slopes uphill and downhill of roadways that are cut into hillsides, or on bare dirt driveways and roads that run downslope (fig. A1.8). Rills are an early stage of the type of channel erosion that occurs upslope of small headcuts, and tend to become tributaries to gullies.

RESPONSE: Rills are a precursor to more erosive gullies. To prevent further degradation of the system, first spread and infiltrate sheet flow above the rill or runnel. Berm 'n basins—particularly their sheet-flow spreader variation—along with increased vegetation and the addition of mulch can all effectively break the continuity of flow from watershed to rill. Next, use brush to spread and infiltrate the flow within the rill itself. Fill the rill with brush placed lengthwise, parallel with the rill. This avoids creating the scour holes that commonly form when barriers are placed perpendicular to the flow. Place the cut base of the brush pointing upslope and branches pointing downslope so the water's flow will help push and anchor the branches into the bed and banks of the rill. This anchoring effect can be further enhanced by pruning those branches and stems that will lay against the banks and basin of the rill to form rigid 1- to 4-inch (2.5- to 10-cm) "hooks" that anchor the brush when it is pushed into place. Avoid layering the brush too thick. You want some direct sunlight to reach the soil beneath the brush to germinate seeds and enable

Fig. A1.9. A deepening gully just below a headcut. Avra Valley, Arizona

vegetation to grow up through the brush to trap sediment and stabilize the land.

Gullies

PATTERN: Gullies are formed as a result of channelized, concentrated, and accelerated water flow. Gullies are large eroded drainages that result from extensive head-cutting (fig. A1.9). A gully is actively eroding if it is downcutting and getting deeper (fig. A1.10). A gully is likely stabilizing if it is getting wider rather than deeper, flow is slowing and becoming shallower, and vegetation is growing within the channel and on its banks (fig. A1.10). This is how nature heals a gully and our response should mimic nature's.

RESPONSE: Gullies are often precursors to even more severe, deeply eroded ravines. The overland flow draining toward the channel should be slowed, spread, and infiltrated into the soil as much as possible before reaching the channel. Where flow paths have been diverted by steeper ditches, trails, or roads, look for

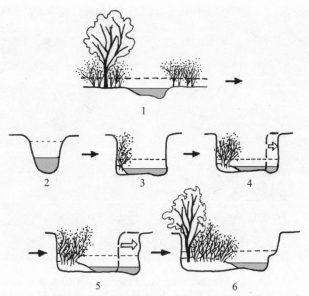

Fig. A1.10. Stable conditions to unstable to stable again. Formation and evolution of a deepening erosive gully with no access to floodplain (2), into a calmer widening water channel that creates a new accessible, vegetated, sediment-trapping floodplain by beneficially eroding its own banks (6). The dark area represents *bankfull* or normal full channel flows just before they would begin to spill over the channel's banks onto the floodplain, while the dotted horizontal line represents depth of typical flood flows. Adapted from *Let the Water Do the Work* by Bill Zeedyk and Van Clothier, Quivira Coalition, 2009

Fig. A1.11. Healthy, meandering waterway during normal bankfull water flow. In flood flows the water flow will beneficially spill over the banks onto the accessible, well-vegetated floodplain. The grasses on the floodplain will help stabilize the soil and harvest sediment, while calming, cleaning, and infiltrating much of the water flow. Credit: Craig Sponholtz

opportunities to redirect the flow back to the more gradual, historic flow path. Changing land management and installing sheet flow spreaders, imprinting, vegetation, mulch, and infiltration basins may all be appropriate above the gully.

Monitor the gully to see whether it is cutting deeper or is widening and stabilizing. If it is cutting deeper, carefully place a series of one-rock check dams constructed perpendicular to the flow at *crossover riffle* locations to help stabilize the drainage. At the same time, repair the broader landscape. See Headcuts above for headcut response, Crossover Riffles below for one-rock dam placement, and Floodplains and Incised Channels below for more on the natural process and response.

Floodplains

PATTERN: A floodplain is the adjacent land area over which a channel spills and diffuses excess flow

during flood events. The average depth and erosive force of floodwater is reduced when it rises sufficiently to flow out of the channel and spread out over the wider—usually vegetated—area of the floodplain (fig. A1.11). In this way, floodwater is transformed into calmer sheet flow that soaks into the ground on the floodplain, reducing the intensity of the flood.

RESPONSE: Floodplains are a water channel's pressure relief valve. We must maintain the floodplains around channels to allow natural over-bank flood flow and its replenishing effects. We must avoid manmade channelizing, narrowing, or straightening of water channels and their floodplains, which result in channel incision and desiccation of the surrounding land. Do not build a house on a floodplain or an alluvial fan, as addressed below. Utilize land management practices and strategies that enhance water-absorbing groundcover and hydrate the watershed, not strategies that denude, pave, compact, and dehydrate the watershed.

Incised Channels

PATTERN: An incised channel is one whose bed has eroded to such a depth that excess flow can no longer spread out over the floodplain. The incised channel often continues to cut deeper because large flows that can no longer spill over and diffuse into the floodplain cut downward instead. Water tables then drop because shallow groundwater adjacent to the channel flows into the deep cut and drains away, rather than being replenished by surface water infiltration as occurs in shallow-bedded streams. As water flow level is lowered in the incised channel, the higher capillary zone on the banks, which used to sustain vegetation along the waterway, dries out and disappears. Riparian plants die out and conditions become more arid and fragile as a result.

RESPONSE: Within incised channels use grade-control strategies like one-rock dams to help stop downward erosion, while allowing the channel to cut laterally to widen its banks in a natural meandering pattern. This recreates a floodplain at a lower level within the widening channel, and enables the channel's flood flows to reconnect with the channel's new floodplain (fig. A1.10). See Meandering Waterways.

Meandering Waterways

PATTERN: In waterways flowing down gradual slopes (less than 4% slope) the energy of the flowing water tends to be diffused by meandering flow patterns (fig. A1.12). This illustrates the fourth water harvesting principle of slowing, spreading, and infiltrating the water (fig. 1.16B in chapter 1). The more the waterway meanders, the greater the distance water must flow to get from the top to the bottom of the waterway. This reduces the gradient or slope of the bed of the waterway, which reduces the erosive speed of the flow while increasing the duration of flow. The water comes into contact with more of the surface area of the channel bed, allowing more water to infiltrate into the subsurface to recharge wells, springs, and aquifers, and to provide moisture for more plants along the waterway. These plants stabilize the channels and adjacent banks, slow flood flows, catch sediment, build soil, and cool and shelter the water, improving fish habitat and water quality.

RESPONSE: Meandering is good. Let it happen. It is a natural process in which the erosive force of water

Fig. A1.12. Meandering waterway with easy access to wide, vegetated floodplain. Northern New Mexico. Credit: Craig Sponholtz

flow in incised waterways is dissipated as the flow cuts into alternate banks of the channel, inducing the redistribution of bed materials, and the gradual evolution to wider waterways with longer flow paths. In the course of this natural process, constricted channels "heal" themselves over time by creating a widened, healthy floodplain. Although it is outside the scope of this book, a strategy to consider in abnormally straight, eroding channels is "induced meandering," which speeds up the widening of narrow incised channels and gullies, lengthening the flow path, and making the overall channel depth shallower as it erodes its banks in a natural meandering pattern. See the book *Let the Water Do the Work: Induced Meandering, an Evolving Method for Restoring Incised Channels* by Bill Zeedyk and Van Clothier for more.

Bank Cutting and Pools

PATTERN: Bank cutting at meander bends occurs where channelized water flowing around a curve cuts the outside bank of the drainage due to the centrifugal force of the water, scouring a basin or pool at the base of the cut. In sandy ephemeral dryland waterways, this pool is hidden when it fills with sand at the end of a runoff event. Bank cutting and pool scouring at meander bends create concave, erosive landforms. The slower moving water on the inside of the curve allows the finer-grained sediment that was cut from the outer bank to deposit on the inside curve, forming a convex *depositional point bar* where vegetation can get a foothold. Immediately downstream of curves, larger sediment is deposited at the *crossover riffles* located in the straight runs of the channel between curves. In figure A1.13, notice the more concave slope of the cutting side of the flow. Then notice the convex shape of the deposition side, usually a gently sloping bank.

The undulating bed of the waterway's lower pools and higher crossover riffles increases the water-holding and water-infiltrating capacity of the waterway, controls flow, and enhances diverse microhabitats—including fish habitat in perennial flows.

RESPONSE: Bank cutting can be very helpful to very harmful depending on the degree of channel incision and relative sinuosity. Where the channel is incised and lacks sinuosity, it can be helpful. Where the channel is not incised and displays appropriate sinuosity, it can be harmful. (See Depositional Bars below for a nonbeneficial example of bank cutting, and the book *Let the Water Do the Work* for more on assessing the health and sinuosity of a channel.)

Bank cutting on a curve and the location of pools at meander bends can help "spread and infiltrate" water by eventually widening and elongating the flow path in an ever more serpentine/meandering channel. This can reduce the slope of a channel and slow the rate of flow. Typically, you should not attempt to control this curving of a channel. Rather, use your recognition of the curving to correctly place one-rock dams if erosion/grade control within the channel is necessary.

One-rock dams are grade-controlling sediment traps, so place them where sediment naturally accumulates. Straight sections of channels at crossover riffles are the ideal sediment-accumulating locations. Do not place them in curved sections of channels where water would likely cut around the outside edges of one-rock dams. One-rock dams placed at the crossover riffles help raise the elevation of the crossover riffles, while increasing the depth of the upstream pools in relation to the riffles, thereby increasing the water-harvesting potential of the waterway. Note that all crossover riffles within a given channel reach should

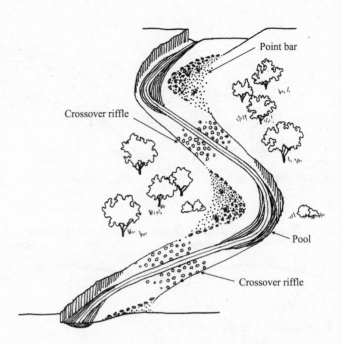

Fig. A1.13. Erosional bank cutting and pools at outer bends, depositional point bars on inside of bends, and depositional crossover riffles between meanders

be raised the same amount of height (if you raise one a foot, raise all the others a foot). Otherwise the slope of that reach of the waterway will be steepened and the flow velocity increased, which would likely increase erosion. See Crossover Riffles for more.

Crossover Riffles

PATTERN: Crossover riffles are the shallow reaches of a channel found in the straight stretches between meander bends and their pools (figs. A1.13, A1.14). Riffles are depositional features formed on the channel bed by large sediment that is scoured out of pools, yet is too heavy to be transported all the way to the next pool. The crossover riffle gets its name because when flowing water is forced to the outside of a meander bend the water must "crossover" the center of the channel on its journey to the outside of the next meander bend. This happens at the crossover riffle location. In the process of crossing over, much of the flow energy needed to transport larger sediment is lost and the material is deposited. This increases the roughness of the channel bed, further slowing the water, and encouraging more deposition.

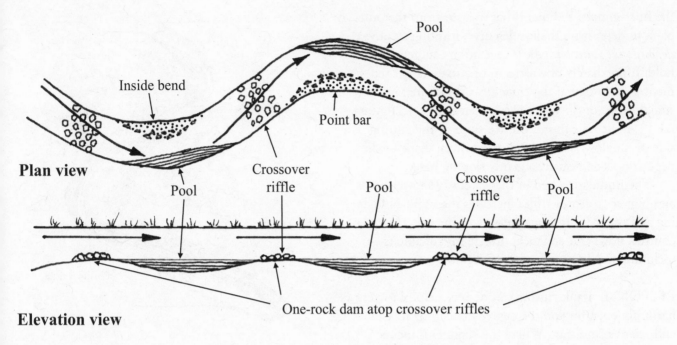

Plan view

Pool

Inside bend

Point bar

Pool

Crossover riffle

Pool

Crossover riffle

Pool

One-rock dam atop crossover riffles

Elevation view

Fig. A1.14. Crossover riffles are locations where water flow crosses over from one meander bend and its pool to another. Sediment naturally accumulates at these riffles; thus they are perfect locations for sediment-trapping one-rock dams. Adapted from *Erosion Control Field Guide*

RESPONSE: Crossover riffles form naturally and function as grade-control structures in channels. The large-sized sediments that make up the riffle are unlikely to be moved during normal water flow (baseflow) and small floods, yet they are transported and replaced during large floods. The result is a dynamic equilibrium in the depth of the channel bed. Identifying crossover riffles is the most critical step in choosing the correct location to properly construct a long-lasting and effective grade-control structure such as a one-rock dam (fig. A1.14). Crossover riffles are generally easier to locate in channels with well-developed meander patterns and variable-sized sediments. The best way to locate the crossover riffle is to find the approximate midway point between two meander bends that curve in opposite directions. If the meander pattern is poorly defined, or the channel bed is composed of sand, it may be necessary to look for larger sediments that are evenly distributed across the bed of the channel. This subtle clue can indicate the location of the riffle. To make certain this is the case, look for deposits of very fine-grained sediments both upstream and downstream of the probable riffle. These fine-grained sediments often indicate locations where water pooled at the end of a flood, allowing these smaller sediments to settle out.

Grade-control structures like one-rock dams (further described and illustrated in chapter 3) mimic the functions of crossover riffles when they are properly placed between the meander bends in a series. Each one-rock dam should be of a similar height, but no more than one foot (30 cm) tall. Placing one-rock dams at every crossover riffle location may effectively raise the bed elevation of a channel for the entire length of the treated segment. One-rock dams placed as such within an eroded, downcut waterway (depending on the depth of the channel bed) can reconnect the stream with its floodplain—dramatically increasing the amount of moisture and sediment the stream can beneficially cycle through the waterway and its stream-side ecology (fig. A1.15).

Depositional Bars

PATTERN: Depositional bars are in-channel sediment deposits formed when relatively slow-moving water drops sediment loads that are too heavy to be moved with the amount of energy available.

Fig. A1.15. A series of one-rock dams just after
construction and hydroseeding of banks.
Arrow denotes water flow.
Santa Fe Botanical Gardens, Santa Fe, New Mexico.
Design by Steve Vrooman of Restoration Ecology.
Rock work and photo by Craig Sponholtz.
Hydroseeding by San Isidro Permaculture

Depositional bars have a convex shape created by the
natural sorting of various-sized sediments. The small-
est particles are deposited on the tops of the bars and
nearest the banks where water is shallow and moves
the slowest. The largest materials are deposited on
the edges of the bars where the deepest water has the
greatest velocity. There are numerous types of depositional bars, each the product of slightly different
hydraulic and sediment transport processes.

Point bars are formed on the inside banks of
meander bends and indicate a developing or well-
developed meander pattern (see point bars labeled
in figures A1.12, A1.13, and A1.14). *Lateral bars* are
formed along banks and can become point bars if
they continue to expand toward the opposite bank
and create a meander bend. In a relatively straight
channel these can indicate the early stages of an evolving meander pattern. *Mid-channel bars* form in the
channel without any contact with adjacent banks and
can be an indication of excess sediment that a channel is unable to transport. This type of depositional
bar causes the flow in a channel to split around the
bar and often leads to unwanted bank erosion. Bank
erosion caused by mid-channel bars does not create

a stable meander pattern and can be a sign of severe
upstream erosion and land degradation.

RESPONSE: Depositional bars of all types are excellent indicators of upstream conditions in the watershed. They show us that sediment eroding from
upstream locations is being transported by the stream
and deposited farther downstream. A combination of
fresh deposition on numerous point bars—typically
with little or no plant growth—accompanied by bank
erosion on the outside bends, can be a sign that a large
volume of sediment is moving down the system. It
can also indicate the channel is somewhat incised and
has lost floodplain access, and that bank erosion is
widening the banks and creating a new floodplain.

The response depends on the type of depositional
bars we observe and the overall trend in watershed
health affecting the location we are observing. A
fairly straight incised channel in which lateral bars are
forming can indicate a meander pattern is forming
and evolving that will result in natural healing of the
incised channel. Lateral bars can be turned into point
bars by constructing baffles that erode the opposite
bank and encourage the deposition of point bars.
(See the book *Let the Water Do the Work* by Bill
Zeedyk and Van Clothier for baffle placement and
construction.) This will create more space for a floodplain and accelerate the natural healing process. The
presence of mid-channel bars can be a symptom of
active degradation and excess sediment production
caused by erosion higher in the watershed. In this
case it would be desirable to find the location of the
erosion that is releasing the sediment and deal with
the problem at its source.

Stepped Pools in Steep Drainages

PATTERN: In general, streams with a slope steeper
than 4% dissipate energy vertically through step
pools as opposed to laterally through meandering and
overflow of flood flows into floodplains. The falling-
pooling-slowing-falling-pooling-slowing effect of multiple step pools prevents water from attaining extreme
speed. Increased channel roughness—created for
example by rocks of various sizes protruding from
the streambed—can also dissipate stream energy.

Fig. A1.16. Step pools constructed by Dr. David Rosgen and associates on a tributary of Three Forks River, Colorado. Reproduced with permission from *Let the Water Do the Work* by Bill Zeedyk and Van Clothier. Credit: Van Clothier

RESPONSE: Stepped pools create a natural pattern that diffuses the energy of steeper flows of water. If you have a steep section of an eroding channel, and rock is available on-site, mimic the stepped-pool pattern (fig. A1.16). Rock-lined plunge pools and one-rock dams can be used to create a series of stepped pools (fig. A1.17).

NOTE: If the bed of an ephemeral channel is bedrock, it might be appropriate to create the conditions for an ephemeral spring by constructing well-built one-rock dams or even loose-rock or dry-stacked-rock check dams that are more than one rock high—though this height can weaken the structure. The one-rock dam or a non-wire-wrapped rock check dam can accumulate porous coarse sediments on its upstream side into which water rapidly infiltrates. This water is then slowly released as a seep spring that exits the sediment at the base of the dam. A structure more than one-rock high might be appropriate in this case only because the bedrock can endure the force of water falling over the structure without a scour hole forming that would undermine the dam (figs. A1.18, A1.19). A series of such sediment-accumulating structures could raise the sediment bed throughout the channel and keep the infiltrated water moving

Fig. A1.17. Looking downstream at water pooled within a rock-lined plunge pool, along with pools behind two one-rock dams further downstream. One year after this photo was taken there were abundant native grasses growing through the rock of the structures and in the soils above and below them. Santa Fe Botanical Gardens, Santa Fe, NM. Design by Steve Vrooman of Restoration Ecology. Installation by Craig Sponholtz

downstream below the surface, reducing water loss to evaporation. The ideal location for these structures is typically at the crossover riffle.

Exposed Roots

PATTERN: The exposure of tree and shrub roots along large or small drainages (fig. A1.20) indicates significant erosion has occurred. The exposed root crown—the point where the trunk once emerged from the ground and under which the roots began to grow—reveals the original soil level before erosion occurred and exposed the roots. Exposed hammock-

Fig. A1.19. A loose-rock check dam built atop bedrock by the Civilian Conservation Corps in the 1930s with resulting spring. Note how progressively smaller material, from small rocks to gravel, was placed against the upslope side of the dam's large rocks to ensure smaller sediment does not erosively wash out between the large rocks. Tucson Mountains west of Tucson, Arizona

Fig. A1.18. A one-rock dam on bedrock, sediment accumulation above, and ephemeral seep below. Northern New Mexico. Work by Craig Sponholtz

shaped roots spreading across the bed of the drainage are telltale tracks of the *hammock effect* indicating that, prior to erosion, the channel level was above this stabilizing fabric of woven roots growing from the trees on the banks (fig. A1.21).

RESPONSE: The presence and extent of exposed roots clues you in to the degree of channel and bank erosion occurring in a drainage. What is the nature and composition of the sediment and soil and its resistance to erosion? Is the material somewhat stable clay or less cohesive sand and gravel? The less cohesive and the more erosion-prone the material, the more important the roots of trees and other plants are for stability. If you are seeing the hammock effect in exposed roots, erosion conditions are bad and the need for repair is urgent. Arrest downcutting by constructing grade-

Fig. A1.20. Bill Zeedyk in a severely eroded drainage in southern Arizona. Note the exposed root crown of the mesquite tree just to the left of Bill's head, and the dangling roots below. A white-colored rock baffle was just constructed where the person is sitting to induce more of a meander into the opposite bank.

Fig. A1.21. Hammock of tree roots exposed in southeast Arizona. Reproduced with permission from *Let the Water Do the Work* by Bill Zeedyk and Van Clothier. Credit: Van Clothier

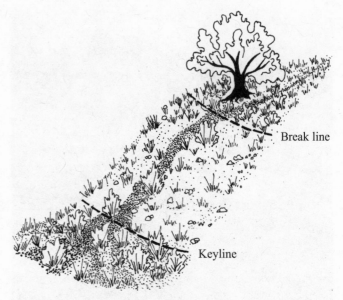

Fig. A1.22. Break line and keyline on a slope

control structures such as one-rock dams interwoven among the hammock-like roots and seeded with native grasses to add more root mass.

Look for the cause of the downcutting and address it. If a road crossing has lowered the bed of the drainage, raise the bed again by constructing a grade-control structure such as a check dam on the downslope side of the road. Rock cobble can be used to fill in upgradient of the dam so the raised road surface is hard and drivable in wet conditions. Raise incised channels to reconnect them with their floodplains. If the watershed draining into the channel has lost its cover of perennial vegetation, take steps to bring that sponge-like vegetation back by planting pole cuttings of fast-rooting plants like willow, sod wads, or seeds.

GENERAL PATTERNS OF WATER, SLOPE, AND FLOW

The following patterns are not limited to sheet flow or channel flow alone. They are caused by various flows of water and sediment, the life forms they support, and the slopes they help shape. These patterns also help you read the landscape and its flows. Look for break lines and keylines, high water marks, vegetation, and animal signs. They are described below.

Break Lines and Keylines

PATTERN: *Break lines* are contour lines at locations in landscapes where grades change from gradual convex slopes where runoff more readily infiltrates and sediments deposit, to steeper concave slopes from which water more readily runs off and sediments erode (fig. A1.22). These sediments can be transported by gravity (colluvium) and/or water flow (alluvium).

Keylines, in contrast, are those contour lines at locations in landscapes where grades change from steeper concave slopes from which water quickly runs off and sediments erode, to more gradual convex slopes where water more readily infiltrates and sediments deposit.

On a micro-level, a slope covered with vegetation, leaves, and silt may change to a steeper patch of naked sloping dirt. The keyline would be downhill from there, where that naked slope changes to a more gradual slope on which fines, silts, organic matter, and more dense vegetation can accumulate.

On a macro-level, a mountain, hilltop, or channel bank may slope down to a breakline where the slope steepens and erosion increases. Below that lies the keyline where the slope lessens again at the top of an alluvial fan, terrace, or bench composed of deposited or accumulated sediments.

RESPONSE: Break lines and keylines help you identify where in the landscape sediments are eroding and depositing. Once you identify these breaklines and keylines, concentrate initial water-harvesting and revegetation efforts above break lines and below keylines. These are the locations where water more readily infiltrates, sediment more readily accumulates, and vegetation more readily germinates. The vegetation you establish and the water you infiltrate above breaklines can help seed and vegetate the slopes below the breakline, though be very cautious about excessively hydrating steep slopes in areas prone to mudslides. In some instances, you may want to focus your water harvesting and revegetation efforts exclusively below the keyline. Here water and eroded sediment that have crossed the break line above then slowly erode the steeper section of the slope between the break line and keyline, and can easily be harvested below the keyline to create a more gradual, stable angle of repose with higher moisture content. If you build a small earthen dam or pond, make sure that both the structure and the water it collects are located below the keyline to minimize earth moving and ensure overflow crosses gentle, easily managed slopes.

Alluvial Fans

PATTERN: The slopes of tributary channels are always steeper than the slopes of the streams they contribute to. Alluvial fans are depositional landforms created when sediment-laden tributary channels widen and their slope is significantly reduced, such as occurs when channels intersect broad gently sloping valley bottoms. As the water slows and spreads, the heaviest sediments are deposited first. The deposition of the heaviest sediments at the apex (upslope side) of the alluvial fan causes the runoff to slow and spread even more. As the runoff continues to slow and spread, finer and finer sediment sizes are deposited. The result is a broad, convex, fan-shaped landform that is composed of deep alluvial deposits (figures A1.23 and A1.24).

Alluvial fans provide important sediment- and water-harvesting functions in watersheds. They are reservoirs of stored sediment and moisture that buffer the amount of sediment and runoff that downstream channels receive. The natural sediment-sorting mechanism

that deposits the largest sediments on the upslope side of the fan and the finest particles on the downslope side creates a soil profile that readily infiltrates large amounts of runoff. Often runoff infiltrates deep into the soil strata, minimizing the availability of surface moisture and consequent evaporation loss. However, the deeply stored moisture and rich alluvial soils create an ideal environment for the growth of deep-rooted plants and trees. Alluvial fans have long been used by indigenous cultures as areas of intense cultivation. The regular input of sediment and nutrient-rich runoff constantly rejuvenates the fertility of the alluvial fan.

RESPONSE: Alluvial fans are created and maintained by the slowing and spreading of regular inputs of sediment-laden runoff water. Alluvial fans with large sediment supplies can be quite steep, but the spreading and slowing of runoff and the resulting deposition of sediment maintain sheet flow despite the relative steepness. Alluvial fans can be fragile landforms since they are composed of loose sediment deposits. Anything that reduces the ability of runoff to slow and spread across the fan will concentrate and accelerate runoff and cause severe erosion and gully formation.

Runoff should be slowed and spread as close to the apex of the fan as possible, especially if there is any gully formation occurring on the fan. The higher upslope the slowing and spreading occurs, the more sediment- and moisture-storage capacity the fan will have. Media lunas (sheet flow spreaders) constructed using rock mulch or brush are ideal to establish and maintain the slowing and spreading of runoff (see figures in the *sheet flow spreader* section of chapter 3). Alluvial fans dissected by one or more channels will actively erode and degrade. In this case it is important to spread runoff out of each channel as close to the apex of the fan as possible. One-rock dams can be used where the channel is deepest and media lunas can be used where the channel is very shallow. Take care not to concentrate runoff that has been spread out of an actively eroding channel.

Alluvial fans can be constructed in locations where a low-banked channel is transporting excess sediment across a gentle slope. A rock or brush sheet flow spreader can be used to slow and spread runoff onto

Fig. A1.23. An alluvial fan forming where a steep slope becomes more gradual and flow spreads. Arrows denote water and sediment flow. Credit: Craig Sponholtz

Fig. A1.24. An alluvial fan where steeper, channelized flow spreads out over a more gradual slope. Dotted arrows and shaded ovals denote changing, meandering flow of water and sediment. Dotted arced lines denote contour lines of the fan (same shape as a sheet flow spreader). Credit: Craig Sponholtz

a relatively broad and flat landform adjacent to the channel. Locate the intended apex of the future fan where the slope and land contours provide the first opportunity to spread runoff out of the shallow channel. A sheet flow spreader constructed at this location will spread and slow the runoff to create sheet flow and assist the sedimentation of the channel. Over time the newly formed sediment deposit will continue to slow and spread runoff to create sheet flow.

High Water Marks

PATTERN: High water marks are the highest points in drainages or floodplains where you can see evidence of past water and sediment flow. Look for lines of discoloration on rocks and vegetation along with snags or accumulations of branches, twigs, grass, and other detritus that indicate the greatest height water reached during flooding. Sometimes these accumulations can be seen surprisingly high on a fence or in the branches of a tree (fig. A1.25).

RESPONSE: High water marks tell you the potential for future high-water flow events in drainages and floodplains. Do not build within the area of potential flooding, and make sure no high water marks appear above or near the level of your existing buildings' foundations. If you find evidence of water backing up to a building's foundation, try to divert and harvest that water with earthworks before it reaches the building. In addition, make sure the grade around the building drains water to a point 10 feet (3 m) away from the building. Flood peaks can become more intense as vegetated, sponge-like land in the watershed is replaced with roofs, streets, and parking lots, or as land is denuded by other disturbances such as overgrazing, overcutting, or excessive all-terrain vehicle use. It is not wise to buy or build a home on flood-prone land, nor to manage the watershed in a way that increases peak flood flows.

Vegetation

PATTERN: Vegetation generally increases in density when more water is present. The presence of different types of plants indicate how long water is available for

Fig. A1.25. High water flow detritus on young cottonwood tree

plant growth—the *hydroperiod*. Perennial hydric species (hydroriparian vegetation) are long-lived plants that are adapted to waterlogging, and their presence indicates that consistent water supplies are close to land surface. In the Southwest, very high-water-use, broad-leafed cottonwoods (*Populus fremontii*), sycamore trees (*Platanus wrightii*), willows (*Salix spp.*), and cattails (*Typha latifolia*) typically indicate the presence of springs, perennial water flow, or shallow groundwater. In contrast, hardy triangle-leaf bursage shrubs (*Ambrosia deltoidea*) are found in well-drained arid upland zones. However, everything is relative. Sometimes even bursage cannot grow in extremely arid zones of the broad landscape, and instead such plants are found along drainages and other areas of greater water concentration. Indicators of increased short-term soil-moisture availability, such as at the bottom of a slope or along with fringes of riparian areas, include the presence of native annuals such as peppergrass (*Lepidium thurberi*) and native perennial shrubs such as saltbush (*Atriplex spp.*) that have grown oversized compared to their upland counterparts.

RESPONSE: Different plant species are indicators of the timing or duration of soil moisture availability. Familiarize yourself with local plants and their water tolerances and needs. Take note of what grows in association with perennial, intermittent, and ephemeral water flows, and drier uplands. What plants grow in hot, dry, exposed microclimates such as equator-facing slopes? What plants grow on cool, moist, protected slopes on the winter-shade-side of ridgelines? What grows in deep, organic-matter rich soils with high moisture-storing capacity? What grows in shallow, less fertile soils with lower moisture-storing capacity? Plants will tell you how much water is in the soil and are an indicator of what other vegetation types with similar water needs and tolerances can grow in various areas of the landscape. Changes in the number, vigor, and density of plants can also alert you to the improving or declining health of the watershed and its ecosystems.

Animals

PATTERN: Animals, insects, and birds that need readily available water can signify the proximity or dependability of a water source. Dragonflies (*Odonata*) are found near open bodies of water but are highly mobile. Less mobile, water-dependent indicator species include stoneflies (*Plecoptera*) and caddisflies (*Trichoptera*), whose presence often also indicates good water quality. A high number of toads probably indicates a water source that is ephemeral or is too small to support predatory fish.

RESPONSE: Familiarize yourself with the water needs of local animals, insects, and birds. When assessing a site, use direct sightings or other signs of animals, insects, and birds to indicate the nature of local water sources. As with different plant species, the increasing or decreasing numbers and varieties of animals, insects, and birds can alert you to improving or declining health of the watershed and ecosystems. However, in urban settings with artificial fountains and pools, the presence of more water-dependent creatures may not indicate the presence of natural water supplies and healthy watershed conditions.

Appendix 2

Water-Harvesting Traditions in the Desert Southwest

By Joel Glanzberg
Illustrated by Roxanne Swentzell

This article first appeared in the August 1994 Permaculture Drylands Journal
and is reprinted with the permission of the Permaculture Drylands Institute.

Before the advent of modern irrigation technology, peoples of the American Southwest relied upon an array of water-harvesting and water-conserving techniques to grow their food. Not only are these techniques still appropriate, but their use, scale, and at times, failures have much to teach us. Several of the systems used by traditional peoples are described and illustrated below.

CHECK DAMS

Check dams are built across drainages that flow only periodically. They are constructed of rock and can range in size from small to large. These rock dams catch soil and water, and were often built higher as more soil accumulated behind them. They provided an excellent way to fertilize soil and stabilize drainages,

and were used for all kinds of crops. There are good examples in Colorado, Utah, and in New Mexico—at Mesa Verde and Hovenweep National Parks, at numerous small dams in the upper Rio Grande and Chama drainages, and throughout the Pajarito Plateau (fig. A2.1).

TERRACES OR LINEAR BORDERS

Terraces themselves were occasionally built, but linear borders or low lines of stone across slopes of hills were more common. At Point of the Pines in Arizona, hilltop pueblos were surrounded by concentric rings of rocks gathered from the entire hillside and laid along contour lines across the slopes. Soil washing down the bare hillsides caught behind the stone walls, accumulating up to 16 inches (406 mm) deep. This loose soil would have been highly fertile and water absorbing (fig. A2.2).

WAFFLE GARDENS

Waffle gardens can be either sunken beds with ground-level berms, or ground-level beds surrounded by raised berms of earth. The bermed beds catch and hold rainwater as well as retain water brought by hand. Waffle beds were built on a very small scale for especially valuable crops. They are best known historically at Zuni (fig. A2.3).

Fig. A2.1. Check dams

Fig. A2.2. Linear borders

Fig. A2.4. Grid gardens

Fig. A2.3. Waffle gardens

GRID GARDENS

Grid gardens are similar to waffle gardens, but have walls made of stone rather than earth. They usually have much larger beds than waffle gardens. The walled beds help hold soil and were often placed to catch water runoff from moderate slopes. They were used extensively during prehistoric times throughout the upper Rio Grande, the Chama, and the Ojo Caliente drainages. They were probably not hand-watered, and it appears likely that they were used to grow major crops such as corn and beans (fig. A2.4).

GRAVEL- AND ROCK-MULCHED FIELDS

Mulch of any kind slows evaporation by sheltering the soil surface. The Anasazi clearly knew this. Throughout the upper Rio Grande and Chama Drainages, vast areas were mulched with gravel. Grid gardens were often covered with mulch. Gravel mulches not only conserve moisture, they also reduce wind and water erosion. Dark gravel mulches increase soil and air temperatures, reducing the threat of early and late frosts. At Wupatki in northern Arizona, the Sinagua people were able to grow food without supplemental water largely because of the natural covering of cinders created by the eruption of Sunset Crater. This eruption, and the resultant layer of cinders that covered highly fertile, water-holding volcanic ash, was responsible for Anasazi, Hohokam, and Mogollon people moving into this area to live, creating the Sinagua culture around A.D. 1000 (fig. A2.5).

CLIFFBASE PLANTINGS

Often the water-collecting surfaces of cliffs were used to provide water for crops. By planting where water would run off and be concentrated, available moisture and fertility could be increased. At

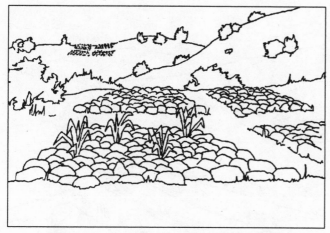

Fig. A2.5. Gravel and rock-mulched fields

Chaco Canyon, this technique was used extensively. A complex irrigation system was developed using cliff runoff. Grid gardens, check dams, and terraces were located in various places to catch this runoff (fig. A2.6).

FLOODPLAIN FARMING

Soil located in or near a channel of flowing water is usually moist and fertile. For this reason, floodplain fields were situated along the margins of permanent or ephemeral streams, the low terraces of arroyos, or within the bottoms of arroyos (fig. A2.7). The principle is similar to the moisture and fertility enhancement utilized by check dam agriculture located in steeper arroyos. In this case, flatter areas in drainages were planted, where a raised water table was also useful to the cultivator.

One disadvantage to such sites is cold air drainage into these canyon bottoms. The accumulation of cold air makes these sites susceptible to late spring and early fall frosts, limiting the length of the growing season. Other disadvantages limiting use of these areas are the danger of floods wiping out fields and the difficulty of clearing thick riparian growth.

Often brush weirs or earthen walls were used to slow or spread water across the fields. This led to irrigation, which spread the water over more land, enabling more crops to be grown with more control.

Fig. A2.6. Cliffbase plantings

Fig. A2.7. Floodplain farming

Fig. A2.8. Floodwater farming

Fig. A2.9. Irrigation

FLOODWATER FARMING

Often the fans of soil below arroyos or small canyons were planted to utilize the flood waters coming down these drainages. This is sometimes known as ak-chin farming and is still practiced by the O'odham and Hopi. In some cases, the field is sited beside the path of the arroyo. Brush weirs are then constructed across the bed of the arroyo to direct water out of the arroyo and onto the field. A destructive flood will blow out the weir but will not destroy the field. Destruction of the field by a flood is a very real problem with ak-chin farming for fields located directly in the paths of arroyos (fig. A2.8).

IRRIGATION

That there was irrigation before the arrival of the Spanish is clear. Its exact extent and character is not. What we do know is that it was not as universal a trait of Pueblo agriculture in the past as it is now, and was only one of a wide range of farming techniques used.

It is not clear to what extent the Rio Grande and other large rivers in northern New Mexico were used prehistorically for irrigation. Modern farming eliminates the evidence of previous irrigation systems, and scholars do not agree on the documentary evidence. At Santa Clara Pueblo, the looser soils in the canyon and immediately surrounding the village seem to have been preferred sites for irrigation. These areas were watered by Santa Clara Creek.

As late as 1940, almost all irrigated fields were located in the vicinity of Santa Clara Creek. There

appears to have been only a gradual shift of landholding from the area around the pueblo to the river bottom, indicating that at Santa Clara, river bottom cultivation was one small part of Pueblo agriculture. Typically, fields were scattered across the landscape at different elevations and in different environments to prevent a disaster from destroying all of the crops. The Hohokam of southern Arizona built substantial irrigation canals to divert water from large rivers, but the eventual failure of these projects contributed to the destruction of their "advanced" civilization. In the Anasazi area, dry farming seems to have been the rule, and irrigation was small in scale compared to the Hohokam (fig. A2.9).

FOLLOWING THE ANASAZI THROUGH TIME

The appearance of the various farming techniques through time gives us a view into the environmental effects of Anasazi agriculture and their responses to these effects. Originally simple swidden agriculture was practiced. Land was cleared, burned, and planted. As it was exhausted, new land was cleared. Eventually the original plot recovered and could be replanted. This extensive clearing increased erosion. Check dams and linear borders were being constructed late in the occupations of Chaco, the Mimbres area, the San Juan Basin, and other sites, such as Pot Creek Pueblo near Taos. These structures were apparently an attempt to halt the serious erosion caused by deforestation and clearing, overuse of wild plants, and foot traffic. Despite these conservation attempts, these areas were ultimately abandoned. When a prolonged drought struck in A.D. 1276–1299, the food-production systems were already under stress from the high population densities. The combination of the drought with the environmental degradation caused by heavy farming and residential use probably led to final abandonment of settlements.

Refugees from these areas built grid gardens, techniques intended to prevent the start of erosion. Rather than waiting for erosion to begin, farmers were now attempting to stop the process at its source before it began, an example of Anasazi farmers learning from past mistakes and adapting to their environments.

As noted above, at an earlier time, the Anasazi had rotated fields, farming one until it was exhausted, and then clearing another, and so on until they returned to the first, many years later. As populations increased, however, the inhabitants were forced to use all available farmland in proximity to their villages. This forced the Anasazi to be somewhat nomadic, moving every 60–100 years when they had depleted a site's soil and other natural resources. After a period of time, the original group or another group could

reinhabit an area, the soil fertility and natural resources having recovered from previous usage. This is a key part of pre-Colombian, Pueblo land use patterns. Even after the adoption of corn, squash, and beans led them to a sedentary lifestyle in villages, they continued to be semi-nomadic at a much slower pace.

It was continuous habitation and the associated large-scale populations, irrigation systems, building projects, deforestation, and soil depletion that contributed to the forced migrations of the 1300s. The land taught the Anasazi to keep things small and to move occasionally to allow it to rest. By remaining fluid within their environment, by using many techniques in various locations, microclimates, and elevations, and by maintaining an appropriately small scale, the Anasazi were able to survive where earlier growth and urbanization had failed. It is ironic that rather than learning from what has failed before and adopting what has succeeded, we have done the opposite. Like the Chacoans and Hohokam, we believe that our technical "advances," power, and grandeur make us exceptions to the constraints of our environment. And just like them, our failure will come to us as a surprise.

Joel Glanzberg is a master permaculture teacher and designer living in northern New Mexico and working with the Regenesis Group (www.regenesisgroup.com). He is the author of The Permaculture Mind *(www.patternmind.org).*

Roxanne Swentzel (www.RoxanneSwentzell.net) is an exceptional artist and co-creator (with Joel) of the thriving Flowering Tree Permaculture Site (www.FloweringTreePermaculture.org) in the Santa Clara Pueblo, New Mexico.

Appendix 3
Water-Harvesting Calculations

List of Equations and Other Information

Box A3.1. Abbreviations, Conversions, and Constants for English and Metric Measurement Units

Note: * items are approximate or rounded off
   † Water density changes slightly with temperature.

ABBREVIATIONS FOR ENGLISH UNITS

inches = in
feet = ft
square feet = ft^2
cubic feet = ft^3
gallons = gal

pounds = lb
pounds per square inch of pressure = psi
acre = a
acre feet = AF

CONVERSIONS FOR ENGLISH UNITS

To convert cubic feet to gallons, multiply cubic feet by 7.48 gal/ft^3 *
To convert inches to feet, divide inches by 12 in/ft
To convert gallons of water to pounds of water, multiply gallons by 8.32 lb/gal *†
To convert cubic feet of water to pounds, multiply cubic feet by 62.23 lb/ft^3 *†
To convert acre feet of water to gallons, multiply acre feet by 325,851 gal/AF
To convert acre feet of water to cubic feet, multiply acre feet by 43,560 cubic feet/AF

CONSTANTS

Pounds of pressure per square inch of water per foot of height = 0.43 psi/ft *†
Ratio between a circle's diameter and its circumference is expressed as π = 3.14 *

ABBREVIATIONS FOR METRIC UNITS

millimeters = mm
centimeters = cm
meters = m
liters = l
kilograms = kg
hectare = ha

To convert Fahrenheit (F) to Celsius (C) for *actual* indoor/outdoor temperature measure ("it's 70°F outside today"), subtract 32 from Fahrenheit temperature. Then multiply result by 5, and divide by 9. To convert °C to °F, multiply by 9, then divide by 5, then add 32.

To convert Fahrenheit (F) to Celsius (C) for temperature *difference* ("it's 20°F hotter today than yesterday"), multiply Fahrenheit by 5, then divide by 9. To convert °C to °F multiply by 9, then divide by 5.

CONVERSIONS FOR METRIC UNITS

1 liter of water weighs 1 kilogram
1 cubic meter = 1,000 liters

CONVERTING BETWEEN ENGLISH UNITS AND METRIC UNITS

To convert inches to millimeters, multiply inches by 25.4 mm/in
To convert inches to centimeters, multiply inches by 2.54 cm/in
To convert feet to meters, multiply feet by 0.30 m/ft *
To convert square feet to square meters, multiply square feet by 0.093 m^2/ft^2 *
To convert cubic feet to cubic meters, multiply cubic feet by 0.028 m^3/ft^3 *
To convert gallons to liters, multiply gallons by 3.79 liter/gal *
To convert pounds to kilograms, multiply pounds by 0.45 kg/lb *
To convert acres to hectares, multiply acres by 0.405 ha/a *
To convert miles to kilometers, multiply miles by 1.6 km/mi *

Best technique to measure rainfall: Buy a simple rain gauge for $10 or so from a hardware or feed store, plant and garden nursery, or a scientific supply house. A rain gauge that is tapered at the bottom makes reading small amounts of rainfall easier.

For resources documenting local rainfall rates and other climatic information, see appendix 6.

Equation 1A.
Catchment Area of Rectangular Surface (English units)

length (ft) × width (ft) = catchment area (ft^2)

EXAMPLE:

A house measures 47 feet long by 27 feet wide at the drip line of the roof. Note that it does not matter whether the roof is flat or peaked; the roof dimensions at the drip line are the same. It is the "footprint" of the roof's drip line that matters.

47 ft × 27 ft = 1,269 ft^2
1,269 ft^2 = catchment area

If the roof consists of two or more rectangles, calculate the area for each rectangle and add together. Again, take the view of a falling raindrop, and only look at the "footprint" of the roof's drip line. Roof pitch cannot be seen from above and does not matter. With conical, octagonal, or other non-standard roof shapes, again calculate the area based on the *drip line*.

Equation 1B.
Catchment Area of Rectangular Surface (metric units)

length (m) × width (m) = catchment area (m^2)

EXAMPLE:

15 m × 9 m = 135 m^2
135 m^2 = catchment area

Again, all the considerations in Equation 1A will apply.

Equation 2A.
Catchment Area of Triangular Surface (right triangle)

Multiply the lengths of the two shorter sides of the triangle then divide by 2 = catchment area

EXAMPLE:

A triangular section of roof measures 9 feet by 12 feet by 15 feet. This is a right triangle, with the 90-degree angle between the 9-foot and 12-foot sides. Taking the measurements of the two shorter sides:

(9 ft × 12 ft) ÷ 2 = catchment area (ft^2)
108 ft^2 ÷ 2 = 54 ft^2

54 ft^2 = catchment area

Equation 2B.
Catchment Area of Triangular Surface (standard math formula)

Multiply the triangle's base times its height then divide by 2 = catchment area
where the base can be any side, and the height is measured perpendicularly from the base to the opposite vertex.

EXAMPLE:

You want to know the area of a triangular section of patio. The length of the section in front of you is 20 feet (triangle base) and you measure 4 feet perpendicularly to the opposite vertex of the triangle.

$(20 \text{ ft} \times 4 \text{ ft}) \div 2 = $ catchment area (ft^2)
$80 \text{ ft}^2 \div 2 = 40 \text{ ft}^2$

$40 \text{ ft}^2 = $ catchment area

Equation 2C.
Catchment Area of Triangular Surface (Heron's formula)

This formula, attributed to Heron of Alexandria (first century A.D.), involves no trigonometry. It only needs the square root (sqrt) function found on most electronic or computer calculators. It may be useful when dealing with non-right triangles where you can measure (or know) all sides of the triangle.

Step 1: Determine the lengths of the sides of the triangle. These are a, b, c.

Step 2: Calculate s.
$(a + b + c) \div 2 = s$

Step 3: Calculate S, using:
$s \times (s - a) \times (s - b) \times (s - c) = S$

Step 4: Calculate the catchment area, which is the square root of S.
sqrt S = catchment area

Equation 3.
Catchment Area of Circular Surface

$\pi \times r^2 = $ catchment area
Note: r = radius of the circle. A circle's radius is half the circle's diameter.

EXAMPLE:

A circular roof has a 25 foot diameter. Divide the diameter by 2 to get the *radius* of 12.5 feet.

$\pi \times (12.5 \text{ ft} \times 12.5 \text{ ft}) = $ catchment area (ft^2)
$3.14 \times 156.25 \text{ ft}^2 = 490.62 \text{ ft}^2$

$490.62 \text{ ft}^2 = $ catchment area

Equation 4A.
Possible Volume of Runoff from a Roof or Other Impervious Catchment Area (English units)

catchment area (ft^2) × rainfall (ft) × 7.48 gal/ ft^3 = maximum runoff (gal)

Note: For a more realistic and conservative estimate see Equation 5.

EXAMPLE CALCULATING *ANNUAL* RUNOFF:

Calculate the gallons of rain running off the roof in an average year from a home that measures 47 feet long and 27 feet wide at the drip line of the roof. (In the example below, the roof dimensions at the drip line are included in the calculation; the catchment area is the same whether the roof is flat or peaked.) Rainfall in this location averages 10.5 inches per year, so you will divide this by 12 inches of rainfall per foot to convert inches to feet for use in the equation. (Note: You can use the same equation to calculate the runoff from a single storm, by simply using the rainfall from that storm instead of annual average rainfall in the equation.) Since the roof is a rectangular area, use the following calculation for catchment area:

(length (ft) × width (ft)) × rainfall (ft) × 7.48 gal/ft^3 = maximum runoff (gal)
(47 ft × 27 ft) × (10.5 in ÷ 12 in/ft) × 7.48 gal/ft^3 = maximum runoff (gal)
1,269 ft^2 × 0.875 ft × 7.48 gal/ft^3 = 8,306 gal

8,306 gal = runoff

EXAMPLE CALCULATING RUNOFF FROM A *SINGLE RAIN* EVENT:

Calculate the maximum gallons of rain running off the roof in a single rain event from a home that measures 47 feet long and 27 feet wide at the drip line of the roof. It is not unusual for heavy storms in the example area to drop three inches of rain. To determine the runoff from such a rain event you will divide the 3 inches of rainfall by 12 inches of rainfall per foot to convert inches to feet for use in the equation. Since the roof is a rectangular area, use the following calculation for catchment area:

(length (ft) × width (ft)) × rainfall (ft) × 7.48 gal/ft^3 = maximum runoff (gal)
(47 ft × 27 ft) × (3 in ÷ 12 in/ft) × 7.48 gal/ft^3 = maximum runoff (gal)
1,269 ft^2 × 0.25 ft × 7.48 gal/ft^3 = 2,373 gal

2,373 gal = maximum runoff

Equation 4B.
Possible Volume of Runoff from a Roof or Other Impervious Catchment Area (metric units)

catchment area (m^2) × rainfall (mm) = maximum runoff (liters)

Calculations for annual rainfall, a rainy season, or an event would be similar to those for English units.

Box A3.2. Estimating Rainfall Runoff Using Rules of Thumb

Rough rule of thumb for calculating rainfall runoff volume on a catchment surface (English units):
You can collect 600 gallons of water per inch of rain falling on 1,000 square feet of catchment surface.

On the really big scale:
You can collect 27,000 gallons of water per inch of rain falling on 1 acre of catchment surface.

Rule of thumb for calculating rainfall volume on a catchment surface (metric units):
You can collect 1,000 liters of water per each 10 millimeters of rain falling on 100 square meters of catchment surface.

On the really big scale:
You can collect 100,000 liters of water per 10 millimeters of rain falling on one hectacre of catchment surface.

Equation 5A.
Estimated Net Runoff from a Catchment Surface Adjusted by its Runoff Coefficient (English units)

catchment area (ft^2) × rainfall (ft) × 7.48 gal/ft × runoff coefficient = net runoff (gal)

Impervious catchment surfaces such as roofs or non-porous pavement can lose 5% to 20% of the rain falling on them due to evaporation, and minor infiltration into the catchment surface itself. The more porous or rough your roof surface, the more likely it will retain or absorb rainwater. On average, pitched metal roofs lose 5% of rainfall, allowing 95% to flow to the cistern. Concrete or asphalt roofs retain around 10%, while builtup tar and gravel roofs can retain 15% to 20%. However, the percent of retention is a function of the size and intensity of the rain event, so more porous roof surfaces could absorb up to 100% of small, light rain events. To account for this potential loss, determine the runoff coefficient that is appropriate for your area and impervious catchment surface (0.80 to 0.95).

EXAMPLE CALCULATING NET *ANNUAL* RUNOFF FROM A ROOF:

Calculate the net gallons of rain running off the roof in an average year from a home that measures 47 feet long and 27 feet wide at the drip line of the roof. Rainfall in this location averages 10.5 inches per year, so you will divide this by 12 inches of rainfall per foot to convert inches to feet for use in the equation. (Note: You can use the same equation to calculate the runoff from a single storm, by simply using the rainfall from that storm instead of annual average rainfall in the equation.) Assume that the loss of water that occurs on the catchment surface is at the high end of the range so you get a conservative estimate of *net runoff*. This means you select a *runoff coefficient* of 80%, or 0.80. Since the roof is a rectangular area, use the following calculation for catchment area:

(length (ft) × width (ft)) × rainfall (ft) × 7.48 gal/ft^3 × runoff coefficient = net runoff (gal)
(47 ft × 27 ft) × (10.5 in ÷ 12 in/ft) × 7.48 gal/ft^3 × 0.80 = net runoff (gal)
1,269 ft^2 × 0.875 ft × 7.48 gal/ft^3 × 0.80 = 6,644 gal

6,644 gal = net runoff

Based on this, a realistic estimate of the volume of water that could be collected off the 47 foot by 27 foot example roof in an average year is 6,644 gallons.

Pervious surfaces such as earthen surfaces or vegetated landscapes can infiltrate up to 100% of the rain falling on them. Their runoff coefficient is greatly influenced by soil type and vegetation density. Large-grained porous sandy soils tend to have lower runoff coefficients while fine-grained clayey soils allow less water to infiltrate and therefore have higher runoff coefficients. Whatever the soil type, the more vegetation the lower the runoff coefficient since plants enable more water to infiltrate the soil. Below are some runoff coefficients for the southwestern U.S., although these are just rough estimates since runoff rates are also affected by rainfall intensity and duration. The more intense or the longer the rainfall the greater the runoff, since more rain is infiltrated in the soil before the soil becomes saturated. A very light rainfall may just evaporate, and not run off or infiltrate at all.

- Sonoran Desert uplands (healthy indigenous landscape): range 0.20–0.70, average 0.30–0.50
- Bare earth: range 0.20–0.75, average 0.35–0.55
- Grass/lawn: range 0.05–0.35, average 0.10–0.25
- For gravel use the coefficient of the ground below the gravel

EXAMPLE CALCULATING NET *ANNUAL* RUNOFF FROM A *BARE SECTION OF YARD*:

In an area receiving 18 inches of rain in an average year, you want to calculate the runoff from a 12 foot by 12 foot bare section of yard that drains to an adjoining infiltration basin. The soil is clayey and compacted, and you estimate its runoff coefficient to be 60% or 0.60.

catchment area (ft^2) × rainfall (ft) × 7.48 gal/ft × runoff coefficient = net runoff (gal)
12 ft × 12 ft × (18 in ÷ 12 in/ft) × 7.48 gal/ft^3 × 0.60 = net runoff (gal)
144 ft^2 × 1.5 ft × 7.48 gal/ft^3 × 0.60 = 969 gal

969 gal = net runoff

Based on this, a realistic estimate of the volume of runoff that could be collected off the 12 foot by 12 foot section of bare earth adjoining the infiltration basin is 969 gallons in an average year.

EXAMPLE CALCULATING NET RUNOFF FROM A *SINGLE STORM EVENT* ON *ESTABLISHED LAWN* (*GRASS*):

The runoff coefficient for this established lawn is assumed to be 20% or 0.20, and the maximum storm event is 3 inches:

12 ft × 12 ft × (3 in ÷ 12 in/ft) × 7.48 gal/ft^3 × 0.20 = net runoff (gal)
144 ft^2 × 0.25 ft × 7.48 gal/ft^3 × 0.20 = 54 gal

54 gal = net runoff

Equation 5B.
Estimated Net Runoff from an Impervious Catchment Surface Adjusted by its Runoff Coefficient (metric units)

catchment area (m^2) × rainfall (mm) × runoff coefficient = net runoff (liters)

EXAMPLE:

In an area receiving 304 millimeters of rain a year, you have a rooftop catchment surface that is 15 meters long and 9 meters wide, and you want to know how much rainfall can realistically be collected off that roof in an average year. You want a conservative estimate of annual *net runoff*, so you use a *runoff coefficient* of 80% or 0.80. (Since the roof is a rectangular area, use the following calculation for catchment area as in Equation 1B—catchment area (m^2) = length (m) × width (m)—which is figured into the calculation below.)

(length (m) × width (m)) × rainfall (mm) × runoff coefficient = net runoff (liters)
(15 m × 9 m) × 304 mm × 0.80 = net runoff (liters)
135 m^2 × 304 mm × 0.80 = 32,832 liters

32,832 liters = net runoff

A realistic estimate of the volume of water that could be collected off this 15 meter by 9 meter roof in a year of average rainfall is 32,832 liters.

Equation 6.
Cistern Capacity Needed to Harvest the Roof Runoff from a Large Storm Event

catchment area (ft^2) × rainfall expected in a local high volume storm (ft) × 7.48 gal/ft^2 × runoff coeficient = catchment runoff (gal)

EXAMPLE:

A water harvester with a 1,200 ft^2 roof lives in an area where a single storm (or two storms just a few days apart) can unleash 3 inches of rain.

1,200 ft^2 × (3 inches ÷ 12 inches) × 7.48 gal/ft^2 × 0.80 = catchment runoff (gal)
1,200 ft^2 × 0.25 ft × 7.48 gal/ft^2 × 0.80 = 1,795 gal

1,795 gal = catchment runoff

This is the *minimum cistern volume* needed to capture the roof runoff for this size storm.

Note: The above calculation is meant to give a rough estimate of a tank size that will reduce water loss to overflow from the tank and extend the availability of a lot of rainfall long after the rain event only—it is not based on estimated water needs. It is a quick and easy calculation for those simply wanting to supplement their water use with efficient rainwater tank storage. I often recommend beginner water harvesters start with a tank not exceeding a 1,500 gallon capacity. The system can always be expanded later. To start small you don't need to begin with a tank harvesting all the roof's runoff; rather begin by sizing a tank capturing water from just one section of the roof.

Equation 7.
Water Storage Capacity Needed for a Household Committing to Use Harvested Rainwater as the Primary Water Source (English units)

number of people × daily water consumption (gal/person/day) × longest drought period (days) = needed storage capacity (gal)

EXAMPLE:

If three people live in a household, each person consumes an average of about 50 gallons per day (their estimates based on the information in boxes 2.6-2.8 or from h2ouse.org), and the typical dry season in their area lasts 140 days then:

3 people × 50 gal/person/day × 140 days = 21,000 gal

21,000 gal = needed water capacity

If the people in this household are planning to live primarily off rainwater at their current water consumption rate they would be wise to plan for at least 21,000 gallons of water collection and storage capacity to get them through up to 140 days of dry times.

If the needed water capacity (and needed catchment area) seems too large to be feasible, see how much you can realistically reduce your water consumption, then do the calculation again. For example, if the same household could reduce its daily water consumption to 20 gallons/person/day only 8,400 gallons of water collection and storage capacity would be needed.

Note: The above calculation will give a ballpark estimate of minimum tank capacity to meet dry season demand in expected drought. Sufficient catchment directing water to the tank is also needed to ensure the tank is full or close to full on day one of the dry season. See volume 3 of *Rainwater Harvesting for Drylands and Beyond*, for additional calculations and considerations.

Equation 8A. Potential Gravity-Fed Water Pressure from Your Tank (English units)

height of water above its destination (ft) × water pressure per foot of height (psi/ft) = passive water pressure (psi)

For every foot your source of water is above the elevation of the place where it will be used you develop 0.43 psi/ft of passive water pressure (gravity is the only force being used to create that pressure). The source of water may be in a tank, or a gutter and its associated downspout. The place you use the water may be a garden bed, a fruit tree basin, or any other location where supplemental water is needed.

EXAMPLE:

The folks with the new 8-foot-tall tank want to figure out how much passive water pressure will be available to deliver water from the tank to their squash plants placed in basins 6 inches (0.5 ft) below the surrounding land surface. The height of water in the 8-foot tank is 4 inches below the top of the tank due to the presence of an overflow pipe that allows excess rainwater to safely flow out of the tank during large storms. Based on this information the height of water above its destination is around 8.1 ft. Using Equation 8, calculate the passive water pressure as follows:

8.1 ft × 0.43 psi/ft = 3.48 psi

3.48 psi = passive water pressure

As the cistern water is used, the water pressure will drop with the dropping level of water (head) in the tank. Also, keep in mind that friction between water and the walls of a hose, pipe, or irrigation line will cut down on water pressure, so to maintain pressure try to use the water close to the tank, reducing the length of pipe or hose. For example, place a garden on the east side of your tank where the veggies will be shaded from the hot afternoon sun by the bulk of the tank, and you won't need a hose any longer than 25 feet (7.6m).

EXAMPLE:

I often place cisterns so their base is at least 2.5 feet above the garden or basin receiving the stored water. This guarantees me at least 1 psi of gravity-fed pressure even when the tank is nearly empty.

height of water above its destination (ft) × water pressure per foot of height (psi/ft) = passive water pressure (psi)
2.5 ft × 0.43 psi/ft = 1.08 psi

1.08 psi = passive water pressure

Equation 8B.
Potential Gravity-Fed Water Pressure from Your Water Tank (metric units)

height of water above its destination (m) x water pressure per meter of height (9.7 kPa/m) = passive water pressure (kilopascals)

See notes for Equation 8A.

Equation 9A.
Storage Capacity of a Cylinder (Can Apply to Both a Cylindrical Cistern or a Length of First Flush Pipe) (English units)

π × (cylinder radius (ft))^2 × effective cylinder height* (ft) × 7.48 gal/ft^3 = capacity (gal)

Note: r = radius of the circle

*Effective height is the height of water you can get back out of the tank when it's full, as opposed to the total height of water in the tank, which includes several inches of water that can never be drained out due to an outflow pipe above the bottom of the tank.

EXAMPLE:

The householders above are considering using a cylindrical tank to store their rainwater. They want to determine the capacity of a tank with a diameter of 3 feet and a height of 8 feet. The radius of the tank is one half the diameter, so it is 1.5 feet. Since they realize the effective tank storage height is going to be reduced by 4 inches because of the raised outlet 4 inches from the bottom of the tank, and by another 5 inches because of the bottom of the tank overflow pipe being 5 inches below the top of the tank, the effective height is going to be 7.25 feet. Using Equation 9A, they calculate the usable capacity of the tank as follows:

$\pi \times (1.5 \text{ ft})^2 \times 7.25 \text{ ft} \times 7.48 \text{ gal/ft}^3 = \text{capacity (gal)}$
$3.14 \times 2.25 \text{ ft}^2 \times 7.25 \text{ ft} \times 7.48 \text{ gal/ft}^3 = 383 \text{ gal}$

383 gal = capacity

Equation 9B.
Storage Capacity of a Cylinder (Can Apply to Both a Cylindrical Cistern or a Length of First Flush Pipe) (metric units)

$\pi \times (r \text{ (m)})^2 \times \text{effective cylinder height (m)} \times 1{,}000 \text{ liter/m}^3 = \text{capacity (liters)}$

See notes for Equation 9A.

Equation 10A.
Storage Capacity of a Square or Rectangular Tank (English units)

length (ft) × width (ft) × effective height (ft) × 7.48 gal/ft^3 = capacity (gal)

EXAMPLE:

A household decides to install a rectangular tank that has interior dimensions: 8 feet tall, 6 feet long, and 4 feet wide. The tank outlet tap is located 4 inches above the bottom of the tank. The underside of the overflow pipe is located 5 inches below the top of the tank. They calculate the effective height of water as 7.25 ft, so the calculation is as follows:

$6 \text{ ft} \times 4 \text{ ft} \times 7.25 \text{ ft} \times 7.48 \text{ gal/ ft}^3 = 1{,}302 \text{ gal}$

1,302 gal = capacity

Equation 10B.
Storage Capacity of a Square or Rectangular Tank (metric units)

length (m) × width (m) × effective height (m) × 1,000 liter/m^3 = capacity (liters)

See notes with Equation 10A.

Equation 11A.
Cistern's One-Time Dollar Price for Storage Capacity (English units)

price of cistern (dollars) ÷ storage capacity (gal) = price of storage capacity (dollars/gal)

EXAMPLE:

The tank in Equation 10A holds 1,302 gallons of water, and would cost around $1300 to purchase and install:

$1,300 ÷ 1,302 gal = $0.99/gal

$0.99/gal = price of storage capacity

Equation 11B.
Cistern's One-Time Price for Storage Capacity (metric units)

price of cistern ÷ storage capacity (liters) = price of storage capacity (price/liter)

See notes with Equation 11A. For non-USA currencies, substitute the appropriate currency.

Equation 12A.
Weight of Stored Water (English units)

stored water (gal) × 8.32* lb/gal = weight of stored water (lb)

EXAMPLE:

A 55-gallon drum under a rainspout has filled to the very top with water and you need to figure out how much it weighs to decide whether you can move it.

55 gal × 8.32* lb/gal = 457.6 lb

457.6 lb = weight of stored water

Water is extremely heavy. Do not underestimate the force you are dealing with when you store it. Platforms supporting storage tanks must be able to hold the water's weight!

*Note: Water density changes slightly with temperature.

Equation 12B.
Weight of Stored Water (metric units)

1 liter of water weighs 1 kilogram

So:
stored water (liters) × 1 kg/liter = weight of stored water (kg)

Appendix 4
Example Plant Lists and Water Requirement Calculations for Tucson, Arizona plus a Sonoran Desert Foods Harvest Calendar

This appendix contains estimated water needs for vegetable gardens and three multi-use perennial plant lists specific for Tucson, Arizona (water needs will fluctuate depending on planting density, soil type, placement, and exposure). There is a far more diverse array of suitable plants and cultivars available for this area than the lists suggest. These lists are meant simply as both a partial introductory guide for Tucsonans, and as a template for people elsewhere to create plant lists specific to their location and climate. See the Plant Lists & Resources page of www.HarvestingRainwater.com for a growing list of regional rain garden plant lists. See the end of this appendix for the Place-Based Wild and Cultivated Food Plants Harvest Calendar illustrating regional harvest periods for diverse plantings that provide resilient bounty in all seasons.

Estimated annual or monthly water requirements can be easily calculated for plants by looking up their mature size, water needs (low, medium, high), and evergreen or deciduous nature on the plant lists, and then using the simple calculations that follow the lists. These estimates are very helpful in determining what plants, and how many, can be sustained within a Tucson, Arizona site's rainwater budget (calculated in chapter 2) and potential supplementary water from household greywater (estimated from boxes 2.6–2.9).

The vegetation section of the resources appendix of volume 2 of *Rainwater Harvesting for Drylands and Beyond* lists some of the books from which I compiled the information. Local gardening groups, herbalists, primitive skills enthusiasts, native plant societies, locally owned plant nurseries, and my own direct observations then fleshed out the lists, and can help you form your lists too. Chapter 4 in this book, and the chapter on vegetation and the planting section of the chapter on infiltration basins in volume 2 offer still more tips.

Box A4.1. Approximate Annual Water Requirements for Mulched Vegetable Gardens in Tucson, Arizona, Planted in Sunken Basins

Based on "Economic Value of Home Gardens in an Urban Desert Environment" by David A. Cleveland, Thomas V. Orum, and Nancy Ferguson, *HortScience* 20(4):694-696.1985

50 ft²	100 ft²	150 ft²	200 ft²	250 ft²	300 ft²
3,180 gallons	6,360 gallons	9,540 gallons	12,720 gallons	15,900 gallons	19,080 gallons
4.5 m²	9 m²	13.5 m²	18 m²	22.5 m²	27 m²
12,080 liters	24,160 liters	36,250 liters	48,080 liters	60,420 liters	72,500 liters

The table in box A4.1 shows, for various size vegetable gardens (square feet or square meters), approximate yearly water needs. Note that these gardens are mulched and in sunken basins, in conformance with the principles and strategies of water harvesting.

In the plant list tables that follow (boxes A4.2–A4.4), APPROXIMATE WATER NEEDS are listed as:

LW = low water use of 10 to 20 inches of water
    per year

MW = medium water use of 20 to 35 inches of
    water per year

HW = high water use of 35 to 60 inches of water
    per year.

The numbers 1, 2, 3, or 4 in parenthesis signify the approximate irrigation needs of the plants once they become established (this often takes 2 to 3 years).

(1) = no supplemental irrigation,

(2) = irrigation once a month in the growing season,

(3) = irrigation twice a month in the growing season,

(4) = irrigation once a week in the growing season.

Ratings based on Arizona Department of Water Resources Low Water Use/Drought Tolerant Plant Lists and direct observation.

Abbreviations signify: D=deciduous, E=evergreen, EO=essential oil, EPS=earth plaster/pigment stabilizer, F=food, FB=firebreak species, FR=fragrant, FW=fiber/basketry/weaving material, G=glue, H=hardy, HC=

Box A4.2. Native Multi-Use Trees for the Tucson, Arizona Area

Species	Water	Size	Cold Tolerance	Elevation Range	Growth Rate	Type of Tree	Human Uses	Wildlife	Domestic Animals That Use Plant
Desert Ironwood (*Olneya tesota*)	LW (1)	25 × 25'	SH 15°F	2,500' and below	moderate	E	F, M, NF, S, T,	Birds, pollinators, large and small mammals	Chickens, goats
Velvet Mesquite (*Propsopis velutina*)	LW (1)	30 × 30'	H 5°F	1,000–5,000'	fast	SD	F, FW, M, NF, P, S, W	Birds, pollinators, large and small mammals	Chickens, goats, cattle, honey bees, dogs
Screwbean Mesquite (*Prosopis pubescens*)	LW (2–3)	20 × 20'	H 0°F	4,000' and below	moderate	D	F, FW, M, S, W, WB	Birds, pollinators, large and small mammals	Chickens, goats, cattle, honey bees, dogs
Cat Claw Acacia (*Acacia greggii*)	LW (1)	20 × 20'	H 0°F	Below 5,000'	moderate to fast	D	M, P, S, T, W	Birds, pollinators, large and small mammals	Cattle, honey bees
Whitethorn Acacia (*Acacia constricta*)	LW (1)	10–15 × 10–15'	H 5°F	2,500–5,000'	moderate to fast	SD	F, G, M, S	Birds, pollinators, large and small mammals	Cattle
Desert Willow (*Chilopsis linearis*)	LW (2–3)	25 × 25'	H –10°F	1,500–5,000'	fast	D	FR, FW, M, S, W, WB	Birds and pollinators	Cattle, honey bees
Canyon Hackberry (*Celtis reticulata*)	MW (2–3)	Up to 35 × 35'	H –20°F	1,500–6,000'	moderate	D	F, S, W, WB	Birds, pollinators, large and small mammals	Chickens
Foothills Palo Verde (*Parkinsonia microphyllum*)	LW (1)	25 × 25'	H 15°F	500–4,000'	slow to moderate	D	F, S, W	Birds, pollinators, large and small mammals, desert tortoise	Cattle, honey bees
Blue Palo Verde (*Parkinsonia floridum*)	LW (2)	30 × 30'	H 15°F	500–4,000'	fast	D	F, S, W	Birds, pollinators, large and small mammals, desert tortoise	Sheep, honey bees

Box A4.3. Native Multi-Use Shrubs, Cacti, and Groundcover for the Tucson, Arizona Area

Species	Water	Size	Cold Tolerance	Elevation Range	Growth Rate	Type Of Plant	Human Uses	Wildlife	Domestic Animals That Use Plant
Oreganillo (*Aloysia Wrightii*)	LW (2)	5x5'	H 15°F	1,500–6,500'	moderate	D shrub	F, FR	pollinators	Honey bees, livestock
Quail-brush (*Atriplex lentiformis*)	LW (1)	Up to 8x12'	H 15°F	Below 4,000'	fast	E shrub	F, FB, M, NF, SC, SP	Birds, large mammals	Honey bees, livestock
Chiltepine (*Capsicum annum*)	LW (2)	Up to 3'	Frost sensitive	Below 4,000'	Slow to moderate	E shrub, D w/ frost	F, M	Birds	Chickens
Desert Hackberry (*Celtis pallida*)	LW (2)	Up to 10'	H 20°F	1,500–3,500'	Slow to moderate	SD shrub	F, M, SC, W	Birds, pollinators, mammals	Chickens, honey bees, cattle
Brittlebush (*Encelia farinosa*)	LW (1)	3'	SH 28°F	Below 3,000'	fast	E shrub	M, G	Pollinators, birds, large mammals	
Mormon Tea (*Ephedra trifurca*)	LW (2)	3–12'	H	Up to 4,500'	Slow	E shrub	E, M, P, T	Pollinators, birds, large mammals	Honey bees
Ocotillo (*Fouquiera splendens*)	LW (1)	Up to 15' tall	H 10°F	Below 5,000'	slow	D "shrub"	E, M, LF	Pollinators, birds	
Chuparosa (*Justicia californica*)	LW (2–3)	4'	SH 28°F	1,000–2,500'	Moderate to fast	D shrub	F	Birds, pollinators	
Creosote (*Larrea tridentata*)	LW (1)	Up to 11'	H 5°F	Below 4,500'	Slow to moderate	E shrub	G, M, W	Birds, pollinators, mammals	
Wolfberry (*Lycium fremontii*)	LW (1)	3–5'	H	2,500' and below	Moderate to fast	D shrub	F, M, SC	Birds, pollinators	Chickens, honey bees, livestock
Penstemon (*Penstemon parryi*)	LW (1)	Up to 3' tall	H 15°F	1,500–4,500'	Moderate	E ground cover	M	Birds, pollinators	
Jojoba (*Simmondsia chinensis*)	LW (1)	Up to 7'	H 20°F	1,000–5,000'	Slow to moderate	E shrub	FB, M, SC, SP, WB	Large and small mammals	Cattle
Saguaro (*Carnegiea gigantea*)	LW (1)	Up to 40' tall	SH 21°F	600–3,600'	slow	E cactus	F, G, M, W, T	Birds, bats, pollinators	Chickens
Barrel Cactus (*Ferocactus wislizenii*)	LW (1)	4–8' tall	H 15°F	1,000–5,600'	slow	E cactus	F, HC, M, P	Birds, pollinators, mammals	Pigs
Staghorn Cholla (*Opuntia versicolor*)	LW (1)	3–10' tall	H	2,000–3,000'	Moderate to fast	E cactus	F, M, SC	Birds, pollinators, mule deer	Cattle
Prickly Pear (*Opuntia engelmanii*)	LW (1)	Up to 5' tall	H 10°F	1,000–6,500'	moderate	E cactus	EPS, F, LF, M, P	Birds, pollinators, mammals, tortoise	Sheep, cattle (when thorns burned off)

Box A4.4. Exotic Multi-Use Fruit Trees, Vines, and Cacti for the Tucson, Arizona Area

Species	Cultivars	Water	Size	Cold Tolerance or Needs	Growth Rate	Type of Plant	Human Uses	Wildlife	Domestic Animals That Use Plant
Apple (*Malus pumila*)	Anna, Ein Shemer	MW (3)	15–20' X 15–20'	150–250 chill hours	moderate	D tree	F, S	Birds, pollinators, deer	Chickens
Apricot (*Prunus armeniaca*)	Royal or Blenheim, Katy	MW (2–3)	25 X 25'	300–400 chill hours	moderate	D tree	F, FB, S, WB	pollinators	Chickens
Carob (*Ceratonia siliqua*)	Casuda, Santa Fe, Sfax	MW (3)	25 X 25'	SH 23°F	moderate	E tree	F, FB, S, WB,		Honey bees, sheep, goats, pigs, cows, horses
Chinese Jujube (*Ziziphus jujuba*)	Lang, Li	LW (2)	20–30 X 10–20'	H 0°F	moderate	D tree	F, M		Chickens
Citrus – grapefruit	Duncan, Ruby Red, Marsh	HW (3)	14–20'	SH 27°F	moderate	E tree	EO, F, FB, M, S	pollinators	Honey bees
Citrus – lemon	Improved Meyer, Lisbon	HW (3)	Up to 20 X 20'	SH 31°F	moderate	E tree	EO, F, FB, M, S	pollinators	Honey bees
Citrus – Sweet orange	Valencia, Trovita, Marrs, Sanguinelli Blood	HW (3-4)	12–20 X 12–20'	SH 27°F	moderate	E tree	EO, F, FB, FR, M, S	Pollinators, hummingbirds	Honey bees
Date palm (*Phoenix dactylifera*)	Medjool, khadrawy, halawy, zahidi, maktoom. Only females produce fruit	MW (3–4)	Up to 40' tall	SH 22°F	moderate	E tree	F, FW, M, S, W, WB	birds	Chickens, dogs, camels, horses
Grape (*Vitis spp.*)	Flame, Ruby, Lomanto, Black Manukka, Thompson	MW (4)	5–90' long	H 0–10°F	moderate	D vine	F, FW, S (on trellis)	Birds, pollinators, small mammals	Chickens, honey bees
Fig (*Ficus carica*)	Black Mission, Conadria	MW (3) 15–30 X	15–30'	H 15°F >100 chill hours	fast	D tree	F, FB, M, S	Birds, bats, pollinators	Chickens
Loquat (*Eriobotrya japonica*)	Big Jim, Tanaka, Champagne, Gold Nugget	HW (4)	20 X 20'	Tree H 10°F , fruit & flowers SH 28°F	moderate	E tree	F, S, WB		Chickens, honey bees
Nopal (*Opuntia ficus-indica*)	Burbank, Quillota, Papaya, Honey Dew, Florida White	LW (1–2)	Up to 10' tall	H 20°F	moderate–fast	E cactus	EPS, F, FB, LF, M, SC	Pollinators, desert tortoise, javalina	Chickens, pigs, sheep, cattle
Olive (*Olea europaea*) *	Ascolano, Barouni, Haas, Manzanillo, Mission	MW (2)	Up to 30 X 30'	Trees H 15°F, Green fruit SH 28°F	moderate	E tree	M, S, W, WB F, FB,	Birds	Chickens
Peach (*Prunus persica*)	Desert Gold, Mid Pride, Rio Grande	MW (3–4)	15–25'	H –15°F , 250–350 chill hours	moderate to fast	D tree	F, FB, M, S	Birds, pollinators	Chickens, honey bees
Pomegranate (*Punica granatum*)	Wonderful, Fleishman, Papago, Sweet	LW (2–3)	12–15'	H 15°F, 100–200 chill hours	moderate	D shrub to tree	F, FB, M, P, SC, T	Birds	Chickens, honey bees

* Order fruiting olives from Santa Cruz Olive Tree Nursery (www.santacruzolive.com) or Peaceful Valley Farm Supply (www.groworganic.com).

hair conditioner, LF=living fence, M=medicinal, NF=nitrogen-fixing, P=pigment or dye, S=shelter/shade, SC=screen, SD=semi-deciduous, SH=semi-hardy, SP=soap, T=tanning hides, W=wood/timber, WB=windbreak.

"Pollinators" can include: butterflies, native solitary bees, beneficial predatory wasps.

How to Estimate the Water Requirements **in a Given Month** for a Listed Plant in Tucson, Arizona

Based on the "How To Develop A Drip Irrigation Schedule" handout from the LOW 4 Program of the Pima County Cooperative Extension/University of Arizona Water Resource Research Center.

A similar "plant water requirement estimator" can be created for other areas according to local evapotranspiration rates.

For an additional resource, see the Arizona Department of Water Resources for their Drought Tolerant/Low Water Use Plant Lists http://www.azwater.gov/azdwr/WaterManagement/AMAs/LowWaterUsePlantList.htm

They have plant lists specific to Tucson, Phoenix, and the Pinal, Prescott, and Santa Cruz Active Management Areas (AMAs)

1. Identify the plant as evergreen or deciduous, and as high, medium, or low water requirement. *For example, a Velvet Mesquite is deciduous with a low water requirement.*

2. Determine the canopy diameter of the plant (the diameter of the leafy part of the plant). This can be the plant's current canopy or its potential canopy at maturity. *Let's say our example mesquite has a 20-foot canopy.*

3. Determine the plant's water requirement in inches for a given month. See the tables in boxes A4.5A and A4.5B, which show how many INCHES of water the plant needs to receive beneath its canopy to maintain its health. *According to the table in box A4.5B, the June water requirement of our deciduous, low water requirement mesquite is 3 inches.*

4. Convert the plant's water requirement from inches to gallons. Find the plant's canopy diameter in Box A4.5C. Then find the corresponding # of gallons per inch of water beneath the canopy, and multiply it by the number of inches required in June to get the

---

Box A4.5A. Monthly Water Requirement in Inches–Evergreen Plants

Water Requirement	J	F	M	A	M	J	J	A	S	O	N	D	Annual Total
Low	0	0	2"	2"	3"	3"	3"	2"	2"	2"	1"	0	20"
Medium	0	0	3"	4"	5"	5"	5"	4"	4"	3"	2"	0	35"
High	0	3"	5"	6"	8"	9"	7"	6"	6"	5"	3"	0	58"

Box A4.5B. Monthly Water Requirement in Inches–Deciduous Plants

Water Requirement	J	F	M	A	M	J	J	A	S	O	N	D	Annual Total
Low	0	0	0	2"	3"	3"	3"	2"	2"	0	0	0	15"
Medium	0	0	0	4"	5"	5"	5"	4"	4"	0	0	0	27"
High	0	0	0	6"	8"	9"	7"	6"	6"	5"	0	0	47"

Box A4.5C. Conversion Table: Canopy Diameter vs. Gallons/Inch under Canopy												
Canopy Diameter in Feet	2	4	6	8	10	12	14	16	18	20	25	30
# of Gallons per Inch of Water beneath Canopy	2	8	18	31	49	71	96	125	159	196	306	441

total GALLONS of water required in that month. *For example, the number of gallons in an inch of water under a 20-foot diameter Velvet Mesquite is 196 gallons. The tree needs 3 inches of water in June, so multiplying 196 × 3 = a June water requirement of 588 gallons.*

How to Estimate the Annual Water Requirements for a Listed Plant in Tucson, Arizona

Use the tables in box A4.5A or A4.5B to find the plant's estimated ANNUAL water requirement in INCHES. Multiply that number by the number of gallons per inch of water beneath the canopy (table in box A4.5C), and the plant's canopy diameter. *For example, the 20-foot diameter Velvet Mesquite needs 15 inches of water annually, and from Table A4.5C we see that there are 196 gallons per inch of water beneath a 20-foot canopy. So multiplying 15 × 196 = an annual water requirement of 2,940 gallons.*

Note 1: Annual water requirement estimates are likely all you will need to consider when designing a landscape of local native plants based on natural wild plant densities and sizes. Such vegetation is naturally adapted to the local rainfall patterns and, once established, can survive the dry periods between rains.

Monthly water requirement estimates are better suited for designing landscapes of exotics or native plants that are planted at a higher than normal density or are irrigated for larger than normal plant sizes. These estimates give you a better idea of what seasons or months require more water so that you can better plan for needed water storage and the timing of supplemental irrigation with cistern water or greywater.

The water requirements for all plants will increase as they grow, since the amount of water they transpire through their leaves increases with the increase in cumulative leaf surface area. Therefore, it is impor-

tant to plan for the water needs of your plants at their mature size. However, by minimizing the amount of water available to native plants you can reduce their mature size—reducing the need for more water. For example, a Velvet Mesquite receiving approximately 6,600 gallons of water per year can grow to be 30 feet tall and wide, but if only 2,940 gallons of water per year is available to the tree, it will likely not grow to be taller and wider than 20 feet.

Note 2: For another method of estimating landscape water needs and tables of information allowing you to do so for many locations in Arizona see the free publication, *Harvesting Rainwater for Landscape Use*, 2nd Edition by Patricia H. Waterfall, 2004.

Estimating Water Requirements of Plants in New Mexico

Web-search and read the free download "Roof-Reliant Landscaping" to estimate plant water needs anywhere in New Mexico.

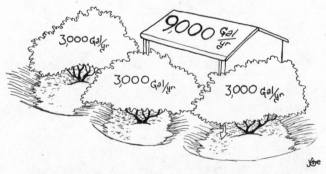

Fig. A4.1A. Annual irrigation demand in Tucson, Arizona, of low-water-use, dry-season-adapted, native mesquite trees met by rainwater supply from roof runoff. Water infiltrated and held in soil—no tanks needed

Fig. A4.1B. Annual irrigation demand in Tucson, Arizona, of higher-water-use, dry-season-susceptible, exotic citrus tree met by rainwater supply from roof runoff, and supplemented with greywater in dry times

Fig. A4.1C. Annual irrigation demand in Tucson, Arizona, of 130 ft² (12 m²) high-water-use vegetable garden in sunken, mulched basins. Cistern needed to dole out roof runoff in long dry times

PLACE-BASED WILD AND CULTIVATED FOOD PLANTS CALENDARS

This calendar builds on the One-Page Place Assessments introduced in chapter 2. They tell another chapter of the story of the challenges and potentials of your place, rooted in the climate and ecosystem where you live.

Figure A4.2 is a Sonoran Desert Foods Harvest Calendar for native and exotic, wild and cultivated, annual and PERRENIAL food plants that grow in the Sonoran Desert. It is not comprehensive, but instead is intended to give you a taste of the wide diversity of climate-appropriate food plants (and their cultivars or sub-varieties) whose different harvest seasons provide resilient year-round harvests. For example, if both blue (*Parkinsonia* [formerly *Cercidium*] *floridum*) and foothills palo verde (*Parkinsonia microphyllum*) trees are planted in a yard in the Sonoran Desert, there can be a three-week difference between the blooming, and

harvesting, season of the two different palo verdes; and thus a three-week extension of the palo verde harvesting period for someone with access to both species as opposed to just one of them. In this location, exotic Marrs oranges are ripe in November through February, while Valencia oranges are ripe February through May and you can bud graft the fruiting stock of both species onto the same tree. The positioning of plants in different microclimates (warmer, cooler, wetter, drier locations), along with the impacts of climate change, affect where plants can grow and their harvest periods.

Plants native to the Sonoran Desert—such as mesquite, ironwood, and palo verde trees—will thrive within or next to passive water-harvesting earthworks or rain gardens. Exotic plants—such as citrus and tomatoes—typically need supplemental irrigation water from rain cisterns, gravity-fed greywater systems, and/or air conditioning condensate. Calculate your plants' water requirements to determine if you have enough on-site water to sustainably meet these supplemental irrigation needs. The plants with an asterisk—such as *pistachio*, *piñon pine*, and the Arizona native *Emory oak* (acorn)—do not grow and produce nuts in the Sonoran Desert, but they do grow in adjacent areas at higher elevations. They remind us to stay conscious of our connections with neighboring communities of plants, people, and whole ecosystems.

I strongly encourage you to prepare a harvest calendar for your location and continually update and expand it. To do this, ask questions of local herbalists, farmers, gardeners, permaculturalists, and localvores; research the local ethnobotanical record; and make your own observations. Find out which plants that can grow and produce in your area are missing from your system and the community. Plant them in locations where on-site, sustainably-harvested water sources will enable them to regeneratively produce. Then care for them, harvest them, and share their bounty.

See www.LorenziniWorks.com for additional, more comprehensive calendars done by the artist Jill A. Lorenzini.

See www.DesertHarvesters.org for native Sonoran Desert food plant planting, harvesting, processing, cooking tips, and resources.

Figure A4.2. Harvest calendar of food plants that grow in the Sonoran Desert

Appendix 5
Worksheets: Your Thinking Sheets

This appendix is meant both as a summary of the design process described in this book and as a checklist of the recommended steps toward creating an integrated conceptual water harvesting design.

Teachers may find these worksheets useful for teaching integrated water harvesting in their classrooms. Note that additional water-harvesting curriculum (designed to be used in conjunction with this book) is available at the Rainwater Harvesting menu tab at www.HarvestingRainwater.com.

This appendix follows the thought-flow in volume 1 of *Rainwater Harvesting for Drylands and Beyond*. It begins with any thoughts you may have on applying the principles of rainwater harvesting to your site, continues with basic observations about your land, and leads to steps you can take to have a more water- and life-abundant home and yard—one more comfortable and beautiful in both winter and summer.

CHAPTER 1/STEP 1: THE RAINWATER-HARVESTING PRINCIPLES

Here's a summary of the "Principles of Successful Rainwater Harvesting." Add any of your thoughts below them.

1. Begin with long and thoughtful observation.

Use all your senses to see where the water flows and how. What is working, what is not? Build on what works.

2. Start at the top (highpoint) of your watershed and work your way down.

Water travels downhill, so collect water at your high points for easy gravity-fed distribution. Start at the top where there is less volume and velocity of water.

3. Start small and simple.

Work at the human scale so you can build and repair everything. Many small strategies are far more effective than one big one when you are trying to infiltrate water into the soil.

4. Spread and infiltrate the flow of water.

Rather than having water erosively runoff the land's surface, encourage it to stick around, "walk" around, and infiltrate into the soil. Slow it, spread it, sink it.

5. Always plan an overflow route, and manage that overflow as a resource.

Always have an overflow route for the water in times of extra heavy rains, and where possible, use that overflow as a resource.

6. Maximize living and organic groundcover.

Create a living sponge so the harvested water is used to create more resources, such as cooling shade and food. As plants grow the soil's ability to infiltrate and hold water steadily improves.

7. Maximize beneficial relationships and efficiency by "stacking functions."

Get your water-harvesting strategies to do more than hold water. Berms can double as high and dry raised paths. Plantings can be placed to cool buildings. Vegetation can be selected to provide food.

8. Continually reassess your system: the "feedback loop."

Observe how your work affects the site—beginning again with the first principle. Make any needed changes, using the principles to guide you.

Are there principles you would like to add or adapt based on your experience? _____

INTERLUDE: YOU AND YOUR WATER

How do you think you currently use your water resources (rough breakdown)?
    Indoors _____
    Outdoors _____

Do you wish you had:
    a cleaner or better source of water? _____
    less expensive water? _____
    a more direct connection with your water source and water use? _____

Have you ever thought of conserving water? _____
    If so, how? _____

If you had a better source of water or less expensive water, how would you use it?
    landscape and garden use _____
    washing and bathing _____
    potable use (drinking and cooking) _____
    other _____

After reading/perusing this book, what might you want to do?
    first? _____
    second? _____
    third? _____

CHAPTER 2/STEP 2: BUILDING ON LONG AND THOUGHTFUL OBSERVATION—ASSESS YOUR SITE'S WATERSHED AND WATER RESOURCES

A. Walk your site's watershed.

Identify its ridgelines/boundaries, and observe how water flows within them. Make any notes below. Refer to pages 42 - 44 for more information.

If runoff flows across your land, pay particular attention to what direction it comes from, its volume, and the force of the water's flow. Look at what surfaces water flows over to estimate the water's quality. Note any observations below. You may also want to look for erosion patterns when it's dry (see appendix 1). Write down your observations.

B. Create a site plan and map your observations.

First, photograph your site so you'll have "before" photos with which to document your progress with future "after" photos.

Now, use grid paper or an aerial photo of your site from Google Earth or Google Maps, and using figure 2.3 as a model create a "to-scale" site plan of your property's boundaries. Leave wide margins to mark the locations where resources—such as runoff from your neighbor's yard—flow on, off, or alongside your site. Draw or identify buildings, driveways, patios, existing vegetation, natural waterways, underground and above-ground utility lines, etc. to scale. Next, draw any catchment surfaces that drain water off your site (for example, a driveway sloping toward the street), and any catchment surfaces draining water onto your site from off-site; indicate the direction and flow of any runoff and runon water. Refer to pages 44 - 49 and figure 2.4 as you do all this. Write down additional notes below.

Additional observations you may want to record at this time:

What vegetation lives solely off on-site water (rainwater), and which depends on pumped water or imported irrigation water? _____

What unirrigated native vegetation do you see growing within a 25-mile (40-km) radius of your site (in similar microclimates as those that exist on your site) that could do well on your site? _____

C. Calculate your site's rainwater resources.

C1. Your Site's Rain "Income"

Determine the "income" side of your site's water budget so you can compare them with the "expense" side. For this section refer to pages 44 - 50 and additional calculations found in appendix 3.

- What is the area's average annual precipitation in inches or mm? _____

- What is the area of your site (land) in square feet, acres, or hectares? _____

- What is the area of the roofs of your house, garage, sheds, and other buildings on your property (see the calculations appendix 3, equations 1–3)? _____, _____, _____

Now, use the calculations in box 2.3, or in appendix 3, to determine your site's rainfall resources. You will want to answer the question: In an average year, how much rain falls on your site? For this you will use your site's area and annual precipitation to get some kind of ballpark figures about your annual rainwater resources in gallons or liters. If you have a difficult time with the math, then just use the "rule of thumb" figures in box A3.2 of appendix 3.

Do your calculations below.

X. Your site's annual rainfall resources in gallons or liters. How much water falls onto your site?
X = _____

C2. Your Site's Rain "Loss"

Now, refer to the calculations in box 2.4 and figure 2.4 to determine how much runoff drains from your impervious catchment surfaces for potential storage/use in adjoining tanks or earthworks. You don't need to be exact; you just want a good estimation.
Roof runoff _____
Driveway runoff _____
Patio runoff _____
_____ runoff _____

Then note how much of that runoff (and additional potential runoff from other built or disturbed surfaces such as mounded sections of the landscape or bare dirt areas) currently drains off your property. Add the estimated total and write it below:

Y. Loss/runoff from your site in gallons or liters. How much runoff runs off your site?
Y = _____

C3. Your Site's Water Gain

Now, you want to estimate how much water (annually) you're gaining from runoff from other properties onto your site. Use the same calculations as above.

Z. Gain/runon in gallons or liters. How much off-site runoff runs onto your site? (Use the same calculations as above.)
Z = _____

C4. Totaling It Up

X (on-site rainfall) - Y (runoff draining off site) + Z (runon to your site) = T (TOTAL: HOW MUCH YOU CURRENTLY DO HARVEST)

X _____ - Y _____ + Z _____

= T _____

T = _____

This equals the total on-site rainwater resources you currently do harvest.

Look to see how much more you can harvest by reducing loss of runoff (Y).

D. Estimate your site's water needs.

This step determines the "expense" side of your water budget by estimating your household and landscape water needs. See page 51.

- What is your annual water consumption based on your water bill? _____
- In which months are your water consumption/needs highest? _____

The next steps are to try to determine how much of your water is used indoors versus outdoors.

- Estimate your average annual *indoor* water consumption using the user-friendly website www.h2ouse.org. _____

- Estimate your average annual *outdoor* water consumption (based on plant water-need requirements; see appendix 4): List some of your larger plants and their water requirements below.

- Or/And: Subtract your estimated *indoor* water consumption from your water bill for a ballpark estimate of your current outdoor needs.

_____

E. Compare your needs and resources.

- How much of your domestic water needs could you meet by harvesting rooftop runoff in one or more tanks? _____

- How much vegetation could you support by simply harvesting rain falling and infiltrating directly in your soil? _____

- How much vegetation could you support if onsite runoff was also directed to the planted areas (this runoff could be diverted directly and passively to planted areas or harvested in a tank and doled out to the planted areas as needed)?

_____

- List what other steps you could take to balance your water budget using harvested rainwater as your primary water source. See the conservation strategy suggestions in box I.7.

Indoors _____

Outdoors _____

F. Greywater sources

Estimate the average volume of accessible household greywater you could reuse within your landscape, using the information in boxes 2.6 to 2.8. *Accessible* means you can access current drain pipes or install new ones to direct the greywater to mulched and vegetated basins within the landscape. You will need to maintain a minimum 1/4-inch drop per linear foot of pipe (2-cm drop per linear meter) for gravity to freely and conveniently distribute your greywater from a point downstream of the P-trap for the greywater source (washing machine, sink, etc.) to the greywater pipe outlet in the landscape.

washing machine _____
shower _____
bathtub _____
bathroom sink _____
kitchen sink (dark greywater) _____
other _____
Total _____

G. Additional "wasted" waters that can become harvested waters

See box 2.9 and estimate the average (or seasonal) volume of household wasted waters you could harvest.

    evaporative cooler bleed-off _____

    air-conditioning condensate _____

    reverse-osmosis water-filter discharge

    _____

    other _____

H. "A One-Page Place Assessment" for Your Location

Look at figure 2.7, "A One-Page Place Assessment" for Tucson, Arizona. Create your own "One-Page Place Assessment" with the how-to tips on the One-Page Place Assessment page on my website HarvestingRainwater.com, or see if an assessment for your community already exists on that webpage. If you lack computer access or skills, take figure 2.7 to your local library or cooperative extension agent, and ask for help finding similar information for your locale. Refer to that One-Page Place Assessment when using this section of the worksheets.

H1. Climate

Temperature variability affects vegetation: human comfort, stress, and potential; and animals both on top and within the soil. The U.S. Department of Agriculture (USDA) plant hardiness/climate zone maps reflect this. To find out the USDA plant zone you are in, see http://planthardiness.ars.usda.gov/PHZMWeb/

The USDA generates new maps as climates change. You can learn a lot when you compare past and current maps. For maps that compare and illustrate differences between 1990 and 2012, see http://www.washingtonpost.com/wp-srv/special/local/planthardinesszones/index.html. And for predicted future changes due to climate change see http://epa.gov/climatechange/

Has your climate zone changed? _____

What was it before? _____

How have USDA Hardiness Zone Maps changed due to a changing climate? _____

For the 1960 and 1965 USDA Hardiness Zone Maps see http://www.garden.bsewall.com/topics/hardiness/history.html

How do you *feel* your climate's temperatures at the following times?

	Winter	Spring	Autumn	Summer
Morning	____	____	____	____
Afternoon	____	____	____	____
Evening	____	____	____	____
Night	____	____	____	____

What are comfortable indoor temperatures and how can we adapt to feel comfortable in an expanded indoor temperature range? In the U.S., 72° F (22.2° C) is often considered a comfortable mean indoor temperature, yet authors David Bainbridge and Ken Haggard, in *Passive Solar Architecture* state that in China "rural residents found indoor winter temperatures of 52.7° F (11.5° C) comfortable, while urban Chinese felt 57° F (13.8° C) was better; … many people when sleeping prefer cooler temperatures and a good comforter, … and can be quite comfortable at 60° F or less; … and in summer 80% of the people may feel comfortable at 86° F (30° C) in tropical climates with outdoor mean monthly air temperatures of 95° F (35° C)."[1]

How could a change of attitudes or practices change your perception of hot and cold? _____

What changes in your clothing could help provide free heating (sweaters or long underwear, for instance) and cooling (shorts and short sleeves)? _____

What do you observe about the seasonal difference between daily outdoor high temperatures and nightly low temperatures? _____

At your location, are hot daily high temperatures offset by cool nights? _____

Are cold nightly temperatures offset by warm daytime temperatures? _____

If so, when _____ ?

In your area, how do average high and low temperatures compare to the record high and low? _____

How could you moderate such extreme temperatures with the addition of on-site water harvesting, shading, solar access, and other passive strategies? See chapters 3 and 4 for ideas. Write down your thoughts. _____

H2. Sun

Where does the sun rise and set on the *summer* solstice at your location? _____

Where does the sun rise and set on the *winter* solstice? _____

Where and how is the sun's rising and setting in summer different from that in winter? _____

Can you visualize the sun's rising and setting, during winter and summer? Spring and fall? Sketch out a site map for your location showing the part of the sky the sun rises and sets in during these seasons. Take a look at the sun path diagram for your latitude in appendix 7 to help you fill out this section. Write any observations. _____

How high is the sun at noon in different seasons of the year (winter and summer solstices, spring and fall equinoxes)? _____

What times of year and where might you want to deflect that sun? When and where might you want to harvest that sun?

How does the sun's seasonal path at your site compare to seasonal sun paths at other latitudes where you, friends, and/or family have lived? _____

Now, look at the monthly average temperatures (in the first section of the "One-Pager") for these different times of the year. What correlation do you observe in your location between the sun's position and average monthly temperatures? _____

What times of year do you have more sunny days, and when do you have more cloudy days (look at the water section)? _____

How might the number of sunny and cloudy days affect the potential to produce renewable power at your site? _____

In the coastal northwest U.S. sunny summers make solar power viable in that season, while overcast rainy falls, winters, and springs make micro-hydro power viable in those seasons. Look at the wind section for wind power potential.

Note how the insulating, blanket-like effect of cloudy high-humidity weather reduces temperature fluctuations between day and night. In contrast, cloudless, low-humidity weather allows radiant heat to quickly dissipate into the atmosphere at night, resulting in lower nighttime temperatures and greater temperature fluctuations between day and night.

What is your site's elevation and how does that affect temperature? _____

On the average, temperatures drop 3.5° F (1.94° C) per 1,000 feet (304 m) of altitude.[2] However, this temperature drop is greater in sunny weather (5.4° F/1,000 ft or 9.8° C/1,000 m) and less in rainy, snowy, overcast weather (3.3° F/1,000 ft or 6° C/1,000 m).[3]

H3. Wind

Note how the wind's speed and direction varies when affected by landforms, buildings, and vegetation. As with water, the wind's velocity increases when its path is constricted and straightened. Thus, when straight roads and paths—and the trees and shrubs planted along them—are parallel with the wind's direction, these locations may become wind tunnels in high-speed winds. Meandering roads and paths, along with their adjoining vegetation, can lessen this effect. See appendix 8 for more on wind patterns and strategies to harvest or deflect these winds, and see chapter 4 for ideas on how to integrate these wind strategies with sun and shade harvesting. Write down any thoughts and/or observations on your local wind patterns. _____

H4. Water

Based on your location's precipitation data, what can you conclude about rainfall distribution? How many "rainy seasons" are there per year, and when do they occur? _____

Compare average monthly temperatures to average monthly rainfall. Does rain fall in warmer months when growing plants need more water? _____

If not, how could you increase water availability and storage when rain does fall to get you through dry months and droughts? _____

See chapter 3 for ideas.

If you have a rainwater tank/cistern, at what time(s) of the year would you be filling it? _____

How often could you fill it, use the rainwater, and refill it in the course of a year? _____

How much of your precipitation falls as rain and how much as snow? _____

If you have a tank, is there a winter-time risk in your location of frozen pipes and tank, or iced-over downspouts and rainhead screens? _____ If so, how would you deal with these issues? _____

Your freeze-protection strategies might include placing the tank, downspouts, and rainhead screens where they would be warmed by winter sun. You could also disconnect, shut off, and/or drain water-holding pipes during months when freezing could occur. In addition, you could install your plumbing underground below the frost line.

How much of your community's utility water consumption could be met by harvesting rainfall community wide? _____

What other conclusions can you draw?

_____

H5. Watergy

See appendix 9 for information about the Water-Energy connection.

What data can you find for your city? _____

If none, use the data provided in the *Water* Costs of Energy, *Energy* Costs of Water, and *Carbon* Costs of Energy charts in appendix 9.

What are some ways you can reduce your water, energy, and carbon footprint, and your community's footprint? _____

H6. Totem Species

Ask staff at a local nature center, natural history museum, or library for information about endangered species, ecosystems, and water sources in your area. What did you learn? _____

Ask long-time residents which animal and plant species they remember being abundant when they first lived here and what the landscape looked like. _____

What species are missing or hardly seen now? _____

Which species are common now, or more common? _____

What contributed to their rise or fall? _____

What can you do to help create or enhance healthy habitat/conditions for indigenous species, ecosystems, and water sources? _____

H7. Taking your One-Page Place Assessment further

What other key indicators (past, present, and future) of the health, challenges, and potential of your location should be captured in your One-Page Place Assessment (these may make your assessment longer than one page)? _____

Consider including:
- Information on your area's geology and soils. For example are you losing, maintaining, or gaining topsoil, organic matter, and fertility?
- Light pollution or its abatement to your assessment, to monitor how many constellations are disappearing or reappearing in your night sky. These constellations more directly connect us to the stories, histories, and knowledge of cultures that have used those stars in the past.

- Culture indicators, including diversity and density of different cultural groups in the past and present, the languages spoken, how long these cultures have been present, what knowledge and strategies supported them then and support them now, and how these have changed.

See the "How to Create Your Own One-Page Place Assessment" subpage on the One-Page Place Assessment page at www.HarvestingRainwater.com for tips on, and resources for, these additional indicators and more.

CHAPTER 3/STEP 3: EARTHWORKS, TANKS, OR BOTH

Refer to the comparisons in box 3.1 and see the overview of strategies later in chapter 3 to decide how you might best harvest the water for your planned uses.

Now that you've (more or less) estimated your on-site water resources and needs (from the previous worksheets), the next step is to answer again the following questions:

How do you currently use your water resources (rough breakdown):
    indoors_____
    outdoors_____

How do you plan to use your water resources?
landscape and garden use_____
washing and bathing_____
potable use_____
other_____

After reviewing chapter 3, and the "principles of successful rainwater harvesting" from chapter 1, what do you think you'd want to do:
    first?_____
    second?_____
    third?_____

Compare the above to your answers in the earlier interlude. What's different, now that you have more information to work from?

CHAPTER 4/STEP 4: INTEGRATED DESIGN

This chapter is intended to heighten your awareness of additional on-site resources and challenges, and to show you how to maximize their potential by integrating their harvest with that of water. The numbers below follow the Integrated Design Patterns and their Action Steps found in chapter four.

If needed, make a new photocopy of your site map and mark the directions of north, south, east, and west.

1. Your site's and buildings' solar orientation (Integrated Design Pattern One)

- What is your site's latitude (web search it or ask your local friendly librarian if you don't know): _____

- How is your site and/or your home oriented? (See figures 4.5A and B). Put this information on your site map as well as writing it below.

On your site map:
- Identify the "winter-sun/equator-facing side," and the "winter-shade side" of your home.
- Map the location of the rising and setting sun on the summer and winter solstice.
- Also mark any of the following, and any additional incoming resources or challenges (see figure 4.4): where you would like more shade or exposure to sun; the direction or location where prevailing winds, noise, or light come from; and the foot traffic patterns of people, pets, or wildlife.
- Where are the warm and cold spots in winter? The hot and cool spots in summer? What areas inside your house get direct sun in the morning and afternoon, in winter and summer? Do you get sun and shade on your garden and outdoor areas when and where you want it? Mark your

site map appropriately and write your observations below.

- Shut off all mechanical heating and cooling systems on a couple of sunny days at least once in each season of the year to observe how direct solar exposure—or the lack of it—affects the comfort of your home and yard. (However, don't do this when there is a possibility your pipes could freeze. And if such pipe freezing could occur then look into how you could retrofit your home so your pipes would not freeze if the power were shut off.) When you shut off all mechanical heating and cooling systems, what are your observations?
winter _____ spring _____
autumn _____ summer _____

2. Equator-facing window overhangs (Integrated Design Pattern Two)

- Do you have window overhangs? _____
- If so, what is their projection length on the winter-sun side of your house? _____
- Use the overhang projection information on pages 99 - 103 to determine appropriate overhang sizes for your winter-sun/equator-facing windows. _____
- Compare existing overhangs to what the calculation or to-scale drawing recommends. _____
- What have you noticed about how overhangs or the lack of them affect your comfort throughout the year? _____
- What can you do (put up awnings or trellises, plant trees, extend overhangs, open up a covered section of winter-sun-facing porch, etc.) to enhance the positive ways sun and shade can affect your building? _____

2B. Windows

Windows are important. Assess where the windows are in your house; how much window area can be opened (casement windows typically let you open the whole window area; sash or slider windows typically let you open half); and whether they have single or double panes, or storm windows (for better insulation), and/or screens (for insect-free summer ventilation)? Write your observations. _____

Windows can affect seasonal comfort. In which rooms of your house is it:

	Winter	Summer
Too warm	_____	_____
Too chilly	_____	_____
Too dark	_____	_____

Which of these rooms have windows, and what direction(s) do they face in each room? _____

How could comfort and energy efficiency be improved by:
Increasing window areas? _____
Adding seasonal shading using an outside overhang or awning? _____
Removing obstructing vegetation outside to allow more sun in? _____
Making inoperable windows operable/openable to increase ventilation? _____

See appendix 8 for how to increase natural ventilation of your rooms and house using different window types and orientations and appropriate placement of outdoor vegetation.

In rooms that are too dark, consider adding windows before adding skylights. Skylights can let in too much direct sun and heat on summer days, and let out valuable heat on winter nights. They can leak when it rains, and may be expensive to replace if damaged.

Prioritize adding windows on the equator-facing side of your house where more winter heating may be needed. Minimize window openings or their access to direct sun on east and west-facing sides if morning and afternoon summer heat gain needs to be reduced.

Consider adding interior operable windows to let light and ventilation pass from one room to the next. High transom windows or opaque glass can be used to maintain privacy between rooms.

Test the winter heat-gain potential of your equator-facing windows on a sunny winter day by putting your hand in the direct sun inside the building window without touching the window glass. If your hand feels warm or hot, it probably is not a low-E window and therefore it is the right kind of window for solar heat gain in winter. If you do not feel significant warmth it is probably a low-E window and will reduce desired passive winter heating. If you need more winter solar gain, you may want to replace low-E windows on the equator-facing side of your house. Write your observations below: _____

Do the same window test when the sun shines through windows on the non-equator-facing sides of your house—especially in the hot months. In these cases, you might not want the added heat from the sun. Depending on your climate and surrounding vegetation, for non-equator-facing windows—including storm windows—you might want to use a low-E type glass that reflects summer heat and insulates interior heat in the cooler seasons. Write your observations below:
_____

In each room with equator-facing windows that receive direct winter sun, what percentage of their floor area equates to the equator-facing windows' area? _____ (See box 4.5 for guidance.)

Is this a big enough window area to keep you warm enough (or largely so) on sunny days in winter?
_____

Is this window area shaded sufficiently by an overhang or awning to keep you cool enough in summer?
_____

How does the percentage of your equator-facing window area compared to your floor area compare to the recommendation(s) for your climate in box 4.5?

3. Solar arcs (Integrated Design Pattern Three)

Make a new site map if needed.

Do you have any elements of a solar arc in place around your home or garden, such as an existing shade tree, covered porch, or building? If so, mark them on your site map and write comments about them below.

- Now, indicate on your site map where missing pieces of a solar arc should be located to complete it and benefit your home or garden.

- Can you use any water-harvesting strategies (earthworks, trees, cisterns) to create or grow a solar arc or a windbreak?

- From where (roof, patio, etc.) could you harvest on-site water to grow shade where you need it? Chapter 3 gives you ideas.

4. Sun & shade traps (Integrated Design Pattern Four)

On your site map, mark where a sun & shade trap might make sense, and indicate any existing elements already in place. Write your observations below.

- Consider the desirability of a fence, new cistern, or plantings within earthworks such as trees, large shrubs, or vines growing on trellises, fences, etc. to create a sun & shade trap. Write any thoughts below.

5. Maintaining winter sun exposure (Integrated Design Pattern Five)

- Where have you noticed winter shadows blocking your sun? Write down observations below, and on your site map indicate features (trees, etc.) that produce long winter shadows.

- What would be the length of a shadow cast by a 20-foot (6-m) tree by the noonday sun on the winter solstice at your latitude (see the shadow-ratio correlation in box 4.9)? _____

- Given this information, based on the tree's height how far to the south (in northern hemisphere) or to the north (in southern hemisphere) would that tree need to be placed from your home if you do not want it to shade your house in winter?

- Solar rights: Are any buildings, trees, or other structures on your site blocking you or any of your neighbors' key winter solar access for equator-facing windows, walls, and roof between the hours of 9 A.M. and 3 P.M. (or between 10 A.M. and 2 P.M. in latitudes above 40°)? _____

What are these obstructions? _____
How much do they block the sun? _____
How can they be altered (for example, pruning or thinning trees or shrubs) to increase solar access? _____

- Now, think about where you might want to add appropriately placed new vegetation, structures, or windbreaks to avoid blocking desired winter sunlight for winter sun-/equator-facing windows, winter gardens, and solar strategies, while providing other benefits. Note any observations below, and pencil in on your site map if necessary.

6. Raised paths, sunken basins (Integrated Design Pattern Six)

- Write your observations of the following: the relative height of paths, patios, sidewalks, driveways and streets compared to adjacent planting areas in your home and community.
- Do you see any "raised path, sunken basin" patterns or a sunken path, raised planting area pattern?

- Is stormwater being directed to vegetation, asphalt, or storm drains?

- Now, identify and map areas where you could develop the raised path, sunken basin pattern at home.

7. Reducing paving and making it permeable (Integrated Design Pattern Seven)

- Below, write examples of pervious and impervious paving around your home and community.

- Write any thoughts about how you can: reduce the paving on your site; either direct the remaining pavement's runoff into adjoining earthworks or make remaining pavement more permeable; turn your driveway into a "park-way" or use porous brick, cobbles, or angular open-graded gravel instead of an impervious material such as concrete; etc.

8. Degenerative, Generative, and Regenerative Investments

Read the section Integrating the Elements and Patterns of Your Site to Create a Regenerative Landscape Design on pages 121 - 123, then answer the following:

What percentage of your investments in your site (and your life) of effort, time, money, material, or labor are degenerative? _____
What are some examples? _____

What percentage are generative? _____
What are some examples? _____

What percentage are regenerative? _____
What are some examples? _____

What degenerative investments can you transform into generative investments? _____
How will you do this? _____

What generative investments can you transform into regenerative investments? _____

How will you do this? _____

9. Tying it all together—creating an integrated conceptual design

Read page 123, and remember that you don't have to implement everything all at once. *Start small, start at the top*… And—if future observations or realizations justify a change in your plan after you've begun implementation, *make a change*. This is all a process based on *long and thoughtful* observation, continuing for the duration of your relationship with the site.

CHAPTER 5

What is your site's regenerative potential?

What is your regenerative potential partnering with the site's?

What is your community's essence?

What is the essence of its Place?

What is its regenerative potential?

How can you partner with, and contribute to that potential?

APPENDIX 1. PATTERNS OF WATER AND SEDIMENT FLOW WITH THEIR POTENTIAL WATER-HARVESTING RESPONSE

After reading this appendix, walk your land and your neighborhood (especially its waterways) and observe water and sediment flow patterns. Which do you see?

APPENDIX 2. WATER-HARVESTING TRADITIONS IN THE DESERT SOUTHWEST

How might (or how do) these practices of the past inform practices in the present?

APPENDIX 4. EXAMPLE PLANT LISTS AND WATER REQUIREMENT CALCULATIONS FOR TUCSON, ARIZONA PLUS A SONORAN DESERT FOODS HARVEST CALENDAR

Are there similar multi-use plant lists and harvest calendars available in your area? If so, what species and information could you add to them? If not, consider generating these lists yourself, or working with others to do so. Then you can share this information and seek out feedback to continue to evolve the lists.

Appendix 6
Resources

To save paper and provide the most up-to-date information and revisions, appendix 6 – Resources now appears in its entirety only at www.HarvestingRainwater.com on the "Books' Resource Appendices" page. The resources provided here are a distillation of those resources covering general rainwater harvesting, water-harvesting earthworks (chapter 3), cisterns (chapter 3), greywater (chapter 3), sun angles and passive solar design (chapter 4), integrated design (chapter 4), building communities' potential (chapter 5), climate change (appendix 9), and soils and vegetation.

General Rainwater Harvesting
Making Water Everybody's Business: Practice and Policy of Water Harvesting, edited by Anil Agarwal, Indira Khurana, and Sunita Narain. Centre for Science and Environment, 2001.

Dying Wisdom: Rise, Fall, and Potential of India's Traditional Water Harvesting Systems, edited by Anil Agarwal and Sunita Narain. Centre for Science and Environment, 1997.

The Negev: The Challenge of a Desert, 2nd ed., by Michael Evenari, Leslie Shanan, and Naphtali Tadmor. Harvard University Press, 1982.

The Collection of Rainfall and Runoff in Rural Areas, by Arnold Pacey and Adrian Cullis. Practical Action Publishing (formerly Intermediate Technology Publications), 1986.

Information on water-harvesting demonstration sites available at HarvestingRainwater.com/rainwater-harvesting-inforesources/water-harvesting-demonstration-sites/

Rainwater Harvesting with Earthworks (chapter 3)
Rainwater Harvesting for Drylands and Beyond, Vol. 2: Water-Harvesting Earthworks, by Brad Lancaster. Rainsource Press, 2008.

A Guide for Desert and Dryland Restoration: New Hope for Arid Lands, by David A. Bainbridge. Island Press, 2007.

Let the Water Do the Work: Induced Meandering, an Evolving Method for Restoring Incised Channels, by Bill Zeedyk and Van Clothier. Quivira Coalition, 2009.

Water Harvesting from Low-Standard Rural Roads, by Bill Zeedyk. Quivira Coalition, 2006.

Erosion Control Field Guide, by Craig Sponholtz and Avery C. Anderson. Quivira Coalition and Drylands Solutions, Inc., 2012.

Rain Gardens: Managing Water Sustainably in the Garden and Designed Landscape, by Nigel Dunnett and Andy Clayden. Timber Press, 2007.

Green Infrastructure for Southwestern Neighborhoods, by Watershed Management Group. Free pdf available at watershedmg.org/sites/default/files/greenstreets/WMG_GISWNH_1.0.pdf

Rainwater Harvesting with Cisterns (chapter 3)
Rainwater Collection for the Mechanically Challenged, by Tank Town. DVD. Tank Town Publishing. www.RainwaterCollection.com

Rainwater Catchment Systems for Domestic Supply: Design, Construction, and Implementation, edited by John Gould and Erik Nissen-Petersen. Practical Action Publishing (formerly Intermediate Technology Publications), 1999.

Rainwater Harvesting: System Planning, by Billy Kniffen, Brent Clayton, Douglas Kingman, and Fouad Jaber. AgriLife Extension, Texas A&M System, 2012.

Water Storage: Tanks, Cisterns, Aquifers and Ponds for Domestic Supply, Fire and Emergency Use, Includes How to Make Ferrocement Water Tanks, by Art Ludwig. Oasis Design, 2005.

Guidelines on Rainwater Catchment Systems for Hawaii, by Patricia S. H. Macomber. College of Tropical Agriculture and Human Resources, University of Hawaii at Manoa, 2001. http://www.ctahr.hawaii.edu/hawaiirain/guidelines.html

Roof-Reliant Landscaping: Rainwater Harvesting with Cistern Systems in New Mexico, Nate Downey, principal author; Randall D. Schultz, editor. New Mexico Office of the State Engineer, 2009. http://www.ose.state.nm.us/wucp_RoofReliantLandscaping.html

Greywater Resources (chapter 3)
Create an Oasis with Greywater: Choosing, Building and Using Greywater Systems, Includes Branched Drains, 5th edition, by Art Ludwig. Oasis Design, copyright Art Ludwig, 1997-2012.

www.OasisDesign.net, www.GreywaterAction.org and www.HarvestingRainwater.com/greywater-harvesting

Sun Angles and Passive Solar Designs (chapter 4)
Sun, Wind & Light: Architectural Design Strategies, 2nd ed., by G. Z. Brown and Mark DeKay. John Wiley & Sons, 2001.

Passive Solar Architecture Pocket Reference, by Ken Haggard, David A. Bainbridge, and Rachel Aljilani. Earthscan Publications, 2009.

Passive Solar Architecture: Heating, Cooling, Ventilation, Daylighting, and More Using Natural Flows, by David A. Bainbridge and Ken Haggard. Chelsea Green Publishing, 2011.

Sun Rhythm Form, by Ralph L. Knowles. MIT Press, 1981.

Ritual House, by Ralph L. Knowles. Island Press, 2006.

Design with Climate: Bioclimatic Approach to Architectural Regionalism, by Victor Olgyay. Princeton University Press, 1963.

A Golden Thread: 2,500 Years of Solar Architecture and Technology, by Ken Butti and John Perlin. Cheshire Books, 1980.

Integrated Design (chapter 4)
Design with Nature, by Ian McHarg. John Wiley & Sons, 1995.

Introduction to Permaculture, by Bill Mollison with Reny Mia Slay. Tagari, 1991.

Gaia's Garden: A Guide to Home-Scale Permaculture, by Toby Hemenway. Chelsea Green Publishing Company, 2001.

Permaculture Mind, by Joel Glanzberg. www.PatternMind.org, 2013.

Living Building Challenge, http://living-future.org/lbc

Building Communities' Potential (chapter 5)
City Repair, www.CityRepair.org

City of Portland, Oregon's Green Street Program, http://www.portlandoregon.gov/bes/44407

www.WalkScore.org and www.LivingStreetsAlliance.org

Desert Harvesters, www.DesertHarvesters.org

The Big Orange Splot, by Daniel Manus Pinkwater. Scholastic, 1977.

Climate Change (appendix 9)
The Weather Makers: How Man Is Changing the Climate and What It Means for Life on Earth by Tim Flannery. Atlantic Monthly Press, 2005.

An Inconvenient Truth: The Planetary Emergency of Global Warming and What We Can Do About It, by Al Gore. Rodale 2006.

Losing Our Cool: Uncomfortable Truths About Our Air-Conditioned World (and Finding New Ways to Get Through the Summer), by Stan Cox. The New Press, 2010.

A Great Aridness: Climate Change and the Future of the American Southwest, by William deBuys. Oxford University Press, 2011.

"Global Warming's Terrifying New Math: Three Simple Numbers That Add Up to Global Catastrophe – and Make Clear Who the Real Enemy Is," by Bill McKibben. *Rolling Stone,* July 19, 2012.

Soils and Vegetation (foundation of the water-harvesting earthworks' living systems)
Teaming with Microbes: The Organic Gardener's Guide to the Soil Food Web, by Jeff Lowenfels and Wayne Lewis. Timber Press, 2010.

The Soil and Health: A Study of Organic Agriculture, by Sir Albert Howard, with a new introduction by Wendell Berry. University Press of Kentucky, 2006.

Food from Dryland Gardens, by David Cleveland and Daniela Soleri. Center for People, Food, and Environment, 1991. CD-ROM, 2002.

Let it Rot!: The Gardener's Guide to Composting, by Stu Campbell. Storey Publishing, 1998.

Worms Eat My Garbage: How to Set Up and Maintain a Worm Composting System, by Mary Appelhof. Flower Press, 2006.

Humanure Handbook: A Guide to Composting Human Manure, by Joseph C. Jenkins. Joseph Jenkins Inc., 2005. www.HumanureHandbook.com

"Green Burials," by Brad Lancaster. http://www.harvestingrain-water.com/2010/03/12/green-burials/

There's a Hair in My Dirt! by Gary Larson. Harper Perennial, 1999.

Appendix 7
Sun Angles and Path

This appendix enables you to determine where the sun will be in relationship to your site any time of any day of the year. Use it as a companion to Integrated Design Patterns One through Five in chapter 4. The goal is to fully understand the seasonally changing play of sun and shadow in order to design passively heated and cooled buildings and landscapes. In addition, this appendix will help you optimally place solar power systems, solar hot water systems, above-ground cisterns, trees, and water-harvesting earthworks in an integrated design that boosts their combined potential in *all* seasons. Summer shade and cooling will be maximized where needed, while winter sun exposure will be retained for free winter heat, light, and solar power. You will reduce consumption, while increasing your sustainable resource production.

This appendix is divided into three sections:

Section one provides basic guidelines for the orientation of buildings and their solar power arrays for all latitudes, and illustrates how sun paths vary by season and latitude across the globe.

Section two gives you sun path diagrams for numerous latitudes to help show where the sun will be at *any time* of *any day* of the year. Use these to design for direct solar access where and when you need sun and for shadows where and when you need shade.

Section three shows you different ways to find True North using the stars, the sun, or a compass.

SECTION ONE

SOLAR POWER GUIDELINES AND SEASONAL SUN PATHS FOR MULTIPLE LATITUDES

Buildings that are oriented correctly to the sun to maximize free/passive on-site solar power, summer cooling, and winter heating save water—as do buildings powered by wind and/or micro-hydro power. That's because, as the WCE: Water Costs of Energy chart in appendix 9 illustrates, from 0.5 to 30 gallons (1.5 to 114 liters) of water are consumed per kilowatt-hour (kWh) of electricity produced at conventional thermo-electric power plants such as coal and nuclear plants, and at hydroelectric plants, which evaporate most of that water from the reservoir behind their dam.[1]

The sun's path and a building's orientation to the sun are used to determine optimal angles for installing solar or photovoltaic (PV) panels that are fixed to the roof. For maximum efficiency, both the long axis of the house and the panels should ideally face True South in the northern hemisphere and True North in the southern hemisphere (see figures 4.5A, B in chapter 4).

True North and True South point directly to the geographic North and South Poles at the Earth's axis points, and thus relate directly to the Earth's movement and the sun's path across the planet. Thus they are more exact solar directions than magnetic/compass north and south, which point to shifting magnetic fields in the general area of the planet's poles. Magnetic/compass north and south vary from place-to-place and over time due to local magnetic anomalies, and they must be corrected by place-specific magnetic declinations. The last section of this appendix shows you different ways to find True North and South.

Panels that are fixed to a roof, facing the sun, are among the least expensive ways to utilize solar panels. A more expensive option is mounting solar panels on a moving tracker that tracks the sun to keep the panels pointing directly at the sun throughout the day. Trackers increase sunlight hitting the panels by about 10% in winter and about 40% in summer[2] when the path of the summer sun traverses more of the sky over a longer period of time.

If using fixed, roof-mounted solar panels, your panels should be angled at the same degree as the angle of your latitude. For example, if you live at 44° latitude, your panels should be angled at 44° facing the winter sun. You can increase the efficiency of the panels by 5 to 10% if you mount them to the roof with a ratchet or other mechanism that allows you to change their angle seasonally. The summer angle should then be set about 15° less than the latitude angle, and the winter angle should be set about 15° greater.[3] Thus, at 44° latitude, the summer angle of the panels will be about 29° and the winter angle will be about 59°.

In some instances you may want to modify the angle of fixed panels to accommodate your climate, seasonal weather, and electrical use patterns. For example, in coastal Oregon the winters are very wet and cloudy resulting in poor solar power production.

The summers are great with lots of sun, so it may be wise to mount the panels to maximize exposure to the summer sun angle rather than to the sparse winter sun. High seasonal electrical needs may make it beneficial to mount panels at the desired angle to favor the season with the highest power usage.

Always mount your panels at angles of 15° or steeper regardless of your latitude, weather, or power usage, so that rainfall and gravity can assist the panels in self-cleaning. Flatter-angled panels tend to accumulate more dust and other materials. For example, in figures A7.1A and B the panels are set at a seasonal angle of 15° rather than 0° to facilitate self-cleaning.

Don't worry about trying to collect all available solar energy during the day. Instead, angle panels to collect what you can in the optimal hours of sunlight between the *solar time* hours of 8 A.M. and 4 P.M. in summer and 9 A.M. and 3 P.M. in winter. Solar time is based on the sun's actual location in the sky and changes throughout the year, while *clock time* divides time into equal 24-hour days throughout the year and does not reflect the sun's location in the sky.

Figures A7.2A – A7.7B show conditions for the *northern* hemisphere; shaded areas denote where the sun does not traverse. In the southern hemisphere the panels would be directed north toward the winter-sun-side, rather than toward the south. The mounted panel angles in the figures have all been set to their optimal seasonal and self-cleaning angle. In a simpler but somewhat less efficient installation, the mounted panels would be fixed in place throughout the year at the same angle as the latitude of the site (unless there is a need to maintain a minimum 15° angle for self-cleaning).

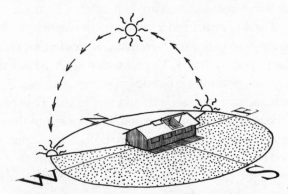

Fig. A7.1A. Sun's path at 0° latitude (the Equator) on the *summer* solstice. Panels set at 15°.
Sun rises and sets 23° N of due east and west.
Noonday sun 67° above *northern* horizon

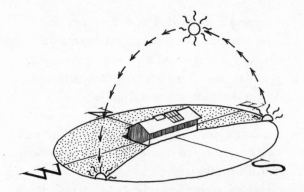

Fig. A7.1B. Sun's path at 0° latitude (the Equator) on the *winter* solstice. Panels set at 15°.
Sun rises and sets 23° S of due east and west.
Noonday sun 67° above *southern* horizon

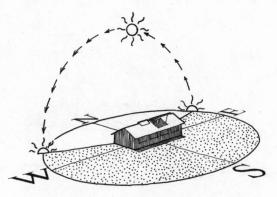

Fig. A7.2A. Sun's path at 12° N latitude on the *summer* solstice. Panels set at 15°. Sun rises and sets 24° N of due east and west. Noonday sun 79° above *northern* horizon

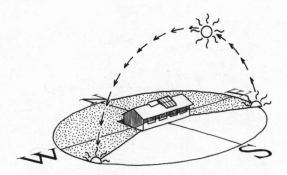

Fig. A7.2B. Sun's path at 12° N latitude on the *winter* solstice. Panels set at 15°. Sun rises and sets 24° S of due east and west. Noonday sun 55° above *southern* horizon

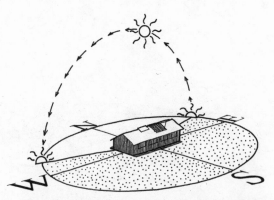

Fig. A7.3A. Sun's path at 23.5° N latitude (Tropic of Cancer) on the *summer* solstice. Panels set at 15°. Sun rises and sets 26° N of due east and west. Noonday sun 90° above *northern* horizon, directly overhead

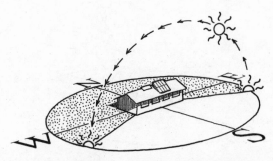

Fig. A7.3B. Sun's path at 23.5° N latitude (Tropic of Cancer) on the *winter* solstice. Panels set at 39°. Sun rises and sets 26° S of due east and west. Noonday sun 43° above *southern* horizon

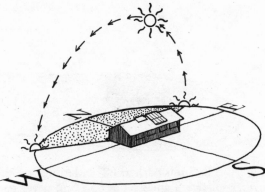

Fig. A7.4A. Sun's path at 32° N latitude on the *summer* solstice. Panels set at 20° (flush with the 20° angle of the roof). Sun rises and sets 28° N of due east and west. Noonday sun 81° above *southern* horizon

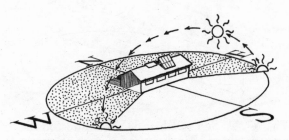

Fig. A7.4B. Sun's path at 32° N latitude on the *winter* solstice. Panels set at 47°. Sun rises and sets 28° S of due east and west. Noonday sun 35° above *southern* horizon

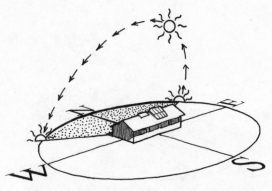

Fig. A7.5A. Sun's path at 44° N latitude
on the *summer* solstice. Panels set at 29°.
Sun rises and sets 33° N of due east and west.
Noonday sun 69° above *southern* horizon

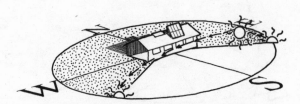

Fig. A7.5B. Sun's path at 44° N latitude
on the *winter* solstice. Panels set at 59°.
Sun rises and sets 33° S of due east and west.
Noonday sun 23° above *southern* horizon

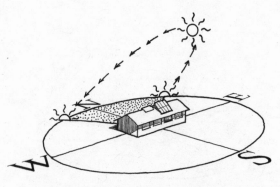

Fig. A7.6A. Sun's path at 56° N latitude
on the *summer* solstice. Panels set at 41°.
Sun rises and sets 44° N of due east and west.
Noonday sun 57° above *southern* horizon

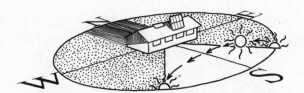

Fig. A7.6B. Sun's path at 56° N latitude on the
winter solstice. Panels set at 71°.
Sun rises and sets 44° S of due east and west.
Noonday sun 11° above *southern* horizon

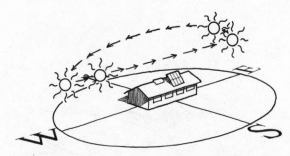

Fig. A7.7A. Sun's path at 68° N latitude on the
summer solstice. Panels set at 53°. Sun never rises
or sets, but is always up on the summer solstice.
Noonday sun 46° above *southern* horizon

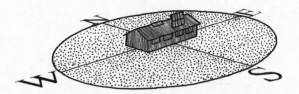

Fig. A7.7B. Sun's path at 68° N latitude on the
winter solstice. Panels set at 83°. Sun never rises
or sets, but is always below the southern horizon on
the winter solstice due to being within the Arctic
Circle, which begins at 66.5° N. Noonday sun is
approximately 1° *below* the *southern* horizon.

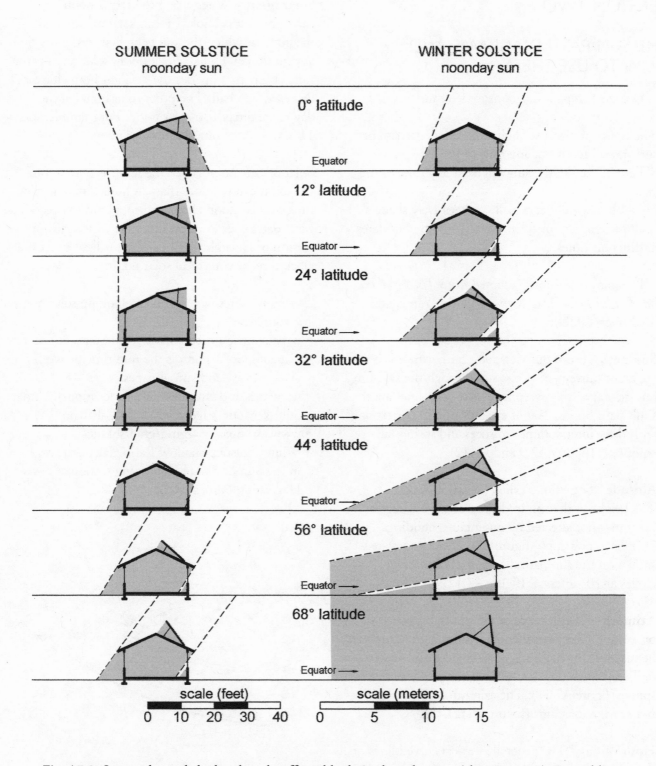

SUMMER SOLSTICE
noonday sun

WINTER SOLSTICE
noonday sun

0° latitude

Equator

12° latitude

Equator ⟶

24° latitude

Equator ⟶

32° latitude

Equator ⟶

44° latitude

Equator ⟶

56° latitude

Equator ⟶

68° latitude

Equator ⟶

scale (feet)

0   10   20   30   40

scale (meters)

0   5   10   15

Fig. A7.8. Sun angles and shadow lengths affected by latitude and season (elevation view). Dotted lines show the angle of the sun as it relates to the shadow cast by the roof and roof-mounted solar panels that are seasonally adjusted to receive more direct sun. Overhangs are the same on all images to highlight the differences in sun angles, but the overhangs could be designed to be retracted where needed for more winter sun for heating, or extended for more summer shade for cooling.

SECTION TWO

THE SUN PATH DIAGRAMS AND HOW TO USE THEM

Use the sun path diagrams to determine:

- The hours of direct sunlight a building, garden, or surface will receive at any time of the year
- Where in the sky the sun will be at any time of the year
- How objects will beneficially or negatively shade a building, garden, or other surface on different dates at different times.

Diagram definitions (adapted from *The Food and Heat Producing Solar Greenhouse* by Bill Yanda and Rick Fisher, 1980[4]):

- **Sun path** - From our viewpoint on Earth, this is the apparent movement of the sun through the sky. On our diagrams, the sweeping east-to-west lines are the sun's path on the 21st of each month, to correspond with the winter/summer solstices and spring/fall equinoxes (figures A7.9 and A7.12).

- **Altitude** - Sometimes called Elevation Angle. The height of the sun in degrees above a true horizon—not affected by mountains, buildings, or other topography (figure A7.10). Altitudes are shown on the sun path diagrams as *concentric* circles at 10° intervals (figure A7.12).

- **Azimuth** - The distance of the sun in degrees east or west of True South (equator-side) in the northern hemisphere, or the distance in degrees east or west of True North (equator-side) in the southern hemisphere (figure A7.11). The azimuth is shown on the *radii* of the diagrams (figure A7.12).

- **Hour of day** - The generally vertical curved lines on the sun-path diagram represent the *solar* (NOT daylight savings or time zone) time of the day. They are marked across the top of the sun-path line (figures A7.9 and A7.12).

- **Solar noon** – When using the term "noon" throughout this book, I mean solar or true noon—the point at which the sun is highest in the sky on any given day. Unlike clock noon, solar noon is the time of day that divides the daylight hours for that day *exactly* in half. The other solar hours of the day relate directly to solar noon. For example, solar 11 A.M. is one hour before solar noon.

Solar or true noon is not the same as *clock* noon. In fact, it can be more than an hour different. The time of solar noon changes throughout the year. The exact date, your location in a time zone, and consideration of Daylight Savings Time will all affect the clock time of your local solar noon.

Several websites where you can look up solar noon for your area:

- www.sunrisesunset.com (United States & Canada). Solar noon is the mid time between the given sunrise and sunset times.
- www.timeanddate.com (World locations). Under Sun & Moon page, click "Sun Calculator."
- www.esrl.noaa.gov/gmd/grad/solcalc/
- Winter-Solstice Shadow Ratio Calculator page in the Sun & Shade Harvesting section at www.HarvestingRainwater.com

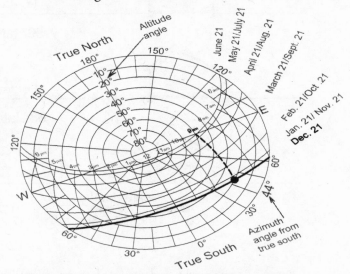

Fig. A7.9. Sun path diagram for 32° N latitude. Line representing the sun's path on December 21 is highlighted in bold. Bold dot represents where the sun will be at 9 A.M. on that date. The dot is the intersection point of the date line (solid bold) and the time line (dashed bold).

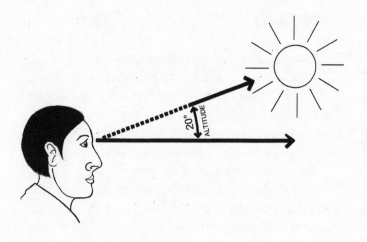

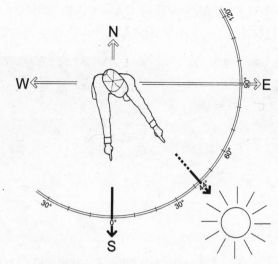

Fig. A7.10. Altitude angle (20°) of the sun above the horizon at 9 A.M. on Dec. 21 at 32° latitude. Adapted from *The Passive Solar Energy Book* by Ed Mazria, Rodale Press, 1979

Fig. A7.11. Azimuth angle (44°) of the sun east of True South at 9 A.M. on Dec. 21 at 32°N latitude. Adapted from *The Passive Solar Energy Book* by Ed Mazria, Rodale Press, 1979

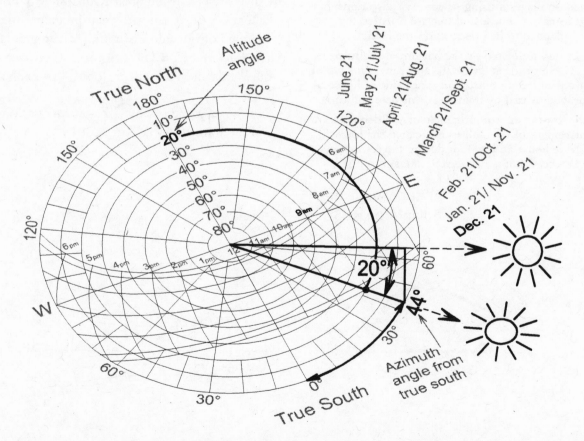

Fig. A7.12. Using the 32°N latitude sun path diagram to find the sun's altitude angle and azimuth angle for 9 A.M. on December 21. Find Dec. 21, and follow the curving line of its sun path. Find 9 A.M., and follow its line to where it intersects with the Dec. 21 date line. Now make a line from the center of the circle to that intersection point and beyond to the outermost ring of the circle to see that at 9 A.M. on Dec. 21 the sun will be 44° (azimuth angle) east of True South. Finally, look again to the date/time intersection point to see it is on the 20° (altitude angle) concentric circle, signifying that on Dec. 21 at 9 A.M. the sun will also be 20° above the horizon.

SOME WAYS YOU CAN USE THE SUN PATH DIAGRAMS

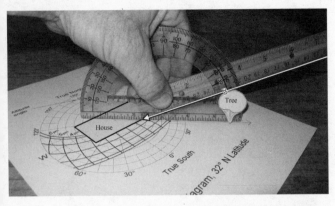

Fig. A7.13. Example of finding length and direction of shadows cast at 9 A.M. on Dec. 21 at 32° N latitude using a cut-out of a tree set in relationship to a cut-out of a house's footprint—all drawn, cut out, and placed to-scale with one another. House footprint and tree oriented on diagram using its compass directions, in the same way house and tree are oriented (or planned to be oriented) on the land.

Sunlight ray, simulated by the bottom of the ruler, is shown at an altitude angle of 20° and runs from top of the cut-out of tree to bottom of south side of house to denote shadow cast by the tree at this date and time.

Protractor resting on azimuth angle of 44° demonstrates that placement of tree and its associated shadow will not shade house at 9 A.M. or during the rest of key solar access hours between 9 A.M. and 3 P.M.

SUN PATH DIAGRAMS

Diagrams in this appendix are given for 0°, 12°, 24°, 32°, 44°, 56°, and 68° of latitude (figures A7.15 – A7.21B). Larger versions of these diagrams are given for every four degrees of latitude on the Sun Path Diagrams page in the Sun & Shade Harvesting section of www.HarvestingRainwater.com. You can also obtain a sun path diagram for your exact latitude at http://solardat.uoregon.edu/PolarSunChartProgram.html, but note that the orientation and layout of the diagram will differ somewhat from those I give you here. To use the figures provided in this appendix, refer to the diagram closest to your latitude.

SUN PATH DIAGRAM REFERENCES:

• University of Oregon Solar Radiation Monitoring Laboratory polar sun path chart program http://solardat.uoregon.edu/PolarSunChartProgram.html
• *The Food and Heat Producing Solar Greenhouse* by Bill Yanda and Rick Fisher, John Muir Publications, 1980
• *Sun, Wind, and Light* by G.Z. Brown and Mark DeKay, John Wiley & Sons, 2001

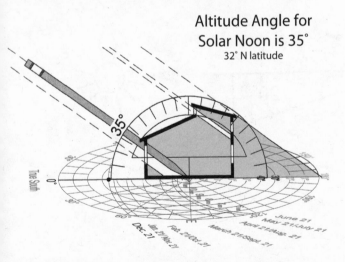

Fig. A7.14. A to-scale model or to-scale drawing of a building can be overlaid on the sun path diagram to see how changing sun angles will interface with the building's windows, walls, and interior. Here an elevation view drawing of a house illustrates the sun's *altitude* angle 35°, *azimuth* angle 0°, and the location of direct sunlight and shadows at solar noon on December 21.

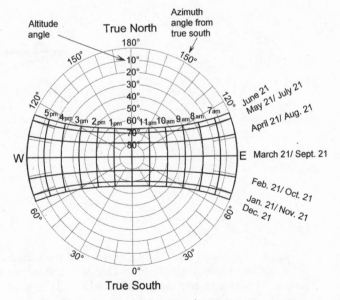

Fig. A7.15. Sun path diagram, 0° latitude (Equator)

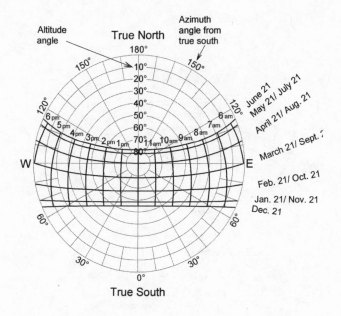

Fig. A7.16A. Sun path diagram, 12° N latitude

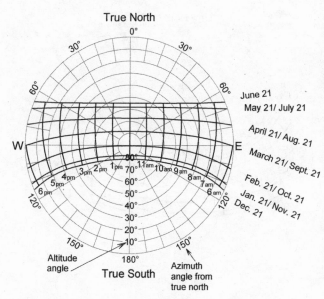

Fig. A7.16B. Sun path diagram, 12° S latitude

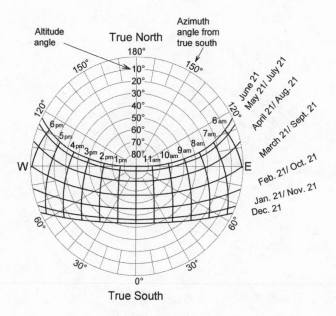

Fig. A7.17A. Sun path diagram, 24° N latitude
(Tropic of Cancer equals 23.5°)

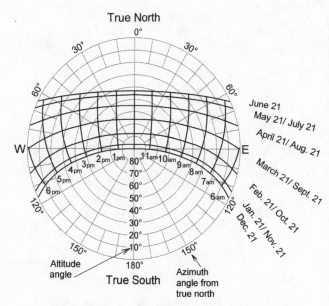

Fig. A7.17B. Sun path diagram, 24° S latitude
(Tropic of Capricorn equals 23.5°)

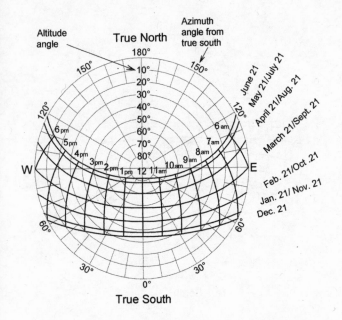

Fig. A7.18A. Sun path diagram, 32° N latitude

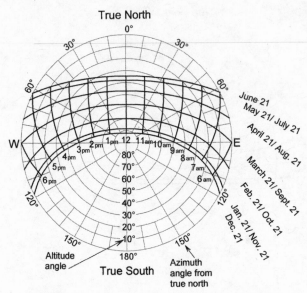

Fig. A7.18B. Sun path diagram, 32° S latitude

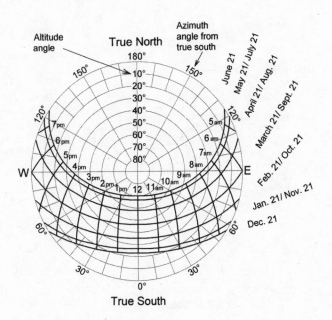

Fig. A7.19A. Sun path diagram, 44° N latitude

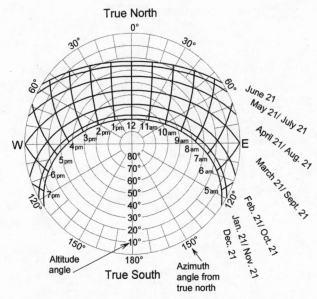

Fig. A7.19B. Sun path diagram, 44° S latitude

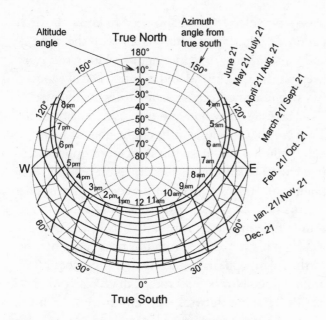

Fig. A7.20A. Sun path diagram, 56° N latitude

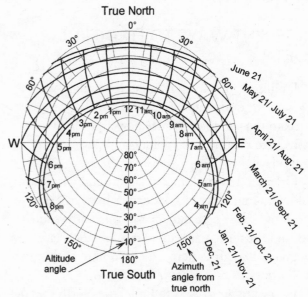

Fig. A7.20B. Sun path diagram, 56° S latitude

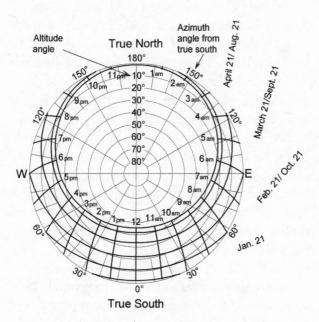

Fig. A7.21A. Sun path diagram, 68° N latitude
(Arctic Circle begins at 66.5° N)

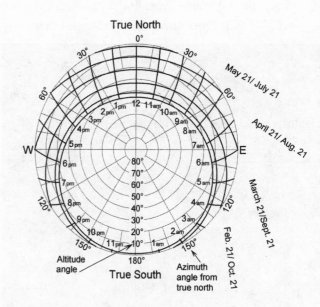

Fig. A7.21B. Sun path diagram, 68° S latitude
(Antarctic Circle begins at 66.5° S)

SECTION THREE

DIFFERENT WAYS TO FIND TRUE NORTH

Finding True North is useful for finding your way, and for optimally orienting a building or solar panels to the sun to maximize power production, passive/free heating and cooling, and passive/free winter daylight.

Orient a new building, an addition to an existing building, or an array of solar panels fixed in place, so their long-axis is perpendicular to this north-south line, and thus it has an east-west orientation (see Integrated Design Pattern One section of chapter 4 for illustrations and more information).

Below are three ways to find True North (besides knowing that at Solar Noon the sun is exactly on True South-True North line).

One: Finding True North using the North Star (*Polaris*)

This is my favorite way to find True North. I find it easy to use and more accurate than using a com-

Fig. A7.22. The North Star is located about midway between the central star of Cassiopeia and the Big Dipper. Extend a line about 5 times as long as the distance between stars Dubhe and Merak in the direction of the Big Dipper's opening to find the North Star.

pass—for which I would also need to know the declination. In addition, it connects me with the night sky and motivates me to reduce light pollution so I can keep on using it. Friend and mentor, David Omick (www.Omick.net) explains the process as follows:

There are a few limitations to using Polaris to find True North. Obviously, it has to be done at night. Bad light pollution can make this difficult or impossible (see www.DarkSky.org for strategies on reducing light pollution). It doesn't work in the southern hemisphere (there's no corresponding star that can be used in the southern hemisphere). It is not precisely accurate, although it's very close. From our perspective here on earth, Polaris appears to revolve in a circle about 0.7° around True North. This means that Polaris will be exactly True North twice a day and twice a day it will be at it's farthest point from True North. At it farthest, it is about 1.5 times the diameter of the moon disc away from True North. That's still very close.

Steps to using Polaris to find True North.

1. Find Polaris. Either use a star guide or this method: Assuming you know where the Big Dipper (scientifically known as *Ursa Major*) is, imagine a line that passes through the two stars (scientifically known as *Dubhe* and *Merak*) that form the far side of the bowl from the handle (the bowl is the part the soup would be in). Now imagine extending this line (in the direction the soup would come out of the bowl, not the other direction). Polaris will be more or less on that line and about 4 or 5 times the distance that Dubhe and Merak are from each other (figure A7.22). Note that Polaris is not as bright as Dubhe or Merak and that although Polaris is visible every night from everywhere in the northern hemisphere (unless there is too much light pollution), the Big Dipper may not be.

2. Now, drive a tall stake into the ground. This can be a wooden stake, piece of rebar, pipe, etc., and should be about 5-6 feet (1.5 m) tall. The stake should be as vertical as possible.

3. Position your eye near the ground on the south side of the stake (the side away from Polaris) and about 10 feet (3 m) away from the stake.

4. Move back and forth until your eye is in line with both Polaris and the stake. Drive another stake

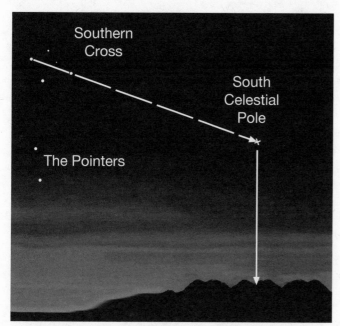

Fig. A7.23. In the southern hemisphere the two stars that make up the long axis of the Southern Cross "point" to an imaginary point, called the South Celestial Pole, which is above the South Pole. Extend an imaginary line four and a half times the length of the cross to the point of the South Celestial Pole. From this point, drop a line vertically down to the horizon. This gives you the direction of True South.

into the ground where your eye is. The two stakes are now on a True North-South line.

Two: Using the shadow cast by a stick to find True North and South

This is the most accurate way to find True North and South. At solar noon, the shadow cast by an object will be its shortest length for the day. To use this fact, follow the steps below:

1. Plant a stick *vertically* in *flat*, level ground that will be in full sun throughout the day. (The stick should stick up above the ground about 18 inches or 45 cm.) This vertical stick will be your *gnomon*, a vertical shaft casting a shadow used as an indicator.

2. On the ground (at least about an hour and a half or so before noon), mark the end of the gnomon's shadow.

3. Every 15 minutes or so, again mark the end of the gnomon's shadow. The shadow will get shorter (though the difference will be slight). At a certain point, the shadow will begin to change direction from

leaning somewhat to the west, to leaning somewhat to the east, and begin to become longer. Mark another several shadow lengths past what you think is the shortest shadow.

4. Draw a line (likely curved) in the dirt connecting all the shadow length marks you made.

5. Now tie a string to the base of your gnomon stick. Make sure the string is long enough, once tied to the gnomon, to cross over the line (connecting the shadow length markers) you just made in the dirt.

6. Tie a smaller pencil-sized stick to the open (non-gnomon) end of the string. Pull the string straight and taut, and mark a circle in the dirt (with your pencil stick kept vertical) around the gnomon.

7. Now find the two points where the circle around your gnomon intersects with the line you made when connecting the ends of the gnomon's shadows. Draw a straight line between these two intersection points—this is your approximate *east-west line* (figure A7.24A).

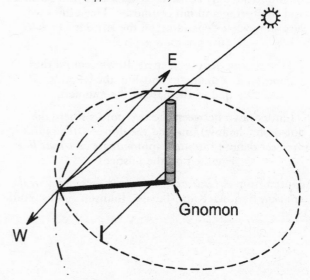

Fig. A7.24A. Finding direction with the shadow cast by a vertical stick gnomon. Dot–dash line denotes the end of the shadow cast by the gnomon, marked throughout the day.

Dotted line denotes a circle drawn around the gnomon at a diameter equaling the length of a string tied to the base of the gnomon.

A line drawn between the two points where the dot-dash (shadow) line and the dotted (string) line intersect is the site's approximate *east-west line*.

Adapted from *A Guide to Prehistoric Astronomy in the Southwest*, by J. McKim Malville, Johnson Books, 2008

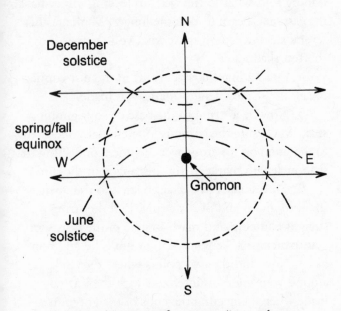

N

December
solstice

spring/fall
equinox

W

Gnomon

E

June
solstice

S

Fig. A7.24B. View of gnomon (in northern hemisphere) from above. Dot-dash lines denote the end of the shadow cast by the gnomon, marked throughout the day, at the December and June solstices and the spring and fall equinoxes. These lines will vary somewhat depending on the latitude of where the gnomon is placed.

Dotted line denotes a circle drawn around the gnomon at a diameter equaling the length of a string tied to the base of the gnomon.

Lines drawn between the two points where the dot-dash (shadow) line and the dotted (string) line intersect denote the site's approximate *east-west line* marked on the solstices.

Adapted from *A Guide to Prehistoric Astronomy in the Southwest*, by J. McKim Malville, Johnson Books, 2008

8. Draw a straight line perpendicular to the east-west line. This is your approximate *north-south line* (figure A7.24B).

Three: Using a compass and magnetic declination to find North

A compass reads magnetic north, so to find True North you'll need to adjust your compass reading by some degrees east or west based on your location's magnetic declination. Magnetic declination is the angle in degrees your compass reading needs to be corrected by in order to find True North. Magnetic declination changes from place-to-place and over time due to magnetic changes in the earth's core.

1. Use the website below to find the magnetic declination for your site.

www.magnetic-declination.com

2. Then make the necessary correction to your compass so it can point to True North. The website below further explains how to adjust your compass to your site's declination.

http://www.compassdude.com/compass-declination.shtml

Appendix 8

Wind Harvesting:
Basic Airflow Relationships, Site Selection, Wind Pumping and Wind Power, Natural Ventilation, Windbreaks, and Snow- and Biomass-Drift Harvesting

Are you considering using wind to pump water and/or generate electricity? Do you want to passively cool your building with ventilation enhanced by placement of room inlets and outlets? How about using windbreaks to cool your building in summer and reduce heat loss from your building in winter? Have you thought about using windbreaks to increase the water gained through the accumulation and melting of snowdrifts? Do you want to protect soils in order to reduce evaporation, and to shelter crops in order to reduce evapotranspiration? Do you want to reduce air-borne dust? If so, your first step is to closely observe conditions at your site and answer the following questions:

- What are the speeds and directions of your prevailing winds?
- What seasons do these prevailing winds occur in?
- How do they vary with season?
- How do they vary between night and day?
- How do they vary in different areas or microclimates of your site? Are they hot, warm, cool, or cold? Are they moist or dry?
- What do they carry? Dust/soil, snow, pollutants, fire?
- Are these winds comfortable or uncomfortable?
- What winds might you want to deflect?
- What winds might you want to harvest?

Answering these questions will start your process of understanding on-site airflow relationships. This will help you determine if and where wind-harvesting or wind-deflecting strategies might be appropriate, and where and how they could be situated to achieve maximum effectiveness at your site.

This appendix builds on your observations and illustrates basic airflow relationships and site selection guidelines. This information will deepen your understanding of the role wind plays in site design, improve you site's efficiency, and alert you to patterns you may want to investigate further in more comprehensive resources. It expands on the Wind section of the One-Page Place Assessment in chapter 2, the integrated passive heating and cooling strategies in chapter 4, and the window-specific tips such as those in box 4.4. "Window Choices Dramatically Affect Passive/Free Heating and Cooling."

Additional resources are at the Wind & Snow Harvesting page at www.HarvestingRainwater.com.

BASIC AIRFLOW RELATIONSHIPS

Work *with* wind flow as you design your site to increase or decrease wind speed, pump water, generate power, ventilate your building, and/or use strategically placed vegetation to deflect wind or harvest snow. To do so, you need to understand the flow. Good news— you're already well underway with your growing understanding of water flow. Air and water are both fluids (flowing materials) that act in similar ways.

The following relationships are drawn in part from pages 15-17 of *Sun, Wind & Light* by G.Z. Brown and Mark DeKay, 2001.[1]

1. Air velocity is slower near the surface of the earth than it is higher in the atmosphere due to friction. The rougher the ground surface—due to terrain, buildings, and vegetation—the slower the wind speed. Wind velocities measured near the ground at your site are likely to be lower than reported wind speeds that are measured perhaps 30 feet (10 m) above land surface at the nearest airport.

2. Air tends to continue flowing in the same direction when it meets an obstruction. Thus air flows like water around objects rather than bouncing off objects in random directions.

3. Wind flow, like water flow, accelerates when constricted. Wind accelerates where its constricted flow passes over and around a building, windbreak, hill, or mountain; through a gap between buildings, walls, or a saddle between two hilltops; and when channelized down a natural canyon or a canyon-like street lined with buildings.

4. Air flows from areas of high pressure to areas of low pressure. A high-pressure zone will be generated on the windward side of an obstruction—the side the wind is blowing toward—while a low-pressure zone with lower velocity will form on the leeward side of the object. This is a key relationship to understand if you want to enhance cross ventilation through a room, house, or outdoor space.

5. Cool air is denser and heavier than warm air, so warm air rises while cool air descends. This is a key relationship to understand if you want to enhance stack ventilation occurring in buildings where hot air escapes out high openings and is replaced by cool air entering through lower ventilation inlets.

SITE SELECTION

A site's climatic region and locally varying microclimates will determine if more or less wind is desirable around a home in that location. Site selection is the first strategy you should employ in working with winds and solar exposure. Figures A8.1A and B are adapted from page 88 of *Sun, Wind, & Light* by G.Z. Brown and Mark DeKay, 2001.[2] They show *general design objectives* and ideal microclimate locations for four climatic regions—cold, temperate, hot-arid, and hot-humid—that will help you in site selection and design.

Cold climate:

• Maximize the warming effects of the sun.
• Reduce the impact of winter wind.
• Ideal house microclimate is low enough on an equator-facing slope to increase solar radiation and to give protection from ridge top winds, but high enough to avoid cold air collection at bottom of valley.

Temperate climate:

• Maximize warming effects of sun in winter.
• Maximize shade in summer.
• Reduce the impact of winter wind, but allow air circulation in summer.
• Ideal house microclimate is in the middle to upper part of the slope with access to both sun and wind, but protected from high winds at top of ridge.

Hot-Arid climate:

• Maximize warming effects of sun in deep winter.
• Maximize shade from early to mid-spring through summer to mid- to late fall.
• Reduce the impact of hot dry winds in any season in which they occur, but allow air circulation at night in hot and warm months if night air is sufficiently cooling.
• Ideal house microclimate is at the bottom of the slope for exposure to cold airflow at night, and on cooler east side of hot afternoon-shading landforms and vegetation.

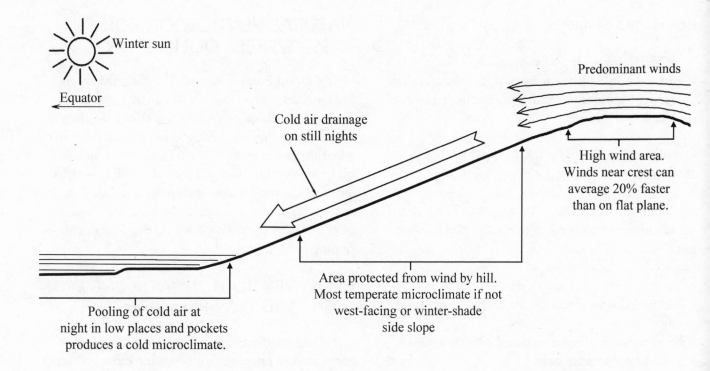

Fig. A8.1A. Basic site microclimates. Dense vegetation, walls, or earth berms can deflect cold air draining down a slope at night when placed above areas you want protected. These same strategies can collect cold air when placed downslope of areas where you want the cold air to "pond." Adapted from the *Passive Solar Architecture Pocket Reference* by D. Yogi Goswami, Ken Haggard, David Bainbridge, and Rachel Aljilani, International Solar Energy Society, 2009

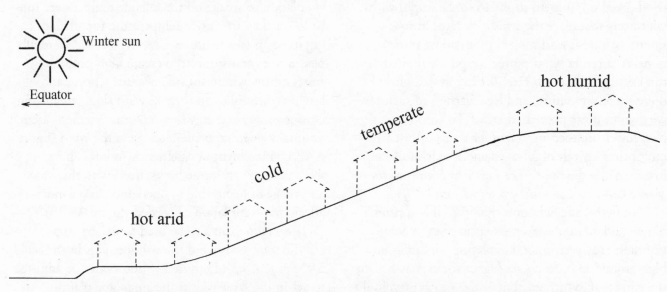

Fig. A8.1B. Basic ideal microclimates for home sites (highlighted by dotted house outlines) for different climatic regions. Adapted from *Sun, Wind & Light: Architectural Design Strategies* by G.Z. Brown and Mark DeKay, John Wiley & Sons, 2001

Hot-Humid climate:

• Maximize shade and ventilation.
• Ideal house microclimate is at the top of a slope for greater exposure to wind, and surrounded by cooling/shading vegetation.

WIND PUMPING AND WIND POWER

If you are considering pumping water or generating electricity using wind power, keep the following in mind:

• Water and electricity don't mix.
• Windmills or wind pumps are for pumping water only.
• Wind turbines or wind-electrical systems are for generating electricity only.

Windmills or wind pumps require a minimum wind speed of 4 to 7 mph (6 to 11 kmph) to begin working[3] while a wind speed of 15 to 20 mph (24 to 32 kmph) is ideal for higher capacity pumping.[4] High production, wind-electrical systems require an average wind speed of 10 to 14 mph (16 to 22 kmph) with tall towers that place the wind generator in strong, smooth winds at least 30 feet (9 m) above obstructions. As a result, wind power is typically limited to rural sites of at least an acre (0.4 ha) in size where there are fewer cultural and legal barriers to such towers than in more populated areas.[5] To construct a simpler, lower-production system for a cost of just $15, built primarily out of salvaged and locally-available materials, see the book, *The Boy Who Harnessed the Wind: Creating Currents of Electricity and Hope.*[6]

Study the Site Selection section of this appendix to get guidance on appropriate placement of wind turbines. However, note that placing windmills on high ground in order to get higher winds may situate the wind pump further from water, which is typically closer to the land surface in low-lying areas.

NATURAL VENTILATION FOR PASSIVE/FREE COOLING

As David Bainbridge and Ken Haggard say in *Passive Solar Architecture*, "Ventilation based on natural forces is quiet, requires no electricity, does not create global climate-altering gases, and works even when the power is off."[7] Ventilation can simultaneously improve comfort and reduce building costs, especially in conjunction with passive summer-shading and winter-sunning strategies, by reducing the size— or eliminating the need—for a mechanical air conditioning system.

CROSS VENTILATION: WORKING WITH HIGH- AND LOW-PRESSURE ZONES

Understanding high-pressure and low-pressure zones are key principles for designing cross-ventilation strategies for buildings. Cross ventilation works best when there are inlet openings perpendicular to the wind (though the openings can be as much as 40° off perpendicular) on the windward, high-pressure side of a building, and outlet openings on the leeward, low-pressure side. Blowing wind is necessary for cross ventilation to work and the outside temperature must be lower than the inside temperature for efficient cooling to occur (see figures A8.2A to A8.7B).[8] Carefully placed trees and shrubs can create high-pressure zones on the windward side of obstructions, and low-pressure zones on their leeward sides, enhancing cross-ventilation even when building openings are not optimally oriented perpendicular to the wind (figure A8.2A). Placement of casement windows and wing walls can have the same basic effect, with the added advantage of being able to open and close windows to affect cross-ventilation as needed (figure A8.2B).

Vegetation can also be used to deflect wind from a building, or focus wind toward openings in the building (figure A8.3). Louvered windows can be adjusted to act in the same way as the landscape plantings to direct air up or down into a living space.

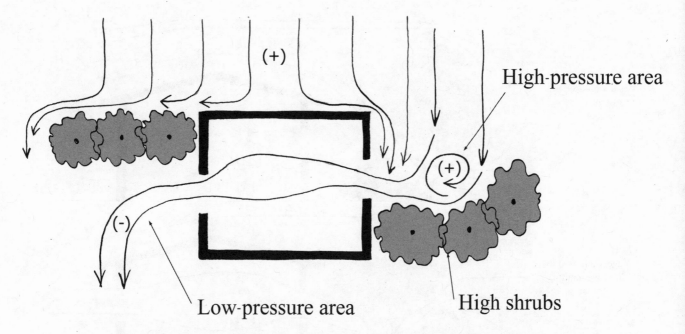

Fig. A8.2A. Enhancing cross-ventilation with trees and shrubs to create high-pressure zones on the windward side, and low-pressure zones on the leeward side, of obstructions. Adapted from the *Passive Solar Architecture Pocket Reference*

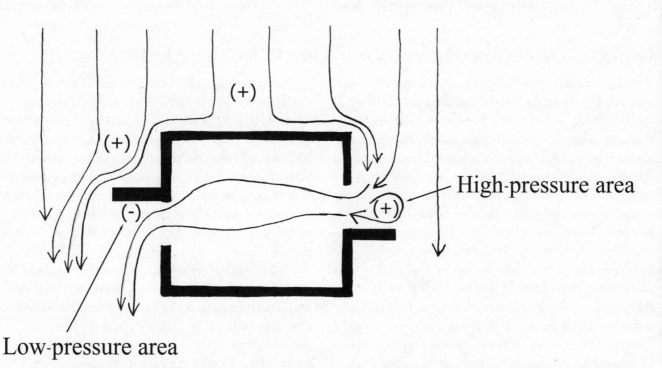

Fig. A8.2B. Enhancing cross-ventilation with casement windows or wing walls to generate the same effect as the vegetation in A8.2A. Windows can open and close to increase or decrease ventilation as needed.

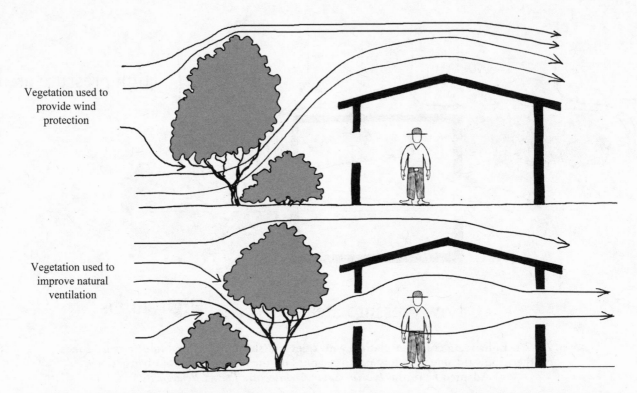

Vegetation used to provide wind protection

Vegetation used to improve natural ventilation

Fig. A8.3. Vegetation can be arranged to deflect wind over a building or to cool and direct airflow through it. Note that interior ventilation is greatly enhanced with more than one wall opening. All other things being equal (wind direction, vegetation, and so forth), the top example with only one opening will be cave-like, with less ventilation, compared to the bottom example. Adapted from the *Passive Solar Architecture Pocket Reference*

WINDOW AND DOOR ORIENTATION

Operable windows and doors with louvered shutters, blinds, or screens help adjust the airflow in individual rooms and whole houses. Healthy ventilation and multi-dimensional natural light that reduces glare are enhanced when rooms have at least two openings on different walls. Multi-directional openings and open floor plans improve cross-ventilation to freshen the air and reduce interior mold.[9] Cave-like rooms with just one opening to the outside can have less than half the ventilation of comparable rooms with openings on two sides (compare upper [only one opening] and lower example in figure A8.3).[10] If the prevailing wind direction is askew to a room, the windows should be on opposite sides of the room to avoid creation of still-air pockets and to increase the interior air speed (figure A8.4). If the room is oriented to face directly into the wind, the windows should be on adjacent walls to get maximum ventilation.

INLET AND OUTLET TREATMENTS

Given a perpendicular alignment of a room to the wind, airflow within the room having windows on opposite walls is largely determined by the inlet location and its relationship to the exterior surfaces of the building, and the resulting high- and low-pressure areas (figure A8.5). If using passive cooling/ventilation to cool a room at night, it is important to bathe the thermal mass of the *interior walls* with cool night air using a technique such as shown in figures A8.2 through A8.8.

Inlet configurations can affect airflow in both horizontal and vertical directions (figures A8.5 and A8.6A and B). The size of outlet openings compared to inlet openings largely determines the speed of airflow in different parts of the room (figure A8.7A). When windows are located at different elevations within a room rather than at the same elevation, the area of the room experiencing airflow increases while the speed of airflow is reduced (figure A8.7B).

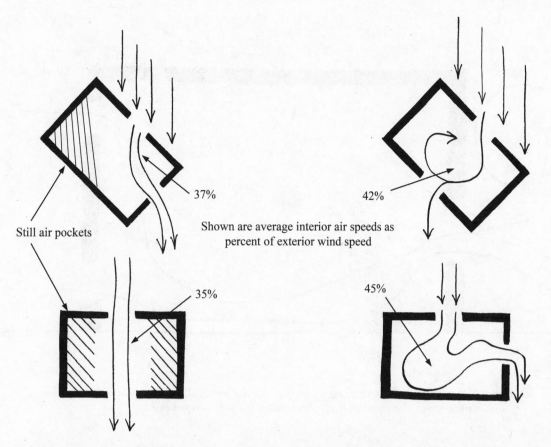

Still air pockets

37%

42%

Shown are average interior air speeds as
percent of exterior wind speed

35%

45%

Fig. A8.4. Variations of window placement in relationship to a building's orientation to the prevailing
wind direction, and the resulting differences in interior wind speed and disbursement.
Adapted from the *Passive Solar Architecture Pocket Reference*

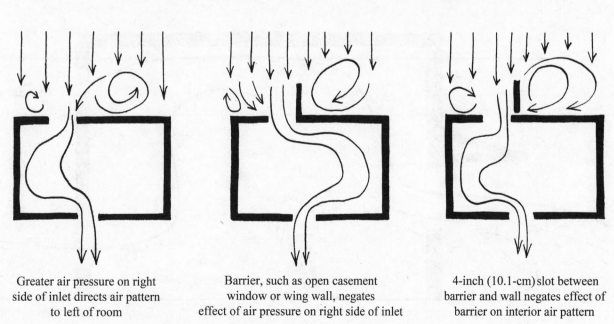

Greater air pressure on right
side of inlet directs air pattern
to left of room

Barrier, such as open casement
window or wing wall, negates
effect of air pressure on right side of inlet

4-inch (10.1-cm) slot between
barrier and wall negates effect of
barrier on interior air pattern

Fig. A8.5. Inlet configurations and exterior building surfaces
that affect airflow in the horizontal dimensions within a room.
Adapted from the *Passive Solar Architecture Pocket Reference*

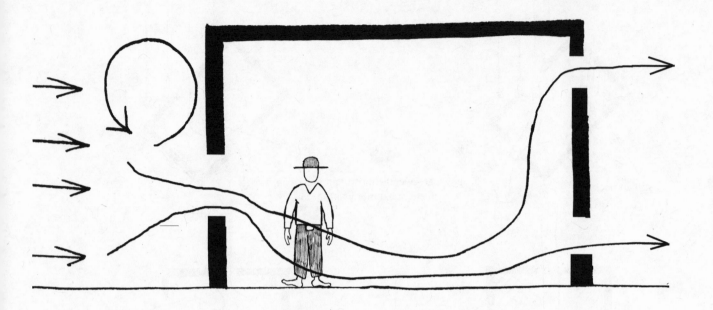

Fig. A8.6A. Inlet and outlet configurations that affect airflow in the vertical dimensions within a room.
Adapted from the *Passive Solar Architecture Pocket Reference*

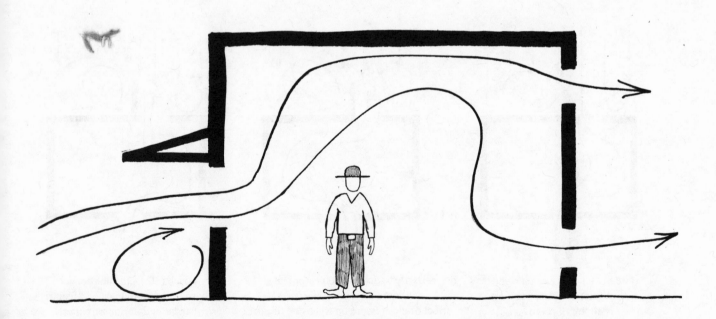

Fig. A8.6B. Overhangs can have the same effect in the vertical dimension that wing walls,
open casement windows, and other barriers have in the horizontal dimension.
Adapted from the *Passive Solar Architecture Pocket Reference*

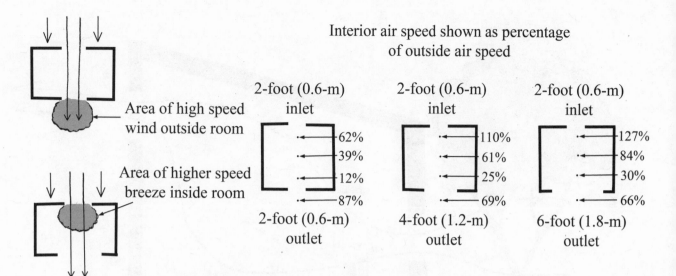

Interior air speed shown as percentage of outside air speed

Area of high speed wind outside room

Area of higher speed breeze inside room

2-foot (0.6-m) inlet
·→ 62%
·→ 39%
·→ 12%
←— 87%
2-foot (0.6-m) outlet

2-foot (0.6-m) inlet
·→ 110%
·→ 61%
·→ 25%
←— 69%
4-foot (1.2-m) outlet

2-foot (0.6-m) inlet
·→ 127%
·→ 84%
·→ 30%
←— 66%
6-foot (1.8-m) outlet

Fig. A8.7A. Effects of inlet and outlet opening sizes on room air speed. Adapted from the *Passive Solar Architecture Pocket Reference*

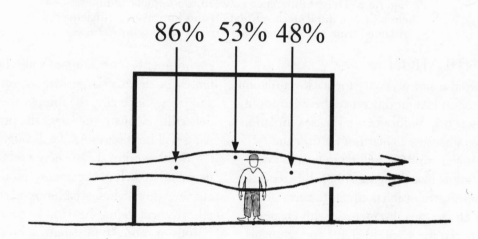

86% 53% 48%

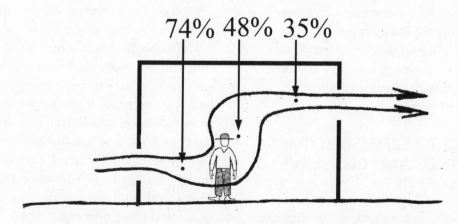

74% 48% 35%

Fig. A8.7B. Changes in location and speed of airflow with differential window heights. Interior air speed is shown as percentage of outside air speed. Adapted from the *Passive Solar Architecture Pocket Reference*

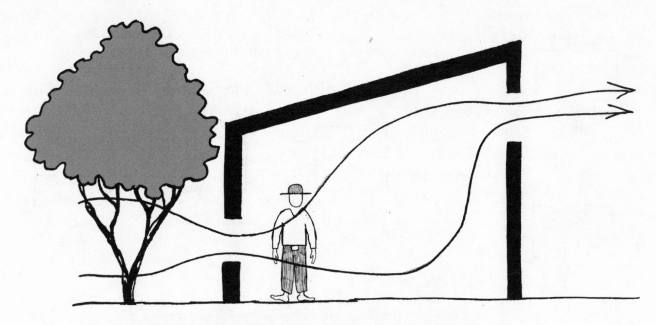

Fig. A8.8. Height difference between room inlets and outlets
help induce natural stack ventilation and improve its efficiency.
Adapted from the *Passive Solar Architecture Pocket Reference*

STACK VENTILATION

Blowing wind is not necessary for stack ventilation to occur in a room. Height differences between room inlets and outlets help induce natural stack ventilation during still times because lighter hot air rises and flows out the upper window, pulling in heavier cool air into the window below (figure A8.8). Efficiencies improve when the area of inlets equals the area of outlets. These efficiencies also increase with greater differences between the vertical heights and temperatures of higher outlets and lower inlets. Stack ventilation works best when the exit air is warmer than the outside air.[11] Black-colored ventilation chimneys heat rising air, improving the stack ventilation process. High ceilings and transom windows, open floor plans, multiple floors, open stairwells, ventilation shafts, and dormers with operable windows all enhance stack ventilation.[12]

WINDBREAKS TO DIFFUSE HOT SUMMER WINDS AND CONSERVE WINTER HEAT

Windbreaks can protect sites from the negative effects of winds that are too hot, too cold, or too strong. Windbreaks can be part of the built environment—consisting of walls, buildings, or fences, or they can be grown—consisting of shrubs and trees. Generally, the thinner the adjacent wind-deflecting element, the larger the protected area down-wind will be (figures A8.9A through F).

Solid windbreaks, such as a solid wall or a very dense shelterbelt of vegetation, produce strong eddy currents downwind, which negate much of their effectiveness (figure A8.10). To prevent this, use a porous or more "transparent" windbreak such as a slatted fence or a more sparsely planted shelterbelt of vegetation instead. Similarly, louvered windows can be opened on the windward side of a building to let in, direct, and diffuse strong winds.

Windbreaks of evapotranspiring vegetation help cool, diffuse, and humidify strong, hot, dry winds in summer. Conversely, by avoiding or deflecting cold, heat-robbing winds in winter, windbreaks and/or protected site placement can save up to 40% in heating costs.[13] For a typical house, winter heat loss can be over twice as great in a 5 mph (8 kmph) wind as in no wind due to air infiltration through attics, crawl spaces, and chimneys and poorly sealed windows, doors, and other openings.[14] The heating load of an unprotected house exposed to a 20 mph (32 kmph) winter wind is over twice as great as the heating load

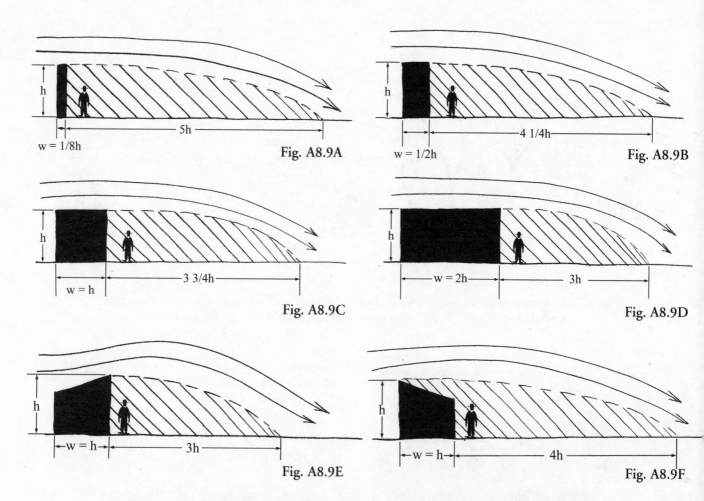

Fig. A8.9 A through F. Wind-deflecting walls, buildings, fences, or vegetation can provide wind protection when located upwind of an impacted site. The thinner the wind-deflecting element, the larger the protected area will be downwind.
Adapted from the *Passive Solar Architecture Pocket Reference*

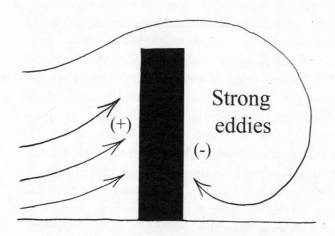

Fig. A8.10A. Strong eddies formed by a solid windbreak. Adapted from the *Passive Solar Architecture Pocket Reference*

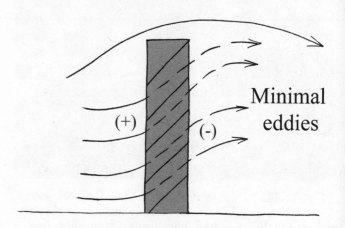

Fig. A8.10B. Minimal eddies formed by a porous windbreak. Adapted from the *Passive Solar Architecture Pocket Reference*

Fig. A8.11. A well-designed windbreak/shelterbelt creates eddies and slows winds to induce snow deposition under the vegetation, to help increase soil moisture there.

Fig. A8.12. Barbed wire fence on Diné lands in northern Arizona sufficiently slows the wind that is blowing perpendicular to the fence (from left to right) for tumbleweed to drop out and accumulate on the leeward side of the fence.

for the same house protected by a windbreak that reduces that 20 mph (32 kmph) wind to just 5 mph (8 kmph).[15]

SNOW- AND BIOMASS-DRIFT HARVESTING

Vegetation can be a highly effective windbreak, especially a combination of trees and shrubs. In the well-designed vegetative windbreak or *shelterbelt* shown in figure A8.11, the wind speed slows as it passes through the trees, dropping the snow it carries within the windbreak/shelterbelt, while deflecting the strongest winds over the house. The resulting snowdrift will accumulate under the trees, melt slowly, and can provide more than double the annual moisture available to the associated vegetation. Yet if this windbreak/shelterbelt was too dense, it could produce similar strong eddy currents downwind, as occur downwind of a solid wall. This could direct the snowdrift onto the house it is meant to protect.

In a similar manner, fences can filter and harvest biomass and mulch carried by the wind that passes through them (figure A8.12). In winter the same process enables the fence to harvest wind-borne snow, creating snowdrifts along the fence. See the Wind & Snow Harvesting page at www.HarvestingRainwater. com for more information on designing shelterbelts that harvest wind-borne snow where you want it.

Appendix 9
Water-Energy-Carbon Nexus

You don't have to believe in climate change to solve it. Everything we do to raise energy efficiency will make money, improve security and health, and stabilize climate.

—Amory Lovins, *Reinventing Fire: Bold Business Solutions for the New Energy Era*

OUR CALL TO ACT

The Water-Energy-Carbon Nexus charts and examples in this appendix show us how the cumulative results of our decisions and actions affect and connect us all. If we strive to understand those connections, and then take informed action, we can make choices that improve life for ourselves and countless others. For example, when we shift to on-site harvested waters and renewable energy, we simultaneously reduce the energy that would otherwise be consumed to pump in imported water, and we decrease the water that would otherwise be consumed to generate our electricity at a distant power plant. This leaves more resources and potential for ourselves, our children, and others—while reducing the emission of carbon dioxide and other pollutants which damage our health, our communities, our ecosystems, our economies, and our futures.

There are many voices calling for us to make better choices. To confirm or correct your path amidst this chorus, keep this in mind: *You will know you are on the right track when domino-style, **positive** impacts beyond what you originally intended, and far exceeding any potential negative impacts, keep springing forth from your choices and actions.* For example, when you harvest your on-site water even if only for conservation purposes, you also equip yourself to irrigate—for free—well-placed climate-appropriate plantings and sponge-like mulched soils. In turn, these plants and soils equip you to passively heat, cool, shelter, and feed yourself and your community. At the same time, they filter out (or bioremediate) environmental contaminants, lure you to be more active outside, improve your health, enable you to see and get to know more of your neighbors, and connect you to a community of interacting people and natural processes. Simultaneously, your water harvesting reduces local flooding, improves stormwater quality, sequesters climate-changing carbon dioxide, and might even help reduce flood- and fire-insurance costs. The impacts don't stop there for they are truly global in scale due to our shared atmosphere and climate.

This appendix focuses on our atmosphere and climate by looking at carbon dioxide (CO_2) emissions from human activities because, by mass (weight), carbon-dioxide emissions represent the greatest proportion of climate-changing gases in our atmosphere. In addition, carbon dioxide is the baseline unit to which all other climate-changing gases are compared in determining their Global-Warming Potential (box A9.1). Carbon-dioxide levels in the atmosphere are rising at an unprecedented rate, contributing to many manifestations of climate change. These include abnormal fluctuations in temperatures around the world; an overall rise in global temperature; more-frequent extreme weather and geologic events; broad-scale changes in precipitation amounts and patterns, and ensuing crop failures; rising sea

To understand the magnitude of a global temperature increase of 3.6° F (2° C), consider this analogy by Anthony Leiserowitz, director of the Yale Project on Climate Change Communication:

Think of the planet as your body, which when healthy maintains a temperature of around 98.7 °F (37° C). When you get sick and get a fever of 1° F (0.5° C) you feel a little off, but you will likely still go to work or school. With a 2° F (1.1° C) fever you feel sick with hot flashes and cold chills, likely keeping you home from work or school. At 3° F (1.6° C) you are starting to get really sick. And with a sustained fever of 4° to 5° F (2.2° to 2.77° C) your brain starts to slip into a coma, and you are close to death.[7]

Fond of life on Earth as we know it? See box A9.2, and read on. Interested in enhancing life? Good news. We currently have the solutions needed to address climate change, but for them to work we must put them into action *now*. You (all of us) have the power to act, demand change, and to show others a better way through your own choices and example!

The combined resource of the charts in this appendix is one tool to use, as are this book's water-harvesting principles, ethics, site-assessment guidelines, integrated-design patterns, and multiple other strategies that turn overlooked or forgotten opportunities and local "wastes" into local zero- to low-carbon, zero- to low-energy, zero- to low-cost resources enhancing your community's and the world's resilience and potential.

THE WATER-ENERGY-CARBON (W-E-C) NEXUS CHARTS

These charts help make the invisible visible. Because the treatment and transport of water is largely unseen by us end-users, we are unaware that when we consume this pumped water, we consume energy. Likewise, the unseen generation of that energy consumes additional water and typically emits polluting, climate-altering carbon dioxide (CO_2). These connections are brought to light by the following:

- The *Water* Costs of Energy (**WCE**) chart
- The *Energy* Costs of Water (**ECW**) chart
- The *Carbon* Costs of Energy (**CCE**) chart

levels and subsequent flooding; accelerated melting of glaciers, which means water shortages for those dependent on their seasonal runoff; melting ice caps, resulting in significant changes in oceanic conditions and death of sea life—and the list goes on.

In 1750, before the start of the Industrial Revolution, total carbon-dioxide levels in the atmosphere were about 280 parts per million (ppm), a level, which according to geologic measurements such as those taken from ice cores, had not been exceeded for over 400,000 years.[3] In 2012, carbon-dioxide levels hit 394 ppm, and these levels are continuing to increase *at an accelerating rate*; the rate in 2012 was about a 2 ppm increase per year.[4] An overwhelming number of climate scientists and environmental thinkers have targeted 350 ppm as a safe concentration of atmospheric carbon dioxide. They warn that unless we rapidly take action to return to levels *below* 350 ppm during the current century, we risk reaching tipping points, such as a global rise in average temperature exceeding 3.6° F (2° C) causing irreversible impacts such as catastrophic climate change.[5, 6]

Box A9.2. Three Simple Numbers That Add Up to Global Catastrophe... or Aversion of That Catastrophe

According to Bill McKibben in his *Rolling Stone* article, "Global Warming's Terrifying New Math: Three simple numbers that add up to global catastrophe – and that make clear who the real enemy is," there are three things we must do if we are to avert irreversibly altering the relative climatic stability to which we humans, our livestock, our crops, and our cultures have adapted.

1. We must keep our cumulative global-temperature increase below **2° C** (3.6° F).
Thus far we've raised global temperatures 0.8° C (1.44° F). Even if we were to stop increasing carbon dioxide emissions now, temperatures would be expected to rise another 0.8° C (1.44° F) due to the effect of the CO_2 already emitted by human activity into the atmosphere.[8]

2. We cannot put more than **565 gigatons** of additional carbon dioxide into the atmosphere by midcentury if we are to have any reasonable hope of keeping our global-temperature increase below two degrees Celsius. Our current carbon emissions are on track to increase 3% per year, and at that rate, "we'll blow through our 565-gigaton allowance in 16 years, around the time today's preschoolers will be graduating from high school."[9] And if we continue this trend beyond that 565-gigaton allowance we will be in line with a temperature increase of about 6° C (10.8° F), "which would create a planet straight out of science fiction."[10]

3. We must take effective action that would keep locked underground 80% of the **2,765 gigatons** of carbon dioxide that would otherwise be released into the atmosphere if we burned the proven coal, oil, and gas reserves of the fossil-fuel companies, and oil-rich countries such as Venezuela and Kuwait. That 2,765 gigatons is five times the 565-gigaton limit climate scientists think is safe to burn.[11]

See McKibben's article and www.350.org for how he recommends we can achieve these three actions above.

Use the information within all the W-E-C nexus charts to approximate your current consumption of hidden water and power, as well as your associated carbon emissions, and then choose:

- energy sources and practices that consume less or no water,
- water sources and practices that consume less or no energy, and
- energy and water sources and practices that emit less or no CO_2.

To orient yourself to the charts, note that each one has three main sections of numbers:

The first section/column on the Water Costs of Energy chart shows how much water is in your energy, on the Energy Costs of Water chart—how much energy is in your water, and on the Carbon Costs of Energy—how much CO_2 is emitted by the energy source(s) used to provide your electricity or used to power the pumping and treatment of your water. Taken one gallon (liter), kilowatt-hour, or

pound (kilogram) at a time, the numbers in each first section might seem small...

However, the second section on each chart scales these numbers up from conceptual to more practical levels, showing how much hidden energy or water it takes, or how much CO_2 is created, to provide an average household's water or power supply for one month. (These charts assume the average U.S. household size is 2.59 people.) If the household numbers still don't cause you concern, remember that just as pennies make dollars, households make communities...

The third section on each chart shows the dramatic cumulative effects of large-scale resource use, multiplying the hidden water for energy, energy for water, and CO_2 emissions attributable to one household by a factor of *100,000*. From there, you can continue to expand up to regional, state, national, and world figures, based on the number of households in your region.

Notes indicated with symbols and letters appear in this appendix; numbered notes are in References.

THE *WATER* COSTS OF ENERGY (WCE) – U.S. units

		How many gallons of water are used[a] to produce **one kilowatt-hour** of electricity? Renewable sources in italics	How many gallons of water are used[a] to produce electricity for **one household** for one month?			How many gallons of water are used[a] to produce electricity for **100,000 households** for one month?	

Average household **kWh/month**:[1,2] **Arizona** = 1,095, **U.S.** = 920, **World** = 240

Ranges of averages are shown: high (darker) & low (lighter). Overall means are not necessarily the means of the given extremes.

Origin	gal/kWh	Arizona	U.S.	World	U.S.	World
Hydroelectric [b,3,4,5,6,7]	30.078	32,935	27,672	7,219	2,767,176,000	721,872,000
	4.500	4,928	4,140	1,080	414,000,000	108,000,000
Geothermal [♦4,6,7]	1.695	1,856	1,559	407	155,940,000	40,680,000
	0.600	657	552	144	55,200,000	14,400,000
Solar Thermal/CSP [♦c,3,4,5,6,7]	0.920	1,007	846	221	84,640,000	22,080,000
	0.750	821	690	180	69,000,000	18,000,000
Nuclear [♦3,4,5,6,7,8]	0.785	860	722	188	72,220,000	18,840,000
	0.400	438	368	96	36,800,000	9,600,000
Biomass [♦5,6,7]	0.665	728	612	160	61,180,000	15,960,000
	0.300	329	276	72	27,600,000	7,200,000
Natural Gas (ST) [♦d,6]	0.645	706	593	155	59,340,000	15,480,000
Coal [♦3,4,6,7,8]	0.560	613	515	134	51,520,000	13,440,000
	0.300	329	276	72	27,600,000	7,200,000
Municipal Solid Waste [♦5]	0.480	526	442	115	44,160,000	11,520,000
	0.300	329	276	72	27,600,000	7,200,000
Oil [♦5,7]	0.480	526	442	115	44,160,000	11,520,000
	0.300	329	276	72	27,600,000	7,200,000
Landfill Gas [e,3,4,9]	0.350	383	322	84	32,200,000	8,400,000
	0.010	11	9	2	920,000	240,000
Natural Gas (CC) [♦d,3,4,5,6,7,8]	0.195	214	179	47	17,940,000	4,680,000
	0.100	110	92	24	9,200,000	2,400,000
Natural Gas (GT) [d,4,6,7]	0.050	55	46	12	4,600,000	1,200,000
	0.010	11	9	2	920,000	240,000
Solar PV [f,3,4,5,6,7,8,9]	0.001	1	1	0	92,000	24,000
	0.000	0	0	0	0	0
Wind [f,4,5,6,7,8,9]	0.001	1	1	0	92,000	24,000
	0.000	0	0	0	0	0
Micro-Hydroelectric [b]	0.000	0	0	0	0	0
Average [4,7]	0.640	701	589	154	58,880,000	15,360,000
	0.570	624	524	137	52,440,000	13,680,000

Available at: HarvestingRainwater.com/water-energy-carbon-nexus

WCE FACTS

39% of fresh water withdrawn in the U.S. is for **thermoelectric power-generation cooling systems**.[5]

Most thermoelectric power plants are only **33% efficient**, which means **2/3 of heat-energy potential is lost**.[5]

Of the **total kWh usage** for the United States, 37% goes to **residential**, 36% to **commercial**, and 27% to **industrial** purposes.[1]

THE *WATER* COSTS OF ENERGY (WCE) – metric units

	How many liters of water are used[a] to produce **one kilowatt-hour** of electricity?	How many liters of water are used[a] to produce electricity for **one household** for one month?			How many liters of water are used[a] to produce electricity for **100,000 households** for one month?	
	Renewable sources in italics	Average household **kWh/month**:[1,2] **Arizona** = 1,095, **U.S.** = 920, **World** = 240				
	Ranges of averages are shown: high (darker) & low (lighter). Overall means are not necessarily the means of the given extremes.					
Origin	liter/kWh	Arizona	U.S.	World	U.S.	World
Hydroelectric [b,3,4,5,6,7]	113.857	124,674	104,749	27,326	10,474,868,030	2,732,574,269
	17.034	18,653	15,672	4,088	1,567,155,600	408,823,200
Geothermal [♦4,6,7]	6.416	7,026	5,903	1,540	590,295,276	153,990,072
	2.271	2,487	2,090	545	208,954,080	54,509,760
Solar Thermal/CSP [♦c,3,4,5,6,7]	3.483	3,813	3,204	836	320,396,256	83,581,632
	2.839	3,109	2,612	681	261,192,600	68,137,200
Nuclear [♦3,4,5,6,7,8]	2.972	3,254	2,734	713	273,381,588	71,316,936
	1.514	1,658	1,393	363	139,302,720	36,339,840
Biomass [♦5,6,7]	2.517	2,756	2,316	604	231,590,772	60,414,984
	1.136	1,244	1,045	273	104,477,040	27,254,880
Natural Gas (ST) [♦d,6]	2.442	2,674	2,246	586	224,625,636	58,597,992
Coal [♦3,4,6,7,8]	2.120	2,321	1,950	509	195,023,808	50,875,776
	1.136	1,244	1,045	273	104,477,040	27,254,880
Municipal Solid Waste [♦5]	1.817	1,990	1,672	436	167,163,264	43,607,808
	1.136	1,244	1,045	273	104,477,040	27,254,880
Oil [♦5,7]	1.817	1,990	1,672	436	167,163,264	43,607,808
	1.136	1,244	1,045	273	104,477,040	27,254,880
Landfill Gas [e,3,4,9]	1.325	1,451	1,219	318	121,889,880	31,797,360
	0.038	41	35	9	3,482,568	908,496
Natural Gas (CC) [♦d,3,4,5,6,7,8]	0.738	808	679	177	67,910,076	17,715,672
	0.379	415	348	91	34,825,680	9,084,960
Natural Gas (GT) [d,4,6,7]	0.189	207	174	45	17,412,840	4,542,480
	0.038	41	35	9	3,482,568	908,496
Solar PV [f,3,4,5,6,7,8,9]	0.004	4	3	1	348,257	90,850
	0.000	0	0	0	0	0
Wind [f,4,5,6,7,8,9]	0.004	4	3	1	348,257	90,850
	0.000	0	0	0	0	0
Micro-Hydroelectric [b]	0.000	0	0	0	0	0
Average [4,7]	2.423	2,653	2,229	581	222,884,352	58,143,744
	2.158	2,363	1,985	518	198,506,376	51,784,272

WCE FACTS (continued)

The country with the **lowest per-capita monthly kWh usage** is **Haïti**: 2 kWh. **Iceland's** is highest: 4,172 kWh. **Jordan**: 174 kWh, **China**: 205 kWh, **France, Germany, & Japan**: ~625 kWh, & **Australia**: 935 kWh.[2]

Available at: HarvestingRainwater.com/water-energy-carbon-nexus

WCE NOTES

Data do not include water used in association with extraction or production of raw energy sources or with lifecycle of power-generating infrastructure (construction of facilities, manufacture & transport of equipment, etc).

◆ These water-for-energy data are for wet-cooled power generation *only*. Wet cooling is a method of transferring waste heat to the atmosphere from water used in power generation. The water is cooled by its reduction to a fine spray, allowing the discharge of heat through evaporation.[10]

a. Regions' monthly water-for-energy quantities are calculated based on U.S. water-for-energy data and region-specific average energy usage. However, each region's actual water-for-energy quantities will vary based on local power-generation specifics, including type of cooling system.

b. Unlike hydroelectric power generation, in which the movement of water flowing over large dams turns turbines to generate power, micro-hydro's turbines are placed in-stream and do not require creation of reservoirs from which large amounts of water are lost to evaporation.

c. CSP = concentrated solar power, a form of solar-thermal energy that uses solar-tracking mirrors or lenses to focus a large area of sunlight onto a small area. The light energy is converted to heat, which is applied to water to create steam to turn turbines, and thus generate electricity via conventional thermoelectric methods.[11]

d. ST = steam turbine. Fuel is combusted to heat water to create steam to turn turbines.

CC = combined cycle. Exhaust of one heat engine is used as heat source for another. This dual use of heat increases system's overall efficiency, but water consumption is higher than in a gas-turbine natural-gas system.[12]

GT = gas turbine (a.k.a., combustion turbine, or single cycle). Force from combustion of fuel turns turbine.

e. As with all power generation, the water costs of landfill gas-generated energy depend on the technology used. We use the same low-end figure of 0.010 gal/kWh (*0.038 liters/kWh*) that is cited for Natural Gas (GT); other technologies use more water. Despite attempts to contact the authors of reference 3, we were unable to confirm the technology behind their Landfill-Gas figure.

f. Solar PV and wind systems consume water if rainfall is not sufficient to wash panels or turbine blades, and if the systems store power in water-filled batteries.

The 0.001 gal/kWh (*0.004 liters/kWh*) figure for Solar PV, given by Kevin Koch, owner of Technicians for Sustainability, was carried over as an estimate for Wind, as the sources that consider wind power to consume water[5,9] state only that it is used in minimal quantities.

CREDITS: Brad Lancaster, Resource concept, oversight | **LeeAnn Lane**, Research | **Megan Hartman**, Research, resource creation | **Brandy Lellou, NV-OC.org**, Research, peer review

Take time now to look over each chart. Note the differences in water, energy, and carbon costs based on water and energy source. Then ask yourself: Where do *your* water, energy, and carbon costs fit on the three charts?

Note 1: Throughout this appendix, I give values in U.S. and metric units, but you may notice some discrepancies if you try to convert directly between the given numbers in the U.S. and metric versions of each chart. This is because initial values on which the metric charts are based (in the first section of each chart), were derived by converting U.S.-unit figures to metric figures; the numbers were then rounded for ease of use. Then, to get from the first to the second and third sections of the charts, metric calculations (in the metric charts) were then carried out parallel to U.S.-unit calculations (in the U.S. unit charts), rather than just converting U.S.-unit calculation results into their metric counterparts. For this reason, the more layers of calculations, the greater one can expect the discrepancies between the metric and U.S.-unit results to be.

Note 2: The intent of the charts, tables, and examples of this appendix is to enable you to compare the relative merits of various water and power sources by looking at the resources potentially consumed and pollutants emitted by their use. It has been a stretch to go as far as we have with our computations, due to the variability of the data. The numbers given are not exact; they are estimates. There are variables we could not incorporate. For example, it is impossible to know in advance what the power-generation-mix numbers will be for any utility, as it might get 15% or more of its power from a multitude of different generating facilities it does not own, either to take advantage of the cheapest power available at any demand moment, or to make up for a power shortage in its system at a time of peak customer power demand. As a result, the exact breakdown of power sources feeding a utility's distributed power can vary from month to month and year to year.

Note 3: In this appendix when we use the term "carbon" we mean carbon dioxide or CO_2.

WATER COSTS OF ENERGY (WCE) CHART

What are the water costs of the energy sources *you* use, and how can you reduce these costs? To get some clues and deepen your personal relation to the Water Costs of Energy, refer to the WCE chart and answer the questions below.

1. What type(s) of energy do you currently use? (If connected to the energy grid, contact your local utility and ask for the source(s) of the power they provide.)

2. How does *your* household's monthly water-for-power consumption compare with the household ranges given for *other* power sources?

3. What are some ways you can reduce your power consumption in order to reduce corresponding water use? (See table A9.1 on page 252 for ideas.)

4. What opportunities do you currently have to shift to less water-consumptive energy sources? (Options may include: installing an on-site solar, wind, or micro-hydro system; contacting your power utility to see if you can purchase renewable/non-nuclear power from them; investing in a regional community solar or wind project, cooperative, or collective; frequenting businesses that do the same).

5. How local is your power source? (On average, about 7% of electricity generated in the U.S. is lost during transmission and distribution. The shorter the transmission distances, the less is lost.)

6. If you were to shift (in whole or in part) to lower water-for-power energy sources, how much water could you save in the process? (Example I in the Integrating the Charts section later in this appendix illustrates the potential of such a shift.)

7. What are your strategies and time frame for making these shifts?

THE ENERGY COSTS OF WATER (ECW) CHART

What are the energy costs of the water(s) *you* consume and/or send down the drain, and how can you reduce these costs? To hone in to the answers to these questions, refer to the ECW chart, and read the notes and answer the questions below as best you can.

Note that two factors dramatically affect the range of energy costs of water:
- The quality of the water source. Remember energy costs are higher to treat lower-quality raw water;
- The distance and elevation increase between the water's source and its end use. Remember the greater the distance and/or elevation increase, the more energy and infrastructure is needed to transport water.

1. What type(s) of water do you currently use? (If you're connected to the water grid, contact your local water utility and ask for the source(s) of the water they provide.)

2. If you're on a well, what is the depth to water? Is the water level consistent, rising, or dropping each year? If you don't know, start monitoring now!

3. How does *your* household's monthly energy-for-water compare with the household ranges given for *other* water sources?

4. What are some ways you can reduce your water consumption in order to reduce corresponding energy use? This book is full of them. (See table A9.1 on page 252 for additional ways you can reduce your water consumption by reducing your energy use.)

5. What opportunities do you currently have to shift to water sources (possibly more than one) that use less or no energy (such as gravity-distributed household roof runoff (directly or via cistern) and/or greywater to irrigate fruit trees planted in or beside water-harvesting earthworks near the house)?

continued on p. 249

THE *ENERGY* COSTS OF WATER (ECW) – U.S. units

	How many kWh of energy are used to source & treat* **one gallon of water?**		How many kWh of energy are used to source & treat* water for **one U.S. household** for one month?		How many kWh of energy are used to source & treat* water for **100,000 U.S. households** for one month?	
			Average U.S. household water use: 7,716 gallons/month[1]			

Ranges of averages are shown from low (lighter) to high (darker). Overall means are not necessarily the means of the given extremes.

Origin	kWh/gal range		kWh/month range		kWh/month range	
On-site rainwater [a] →+	0.0000	0.0007	0	5	0	540,120
On-site greywater [b] →+	0.0000	0.0002	0	2	0	154,320
On-site blackwater [c] →+	0.0000	0.0011	0	8	0	848,760
On-site AC condensate [d] →+	0.0000	360.0000	0	2,777,760	0	277,776,000,000
Stormwater [e] →+	0.0000	0.0034	0	26	0	2,623,440
Surface water [2,3] →+	0.0002	0.0014	2	11	169,752	1,084,870
Central Arizona Project [f] →+	0.0126	0.0152	97	117	9,738,209	11,745,049
Groundwater [2,3] →+	0.0006	0.0020	5	15	478,392	1,543,200
Brackish groundwater [g,2,3,4,5] →+	0.0032	0.0379	25	292	2,469,120	29,243,640
Desalinated seawater [2,3,5] →+	0.0087	0.0882	67	681	6,712,920	68,055,120
Wastewater [2,3] →+	0.0010	0.0030	8	23	771,600	2,314,800
Recycled water [3,6] →+	0.0011	0.0041	8	31	848,760	3,144,642
Average utility water [3] ↻	**0.0013**	**0.0065**	10	50	964,500	5,015,400

ECW FACTS

The **average U.S. residential water usage** is 98 gallons per capita per day (gpcd) (*371 liters per capita per day (lpcd)*).[7]
The **virtual water footprint** of each **U.S.** citizen is 1,146 gallons (*4,337 liters*) per day.[8]
The **virtual water footprint** of each **world** citizen is 366 gallons (*1,386 liters*) per day.[9]

Democratic Republic of Congo's virtual gpcd (*lpcd*) is lowest: 9 (*32*) | **Jordan**: 120 (*434*)| **Germany & China** ~290 (*~1,100*)
France: 371 (*1,404*) | **Japan**: 517 (*1,941*) | **Australia**: 834 (*3,148*)| **Iraq**'s is highest: 1,894 (*6,918*).[8]

Of **all water withdrawn** in 2005 for use in the U.S., **5% was for industry/mining, 12% for public supply, 34% for agriculture, 49% for thermoelectric power generation**.[h,7]

ECW NOTES

***Sourcing** (→) includes pumping from aquifer, surface source, ocean, wastewater facility, etc, to treatment plant only.
Treatment (+) includes raw-water treatment to potable standards, or wastewater to discharge standards.
Lifecycle (↻) means → plus + plus distribution to end-user & wastewater collection, treatment, & discharge.

Energy costs of infrastructure (tank & pump manufacture, canal & building construction, etc.) relevant to water sources are beyond intended scope of this resource, & are not included herein.

Range in kWh/gal (*kWh/liter*) is due to pumping distance, depth, & quality of source water, &/or variations in equipment/processes (e.g., 0.0040–0.0080 kWh is used to lift 1 gallon of water 1,000 feet (*0.0035–0.0069 kWh is used to lift 1 liter of water 1,000 meters*)).[3]

a. Energy use is zero for gravity-fed untreated rainwater systems. High end is calculated with Flotec 3/4-HP (*559W*) shallow-well jet pump lifting water 0–5' (*0–1.5 m*) at 14.4 gpm (*54.5 l/min*)[10] & UV system treating to NSF/EPA standards using a Sterilight Silver S12Q-PA[11] or a Trojan UV Max IHS12-D4.[12]

THE *ENERGY* COSTS OF WATER (ECW) – metric units

	How many kWh of energy are used to source & treat* **one liter of water?**		How many kWh of energy are used to source & treat* water **for one U.S. household** for one month?		How many kWh of energy are used to source & treat* water **for 100,000 U.S. households** for one month?	
			Average U.S. household water use: 29,211 liters/month[1]			

Ranges of averages are shown from low (lighter) to high (darker). Overall means are not necessarily the means of the given extremes.

Origin	kWh/liter range		kWh/month range		kWh/month range	
On-site rainwater [a] →+	0.0000	0.0002	0	5	0	540,173
On-site greywater [b] →+	0.0000	0.0001	0	2	0	154,335
On-site blackwater [c] →+	0.0000	0.0003	0	8	0	848,843
On-site AC condensate [d] →+	0.0000	95.1022	0	2,778,031	0	277,803,138,374
Stormwater [e] →+	0.0000	0.0009	0	26	0	2,623,696
Surface water [2,3] →+	0.0001	0.0004	2	11	169,769	1,084,976
Central Arizona Project [f] →+	0.0033	0.0040	97	117	9,739,160	11,746,197
Groundwater [2,3] →+	0.0002	0.0005	5	15	478,439	1,543,351
Brackish groundwater [g,2,3,4,5] →+	0.0008	0.0100	25	292	2,469,361	29,246,497
Desalinated seawater [2,3,5] →+	0.0023	0.0233	67	681	6,713,576	68,061,769
Wastewater [2,3] →+	0.0003	0.0008	8	23	771,675	2,315,026
Recycled water [3,6] →+	0.0003	0.0011	8	31	848,843	3,144,949
Average utility water [3] ↻	0.0003	0.0017	10	50	964,594	5,015,890

Available at: HarvestingRainwater.com/water-energy-carbon-nexus

ECW NOTES (continued)

b. Energy use is zero for gravity-fed greywater systems. High end was calculated based on EcoVort 650W dirty-water pump lifting water 5' (*1.5 m*) at 56 gpm (*212 l/min*).[13]

c. Energy use is zero for gravity-fed & -discharged septic tanks & leachfields. The high end of range is for lagoons or ponds with oxidation.[14]

d. Energy use is zero for passive harvest (secondary to normal operation of air conditioner (AC)). Cost rises dramatically for active harvest (if AC is installed or run primarily to harvest condensate).

Energy intensity = energy use ÷ condensate yield. For 2- to 3-ton (*7- to 10.5-kW*) central AC system, energy use: 1.4–3.6 kW/hour;[15] condensate yield in dry air: 0.01–0.02 gal/hour (*0.04–0.08 l/h*); in humid air: 0.1–0.2 gal/hour (*0.4–0.8 l/h*).[16]

Range includes dry air: 7–36 kWh/gal (*2–10 kWh/l*), humid air: 70–360 kWh/gal (*18–95 kWh/l*). Values are for chemical-free AC, not cooling tower. Indoor & outdoor humidity & temperature, SEER rating, etc, affect kWh/gal (*kWh/liter*).

e. Zero value is for gravity-fed stormwater in separated storm & sewer systems (MS4). High value is for combined storm & sewer overflow systems (CSO), where stormwater is treated at wastewater treatment plant & often pumped from deep underground storage. Values for MS4 in low-lying areas (prone to flooding & requiring stormwater pumping stations) would fall within given range.[17]

f. Central Arizona Project (CAP) diverts water from Colorado River near Lake Havasu to supply central & southern Arizona. The given statistics for southern Arizona are 4–5 times higher than energy intensity of water delivered to central Arizona, due to increased treatment & pumping.[18] Higher value includes proportionally small kWh usage to distribute treated water to end-users.[19]

g. Definition of brackish groundwater varies by source. Broadly, it is groundwater containing 500–30,000 mg/liter of TDS (total dissolved solids)—more salty than freshwater, less salty than seawater.[20]

h. A large percentage of water withdrawn for power generation is typically returned to its source, but the volume of withdrawal matters: If the quantity of water isn't available, the power plant will have to shut down. Also when water is withdrawn for one use, it is then unavailable for others, such as municipal water supply & environmental needs.[2]

CREDITS: Brad Lancaster, Resource concept, oversight | **LeeAnn Lane**, Research | **Megan Hartman**, Research, resource creation | **Brandy Lellou & Valerie Strassberg**, NV-OC.org, Research, peer review

THE *CARBON* COSTS OF ENERGY (CCE) – U.S. units

| | How many pounds of CO_2 are emitted[a] to generate **a kilowatt-hour** of electricity? | How many pounds of CO_2 are emitted[a] to generate electricity for **one household** for one month? | | | How many pounds of CO_2 are emitted[a] to generate electricity for **100,000 households** for one month? | |
| | Renewable[b] sources in italics | Average household **kWh/month**:[4,5] **Arizona** = 1,095, **U.S.** = 920, **World** = 240 | | | | |
Origin	lb/kWh	Arizona	U.S.	World	U.S.	World
Municipal Solid Waste [c,d,1]	2.988	3,272	2,749	717	274,896,000	71,712,000
Coal [c,1]	2.249	2,463	2,069	540	206,908,000	53,976,000
Oil [c,1]	1.672	1,831	1,538	401	153,824,000	40,128,000
Natural Gas [c,1]	1.135	1,243	1,044	272	104,420,000	27,240,000
Biomass [c,d,2]	0.997	1,092	917	239	91,724,000	23,928,000
Geothermal [2]	0.067	73	62	16	6,164,000	1,608,000
Landfill Gas [c,d,2]	0.043	47	40	10	3,956,000	1,032,000
Solar Thermal/CSP [b,c,e,1]	0.000	0	0	0	0	0
Nuclear [b,c,1]	0.000	0	0	0	0	0
Hydroelectric [1]	0.000	0	0	0	0	0
Solar PV [f,1]	0.000	0	0	0	0	0
Wind [1]	0.000	0	0	0	0	0
Micro-Hydroelectric [1]	0.000	0	0	0	0	0
Average [3]	1.293	1,416	1,190	310	118,965,200	31,034,400

Available at: HarvestingRainwater.com/water-energy-carbon-nexus

CCE FACTS

Approximately **7% of electricity generated in the U.S. is lost** during transmission/distribution (U.S. EIA). In 2007, the CO_2 **emissions** associated with this loss **weighed 188 million tons (*170 million metric tons*)**.[3]

Of the **total United States greenhouse-gas emissions, electric power generation accounts for 34%, transportation 28%, industry/commerce 26%, agriculture 8%,** and **residential use 5%**.[6]

Selected 2009 **per-capita CO_2 emissions**, in tons (*metric tons*):[7] **World average**: 5 (*4.5*)
Afghanistan: 0 | **Jordan**: 3.5 (*3.2*)| **China**: 6.4 (*5.8*) | **France**: 6.9 (*6.3*)| **Japan**: 9.5 (*8.6*)
Germany: 10.2 (*9.3*) | **U.S.**: 19.5 (*17.7*) | **Australia**: 21.6 (*19.6*) | **Qatar**: 84 (*76.4*)

CCE NOTES

CO_2 = Carbon dioxide, at generation only. Data do not include climate-change potential of other greenhouse gases, or emissions associated with extraction of raw energy sources & lifecycle of infrastructure.

a. All regions' emissions are calculated based on U.S. carbon-emissions data and region-specific energy-consumption data. However, each region's actual average emissions will vary based on its own power-generation specifics.

b. Some of the zero-carbon benefits of nuclear and solar-thermal power generation are offset by their high water intensity.

A typical nuclear plant creates 22 tons (*20 metric tons*) of toxic spent nuclear fuel per year, which can take thousands of years to degrade.[8]

Utility-scale concentrated solar power (CSP), if situated in remote locations, requires transmission buildout to bring generated power to populated areas. Such buildout typically costs between $2 million and $4 million per mile[9] (*between $1.2 million and $2.4 million per kilometer*).

In 2007, non-biomass renewable fuel sources generated 7% of U.S. electricity & produced less than 0.1% of U.S. CO_2 emissions.[6]

c. Thermoelectric power plants, which provide 90% of U.S. electricity, heat water to create steam to turn turbines that generate power. The heat comes from burning municipal solid waste, coal, oil, natural gas, biomass, or landfill gas; concentrated sunlight; or a nuclear reactor.[10]

THE *CARBON* COSTS OF ENERGY (CCE) – metric units

		How many kilograms of CO_2 are emitted[a] to generate **a kilowatt-hour** of electricity?	How many kilograms of CO_2 are emitted[a] to generate electricity for **one household** for one month?			How many kilograms of CO_2 are emitted[a] to generate electricity for **100,000 households** for one month?	
	Renewable[b] sources in italics		Average household **kWh/month**:[4,5] *Arizona* = 1,095, **U.S.** = 920, **World** = 240				
Origin		kg/kWh	Arizona	U.S.	World	U.S.	World
Municipal Solid Waste [c,d,1]		1.355	1,484	1,247	325	124,692,826	32,528,563
Coal [c,1]		1.020	1,117	939	245	93,853,469	24,483,514
Oil [c,1]		0.758	830	698	182	69,774,566	18,202,061
Natural Gas [c,1]		0.515	564	474	124	47,364,912	12,356,064
Biomass [c,d,2]		0.452	495	416	109	41,606,006	10,853,741
Geothermal [2]		0.030	33	28	7	2,795,990	729,389
Landfill Gas [c,d,2]		0.020	21	18	5	1,794,442	468,115
Solar Thermal/CSP [b,c,e,1]		0.000	0	0	0	0	0
Nuclear [b,c,1]		0.000	0	0	0	0	0
Hydroelectric [1]		0.000	0	0	0	0	0
Solar PV [f,1]		0.000	0	0	0	0	0
Wind [1]		0.000	0	0	0	0	0
Micro-Hydroelectric [1]		0.000	0	0	0	0	0
Average [3]		0.587	642	540	141	53,962,615	14,077,204

Available at: HarvestingRainwater.com/water-energy-carbon-nexus

CCE NOTES (continued)

d. Biomass is a fuel category whose subtypes include landfill gas, agricultural byproducts, plant-based component of municipal solid waste (estimated at 2/3 of total materials), wood/wood waste, etc. EPA considers biomass to have zero net atmospheric CO_2 impact, as amount of CO_2 used by growing plants is equal to that released upon their combustion.[11] This view does not account for the *rate* at which CO_2 is released via combustion vs. gradual decomposition.

e. CSP = concentrated solar power, a form of solar-thermal energy that uses solar-tracking mirrors or lenses to focus a large area of sunlight onto a small area. The light energy is converted to heat, which is used to generate electricity via conventional thermoelectric methods (see note c, above).[12]

f. Solar PV = on-site photovoltaic solar panels, which use semiconductors to convert solar energy into direct-current electricity.[13]

CREDITS: Brad Lancaster, Resource concept, oversight | LeeAnn Lane, Research | Megan Hartman, Research, resource creation | Brandy Lellou & Valerie Strassberg, NV-OC.org, Research, peer review

6. If you were to shift (in whole or in part) to these other water sources, sourced as close as possible to their end use, how much would you decrease energy costs? (Example II in the Integrating the Charts section later in this appendix illustrates the potential of such a shift.)

7. What are your strategies and time frame for making these shifts?

THE *CARBON* COSTS OF ENERGY (CCE) CHART

What are the carbon costs of the energy source(s) you use, and how can you reduce these costs? The answers to these questions—and those below—can give you some ideas how you can affect climate change. Refer to the CCE chart, boxes A9.3, A9.4, and A9.5 to expand your understanding and options.

1. What type(s) of energy do you currently use? (If

Box A9.3. Converting Monthly Power-Plant CO_2-Emission Units from *Pounds* (or *Kilograms*) to *Cars*

Even if you know how much CO_2 a power plant emits in a certain amount of time, or to generate a certain amount of energy, it can be difficult to imagine what that number really means. So here is a more accessible way to think about it:

Fact: The average U.S. driver logs about 1,000 miles (about 1,600 kilometers) per month.[12]

Fact: The average U.S. car emits 0.916 pounds of CO_2 per mile (0.26 kg per km).[13]

That means that in an average month, the average car emits 0.916 lbs/mile (0.26 kg/km) times 1,000 miles (1,600 km), or 916 lbs (416 kg) of CO_2. To figure out how many cars per month it would take to emit the same amount of CO_2 as, for instance, your power plant, start with the pounds (or kg) of CO_2 the power plant emits and divide it by 916 pounds (416 kg) per car.

For example, a hypothetical coal-fired U.S. power plant generates 206,908,000 pounds (93,853,469 kg)* of carbon emissions to provide the energy for 100,000 U.S. households for one month. Divide 206,908,000 pounds of carbon by 916 pounds (416 kg) of carbon per car per month to get the number of cars that emit an equivalent amount of carbon:

206,908,000 lbs/month CO_2 (power plant) ÷ 916 lbs/month/car = 224,890 cars

or

93,853,469 kg/month CO_2 (power plant) ÷ 416 kg/month/car = 225,609 cars*

*See note on page 244 for why there are discrepancies between results from calculations using U.S. units compared to those using metric units.

---

you are connected to the energy grid, contact your local utility and ask for the source(s) of the power they provide.)

2. How many cars, each driving 1,000 miles (1,600 km) per month, does it take to emit the same amount of carbon as is emitted by your household energy consumption? (To figure this out, see box A9.3.)

3. How many trees does it take to sequester your household's monthly carbon emissions, and do you have enough space on your property for these trees? Is there enough space in your community to for the number of trees needed to sequester your community's monthly household carbon emissions? (See box A9.4 to figure this out.)

4. What are some ways you can reduce your (and your community's) power and fuel consumption in order to reduce corresponding CO_2 emissions? (See boxes A9.5 and A9.6, and table A9.1 in the next section for some ideas.)

5. What opportunities do you currently have to shift to energy sources with lower CO_2 emissions? Are on-site or local solar PV, micro-hydro, and/or wind power possible? What about contacting your power utility to see if you can purchase renewable power from them; investing in a regional community solar or wind project, cooperative, or collective; or frequenting businesses that use renewable power?

6. How local is your power source? (On average, about 7% of electricity generated in the U.S. is lost during transmission and distribution. The shorter the transmission distances, the less is lost.)

7. If you were to shift (in whole or in part) to these other energy sources, how much would you decrease CO_2 emissions? (Example I in the Integrating the Charts section later in this appendix illustrates the potential of such as shift.)

8. What are your strategies and time frame for making these shifts?

Box A9.4. How Many Trees Are Needed to Sequester/Offset Monthly Carbon Emissions?

Fact: The average tree (with a 9.7-inch (24.6-cm) trunk diameter) sequesters 6.08 lbs (2.8 kg) of carbon dioxide per month.[14]

To figure out how many trees it would take to sequester/offset the CO_2 emissions from a given source, start with the number of pounds (kilograms) of CO_2 emitted in a month, then divide that number by 6.08 lbs (2.8 kg) sequestered per tree per month.

For example, to sequester/offset the 206,908,000 pounds per month of carbon-dioxide emissions from the hypothetical coal-burning plant in box A9.4, divide the mass (weight) of the emissions by 6.08 pounds (2.8 kg) per tree per month to get the number of trees needed:

$$206,908,000 \text{ lbs/month } CO_2 \text{ (power plant)} \div 6.08 \text{ lbs/month/tree} = 34,030,921 \text{ trees}$$

How much land would be needed for those trees?

Depending on how much carbon we emit, we may or may not have enough land on our property, or within our community, to grow trees to sequester our household or community emissions. Use the figures below to see if there is enough plantable land on your site or in your area to sequester your or your community's emissions.

In the calculations in the section above, the diameter of the average tree's crown was calculated to be 17.6 feet (5.34 m), based on a crown area of 242 ft² (22.48 m²), based on 180 trees per acre (445 per hectare) or 115,200 trees per square mile (44,479 per km²).[15]

Currently, we do not have enough trees on the planet to sequester the global carbon dioxide being emitted into the atmosphere. If you don't have enough plantable land to sequester emissions on your site or in your area, reduce your emissions. Even if you do have enough plantable land, emission reduction should be your primary strategy. This book has given you numerous strategies to reduce emissions, often emphasizing those that use the strategic planting of on-site water, organic mulches, and trees to reduce water and energy consumption, while also maximizing the carbon sequestration of the soil and those trees. It's time for action. Other than 20 years ago, there is no better time to reduce our carbon emissions, and plant trees and the water and mulch they need, than today.

Box A9.5. Pounds of CO_2 Emitted Per Mile Traveled

This graph was reproduced with permission from the Sightline Institute's post "How Low-Carbon Can You Go: The Green Travel Ranking."

Box A9.6. Water Costs of Fossil Fuel-Fueled Travel

Every gallon (liter) of gasoline requires about 12.9 times that volume of water to process.[16]

Water use for air travel averages 0.3 gallons per airline-passenger mile, or 0.7 liters per airline-passenger kilometer.[17]

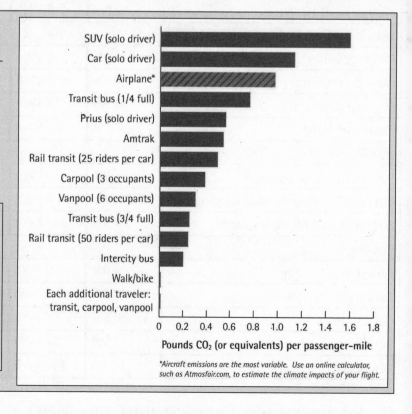

Aircraft emissions are the most variable. Use an online calculator, such as Atmosfair.com, to estimate the climate impacts of your flight.

Table A9.1. One Average U.S. Household's Monthly Energy-Water-Carbon Costs

Region: United States

Region's average _monthly_ energy use (2009)[2]: 920 kWh

Average household size[1]: 2.59 people

Time Frame: 1 _month_

	Percent of total household energy[a]	Energy consumed for this purpose	Water consumed to generate this energy[b,c,3]		Carbon emitted to generate this energy[d,4]	
Clothes drying						
Clothes line	0%	0 kWh	0 gallons	_0 liters_	0 lb	_0 kg_
Electric clothes drier[5]	5.8%	53 kWh	32 gallons	_122 liters_	69 lb	_31 kg_
Water heating						
Solar water heater	0%	0 kWh	0 gallons	_0 liters_	0 lb	_0 kg_
Electric water heater[5]	9.1%	84 kWh	51 gallons	_192 liters_	108 lb	_49 kg_
Food storage & preparation						
Root cellar, dehydration*, fermentation	0%	0 kWh	0 gallons	_0 liters_	0 lb	_0 kg_
Freezer[5]	3.5%	32 kWh	19 gallons	_74 liters_	42 lb	_19 kg_
Refrigerator[5]	13.7%	126 kWh	76 gallons	_289 liters_	163 lb	_74 kg_
Lighting						
Natural daylight[e]	0%	0 kWh	0 gallons	_0 liters_	0 lb	_0 kg_
Electric lighting (indoor & outdoor)[5]	8.8%	81 kWh	49 gallons	_185 liters_	105 lb	_48 kg_
Single light bulb[f] (3 hrs/day)	_Values vary based on light-bulb technology:_					
LED (6W)[6]	n/a	0.5 kWh	0.3 gallons	_1.2 liters_	0.7 lb	_0 kg_
Compact fluorescent (13W)[6,7]	n/a	1.2 kWh	0.7 gallons	_2.7 liters_	1.5 lb	_0.7 kg_
Incandescent (60W)[7]	n/a	5.4 kWh	3.3 gallons	_12 liters_	7.0 lb	_3.2 kg_
Learning & entertainment						
Reading, writing, making music	0%	0 kWh	0 gallons	_0 liters_	0 lb	_0 kg_
Laptop computer[5]	0.1%	0.9 kWh	0.6 gallons	_2.1 liters_	1.2 lb	_0.5 kg_
Desktop computer[5]	1.5%	14 kWh	8.3 gallons	_32 liters_	18 lb	_8 kg_
Color television[5]	2.9%	27 kWh	16 gallons	_61 liters_	34 lb	_16 kg_
Heating						
Passive/free solar heating[g]	0%	0 kWh	0 gallons	_0 liters_	0 lb	_0 kg_
Electric secondary heating equipment[h,5]	0.6%	5.5 kWh	3.3 gallons	_13 liters_	7.1 lb	_3.2 kg_
Electric main heating systems[h,5]	9.6%	88 kWh	53 gallons	_202 liters_	114 lb	_52 kg_
Cooling						
Passive/free cooling[i]	0.0%	0 kWh	0 gallons	_0 liters_	0 lb	_0 kg_
Evaporative cooler[c,5,8]	0.3%	2.8 kWh	1.7 gallons	_6.3 liters_	3.6 lb	_1.6 kg_
Ceiling fan[5]	0.8%	7.4 kWh	4.5 gallons	_17 liters_	10 lb	_4.3 kg_
Room air conditioner[5]	1.9%	17 kWh	11 gallons	_40 liters_	23 lb	_10 kg_
Central air conditioner (AC)[5]	14.1%	130 kWh	78 gallons	_297 liters_	168 lb	_76 kg_

*solar dehydration

INTEGRATING THE CHARTS

Can the WCE, ECW, and CCE charts be used to determine a more complete picture of the connections between water, energy, and carbon? I'm glad you asked—yes! You can integrate the information in the charts to calculate your combined water consumption, energy use, and CO_2 emissions—and discover, for example, how much *more energy* you can save via the *water* you save when you save energy. See table A9.1 for an average U.S. household's combined energy, water, and carbon costs for various uses. Use the options and recommendations presented throughout this book, and particularly in chapters 4 and 5 and the two examples below, for ideas and inspiration on how to maximize the potential savings of your integrated water, energy, and carbon, all the while maximizing the efficiency of passive/free cooling, heating, and daylight strategies.

Example I illustrates a household and community shift from conventional non-renewable utility-supplied power to renewable on-site solar-power generation.

Example II illustrates a household and community shift from utility-supplied water to on-site harvested rainwater and greywater.

Example I: Shift to grid-tied solar PV power generation:

The Dalmar-household example illustrates how making the switch from conventional utility power to on-site grid-tied renewable-energy production triggers conventional-energy savings, water savings, and CO_2-emission reductions at the local power plant. This example's savings calculations combine the use of WCE and CCE charts.

Location: Tucson, Arizona
Scale: The Dalmar household, with 2.63 people[18]
Time frame: One month
Current monthly energy use: 1,095 kWh[19]
***Initial* energy source:** Tucson Electric Power (TEP) generates 71% of its power from coal and 29% from steam-turbine natural gas (NGST) (a simplified approximation).[20]
***Subsequent* energy source *after* shift:** Grid-tied rooftop solar PV system is designed to generate 1,139 kWh per month—or 104% of the Dalmar residence's kWh usage.[21]

Note: Grid-tied household PV systems typically are designed to produce enough solar power during sunny days to slightly exceed (in this case by 4%) the household's average annual electrical consumption. Grid-tied solar PV systems without battery storage are most common in urban settings, while stand-alone solar PV systems with battery storage and no connection to the grid are more common in rural areas far from the grid.

With the grid-tied system in use, 28% (307 kWh) of the Dalmars' monthly 1,095-kWh energy usage would be drawn on sunny days, immediately at the time of power generation, from the household's own

Table A9.2. Integrated Benefits of Grid-Tied Rooftop PV Solar Power

	A. Dalmars' Grid-Tied Solar Household		B. 100,000 Grid-Tied Solar Households	
1. Total monthly TEP utility-energy offset	1,138.8 kWh		113,880,000 kWh	
a. Direct offset	306.6 kWh		30,660,000 kWh	
b. Indirect offset	832.2 kWh		83,220,000 kWh	
	low end	high end	low end	high end
2. Total monthly water savings at TEP	455.8 gallons	665.6 gallons	45,586,320 gallons	66,556,880 gallons
	1,725.0 liters	2,529.2 liters	172,504,800 liters	252,030,480 liters
a. Direct savings	122.7 gallons	179.2 gallons	12,273,240 gallons	17,919,160 gallons
	464.4 liters	687.5 liters	46,443,600 liters	67,854,360 liters
b. Indirect savings	333.1 gallons	486.4 gallons	33,313,080 gallons	48,637,720 gallons
	1,260.6 liters	1,841.7 liters	126,061,200 liters	184,176,120 liters
	U.S. units	metric units	U.S. units	metric units
3. Total monthly CO_2 reductions at TEP	2,193.6 lbs	994.9 kg	219,356,800 lbs	99,489,520 kg
a. Direct reductions	590.6 lbs	267.9 kg	59,057,600 lbs	26,785,640 kg
b. Indirect reductions	1,603.0 lbs	727.0 kg	160,299,200 lbs	72,703,880 kg
4. Monthly CO_2-reduction equivalents	2,193.6 lbs	994.9 kg	219,356,800 lbs	99,489,520 kg
a. Equivalent in cars parked	2.4 cars	2.4 cars	239,472 cars	239,158 cars*
b. Equivalent in trees planted	360.8 trees	360.5 trees*	36,078,421 trees	36,046,928 trees*

*Discrepancies due to rounding of inputs

on-site solar system. This solar energy usage is their *Direct offset*—a reduction of the power and the resulting emissions of carbon dioxide that would otherwise have been generated at the TEP power plant for the Dalmars' use.

The remaining 72% (788 kWh) of their energy usage would be drawn on overcast days and at night from the TEP utility grid (see table A9.2). So why, if 72% of the energy the Dalmars use comes from conventional energy from TEP, would we give them credit for generating *more* renewable on-site energy than they use?

Remember that the Dalmars are literally connected with their neighbors by the utility grid. Even though the Dalmars would use 788 kWh of *utility* energy a month, their solar PV system would kick back into the grid *832 kWh* of solar power—the equivalent of 76% of an average household's monthly usage—which nearby households would use instead of

conventional energy on sunny days. This counts as the Dalmars' *Indirect offset* (a reduction of power and the resulting consumption of water and emissions of carbon dioxide that would otherwise have been generated at the TEP power plant for the Dalmars' neighbors' use). See table A9.2 for the resulting integrated benefits of this conversion.

CONSERVATION FURTHER INCREASES THE BENEFITS

The more a grid-tied household reduces power consumption to an amount less than its PV system's capacity—particularly at night when their solar system is not producing electricity—the more power, water, and carbon-emission savings occur at the utility power plant. For example, charge laptops with direct-use solar power on sunny days and run them on their computer batteries at night and on cloudy days. Run a cooler during summer days but turn it off at *night* and stay comfortable with the accumulated cool stored in the building's interior mass—and/or by using free/passive cooling via natural ventilation. By utilizing such strategies, on an annual basis a more conserving grid-tied household in Tucson, AZ, produces 177%, as opposed to the more-typical 104%, of its electricity needs with its grid-tied rooftop solar PV.[22]

For a more thorough explanation of the benefits in table A9.2 and to see the calculations used to get these answers, see the *Integrating the Charts—Calculations* section on the *Water-Energy-Carbon Nexus* page at www.HarvestingRainwater.com. Check that same webpage for updates on the online *Water-Energy-Carbon Nexus Calculator* we are developing, which will do the integrated calculations for you based on your water and energy source inputs.

Example II: Shift to rainwater and greywater use:

The Flores-household example demonstrates that a household's use of on-site, low-power water-harvesting strategies leads to initial utility-water savings, energy savings, secondary utility-water savings, and reduced carbon emissions. To calculate the savings for this scenario, combine the use of WCE, ECW, and CCE charts.

Location: Tucson, Arizona

Scale: The Flores household, with 2.59 people[23]

Time frame: One month

Water use: 7,716 gallons (29,211 liters) per month[24]

Initial sole **water source:** Tucson Water, a utility that currently gets 60–70% of its water[25] from the energy-intensive (and thus, carbon- and water-intensive) Central Arizona Project (CAP), a source on which Tucson Water is projected to be almost fully dependent in the near future.[26]

Subsequent primary **water source:** Harvested rainwater and greywater is implemented and utilized in a manner similar to that of the Lancaster household in chapter 5.

MORE ABOUT CAP

The CAP canal diverts water from the Colorado River and pumps it uphill to end-users throughout central and southern Arizona, ending just south of Tucson. About 77% of the energy needed to source, treat, and deliver CAP water to end users in Tucson is provided by the coal-powered Navajo Generating Station.[27] Upon delivery, Tucson's allocation of CAP water is used to recharge a groundwater aquifer west of Tucson using constructed recharge basins. In the recharge process, some impurities are naturally filtered out. Next, large production wells down-gradient of the recharge basins pump out a combination of CAP water and

Table A9.3. Integrated Benefits of Integrated Water Harvesting and Reuse

	A. Flores' Integrated Water-Harvesting & -Reusing Household		B. 100,000 Integrated Water-Harvesting & -Reusing Households	
1. Primary monthly CAP-water savings	6,385 gallons	24,172 liters	638,500,000 gallons	2,417,200,000 liters
	low end	high end	low end	high end
2. Monthly utility-energy savings	80 kWh	97 kWh	8,045,100 kWh	9,705,200 kWh
3. Secondary monthly water savings	26 gallons	55 gallons	2,600,000 gallons	5,500,000 gallons
	98 liters	208 liters	9,800,000 liters	20,800,000 liters
4. Monthly CO$_2$-reductions equivalents	174 lbs	211 lbs	17,400,000 lbs	21,100,000 lbs
	78.8 kg	95.5 kg	7,880,000 kg	9,550,000 kg
a. Equivalent in cars parked	0.19 cars	0.23 cars	18,996 cars	23,035 cars
b. Equivalent in trees planted	29 trees	35 trees	2,861,842 trees	3,470,395 trees

groundwater, which is further treated, then delivered to households in Tucson. The recharge, well-pumping, treatment, and final-delivery stages use about 23% of the total energy needed to provide the water supply.[28] These stages are powered by utility company TEP, which generates 71% of its power from coal and 29% from natural gas (see Example I, above).

If the Flores family were to reduce their reliance on CAP/utility water supply from 7,716 gallons (29,211 liters) to 1,331 gallons (5,039 liters) per month by harvesting rainwater and greywater (in a manner similar to the Lancaster-household case study in chapter 5), they would attain a utility-water savings of 6,385 gallons (24,172 liters) per month, which amounts to an 83% *reduction* of utility water consumption from that of a typical household. See table A9.3 for the resulting integrated water, energy, and carbon savings for the Flores household (and 99,999 others).

For more explanation and to see the calculations used to get the figures in table A9.3, see the Integrating the Charts—Calculations section on the Water-Energy-Carbon Nexus page at www. HarvestingRainwater.com. Check that same webpage for updates on the online Water-Energy-Carbon Nexus Calculator we are developing, which will do the integrated calculations for you based on your water- and energy-source inputs.

References

References are listed by book section—introduction, chapter, appendix, etc. At the end of these reference sections, some may also separately list references for boxes, figures, charts, or tables found within those sections of the book.

INTRODUCTION

1. Tucson Water, "Homeowner's Guide to Using Water Wisely," rev. 2009, http://cms3.tucsonaz.gov/files/water/docs/homeowner.pdf (accessed 30 January, 2013).

2. United Nations Environment Programme (UNEP), *World Atlas of Desertification*, 2nd ed. (London: Arnold; New York: John Wiley & Sons, 1997).

3. Ibid.

4. Wendy Price Todd and Gail Vittori, *Texas Guide to Rainwater Harvesting* (Austin: Texas Water Development Board in Cooperation with the Center for Maximum Potential Building Systems, 1997).

5. Brian Barbaris, Senior Research Specialist, Department of Atmospheric Sciences, University of Arizona, personal communication (interview with author), 12 February 2003.

6. Todd and Vittori, *Texas Guide.*

7. Ibid.

8. John Begeman, "Thanks to Storms, Rain Delivers More Than Water to Desert," *Arizona Daily Star*, 2 August 1998, Home Section, p. 1.

9. David Cleveland and Daniela Soleri, *Food From Dryland Gardens* (Tucson: Center for People, Food and Environment, 1991).

10. Ibid.

11. Ibid.

12. Michael Evenari, Leslie Shanan, and Naphtali Tadmor, *The Negev: The Challenge of the Desert* (Cambridge: Harvard University Press, 1971).

13. John Gould and Erik Nissen-Petersen, *Rainwater Catchment Systems for Domestic Supply: Design, Construction, and Implementation* (London: Intermediate Technology Publications, 1999).

14. Charles Bowden, Killing the Hidden Waters (Austin: University of Texas Press, 1977), 119–20.

15. Ibid.

16. Maude Barlow, *Blue Gold: The Global Water Crisis and the Commodification of the World's Water Supply*, A Special Report issued by the International Forum on Globalization, June 1999.

17. Ibid.

18. Elizabeth Shogren, "Sprawl Adds to Drought, Study Says," *Los Angeles Times*, 29 August 2002, p. A12.

19. American Rivers, Natural Resources Defense Council, Smart Growth America, *Report: Paving Our Way to Water Shortages: How Sprawl Aggravates the Effects of Drought*, 28 August 2002, http://www.smartgrowthamerica.org/documents/DroughtSprawlReport09.pdf (accessed 30 January 2013).

20. Frank Sousa, Tucson Department of Transportation and Engineering Division, Stormwater Section, email to author, 7 November 2002.

21. P. Condon, and S. Moriarty, eds., *Second Nature: Adapting LA's Landscape for Sustainable Living* (Los Angeles: Metropolitan Water District of Southern California, 1999).

22. Anil Agarwal, Sunita Narain, and Indira Khurana, *Making Water Everybody's Business* (New Delhi: Centre for Science and Environment, 2001).

23. American Rivers, et al. *Report: Paving Our Way.*

24. Robert Glennon, *Water Follies: Groundwater Pumping and the Fate of America's Fresh Waters* (Washington DC: Island Press, 2002).

25. Barlow, *Blue Gold.*

26. Lester R. Brown and Brian Halweil, "China's Water Shortage Could Shake World Food Security," *Worldwatch*, July/August 1998.

27. Paul Hawken, Armory Lovins, and L. Hunter Lovins, *Natural Capitalism: Creating the Next Industrial Revolution* (New York: Little, Brown and Company, 1999).

28. Barlow, *Blue Gold.*

29. Michael Parfit, "Sharing the Wealth of Water," *National Geographic Special Edition: Water, the Power, Promise, and Turmoil of North America's Fresh Water*, 1993, p. 28.

30. Penn State, College of Education, *Investigations: Lesson 13: Lifestyles and Global Warming—Any Connection?* www.ed.psu.edu/ci/Papers/STS/gac-3/in13.htm.

31. Tom Hansen of Tucson Electric Power (email of 30 April 2003) in which he stated, "The Springerville Generating Station produces electricity with an annual average water consumption of about 0.45 gallon per kWh." Additional information can be found at www.powerscorecard.org.

32. Tim Flannery, *The Weather Makers: How Man is Changing the Climate and What It Means for Life on Earth* (New York: Grove Press, 2001).

33. H2ouse.org, "Save Water, Money, Energy Now!" www.h2ouse.org/action/top5.cfm (accessed 15 January 2013).

34. John Woodwell, Jim Dyer, Richard Pinkham, and Scott Chaplin, *Water Efficiency for Your Home: Products and Advice Which Save Water, Energy, and Money*, 3rd ed. (Snowmass CO: Rocky Mountain Institute, 1995); www.rmi.org/images/other/Water/W95- (accessed 15 January 2013).

35. H2ouse.org, "Save Water, Money, Energy Now!" http://www.h2ouse.org/action/top5.cfm

36. Ibid.

37. Martin M. Karpiscak, Thomas M. Babcock, Glenn W. France, Jeffrey Zauderer, Susan B. Hopf, and Kenneth E. Foster, *Evaporative Cooler Water Use Within the City of Phoenix: Final Report*, Arizona Department of Water Resources, Phoenix Active Management Area, April 1995.

38. Arizona Department of Water Resources, "Outdoor Water Use" pamphlet.

39. Figures determined from calculations from Pima County, Arizona, Cooperative Extension, Water Resources Center, Low 4 Program, "How to Develop a Drip Irrigation Schedule" handout.

40. Richard Heede and Staff of Rocky Mountain Institute, *HOMEmade Money*, (Snowmass CO: Rocky Mountain Institute, 1995).

41. Arizona Department of Water Resources, "Outdoor Water Use."

42. H2ouse.org "Pool and Spa Water Savings," http://www.h2ouse.org/tour/details/element_action_contents.cfm?elementID=d21acae2-1fc4-41d0-bc9a16b993ed790a&actionID=78FA9A8B-2756-4B2E-88D58A48310FAA76 (accessed 31 January 2012).

43. Michael Corbett and Judy Corbett, *Designing Sustainable Communities: Learning from Village Homes* (Washington DC: Island Press, 2000).

44. U.S. Environmental Protection Agency, "The EPA and Food Security," *Pesticides: Topical & Chemical Fact Sheet Sheets*, http://www.epa.gov/pesticides/factsheets/securty.htm (accessed 1 January 2013).

45. Barlow, Blue Gold.

46. National Wild and Scenic Rivers System, "River and Water Facts." http://www.rivers.gov/rivers/kids/funfacts.html (accessed 1 February 2013).

47. Barlow, *Blue Gold.*

48. Amy Vickers, *Handbook of Water Use and Conservation* (Amherst MA: WaterPlow Press, 2001).

49. Ibid.

50. *New Internationalist*, "Factfile on Water," April 2000.

51. Condon and Moriarty, *Second Nature.*

52. New York Times Special Supplement: *Water, Pushing the Limits of an Irreplaceable Resource*, 8 December 1998.

53. Barlow, *Blue Gold.*

54. Ibid.

55. Ibid.

56. Karpiscak et al., *Evaporative Cooler Water Use.*

57. Pima County, Arizona, Cooperative Extension, Water Resources Center, "How to Develop a Drip Irrigation Schedule" handout.

58. Heede et al., *HOMEmade Money.*

59. Vandana Shiva, *Water Wars: Privatization, Pollution, and Profit* (Cambridge MA: South End Press, 2002).

60. United Nations Committee on Economic, Cultural and Social Rights 2002, General Comment No. 15, The right to water (articles 11 and 12 of the International Covenant on Economic, Social and Cultural Rights), Twenty-ninth Session, Geneva E/C. 12-2002-11.

61. Joe Gelt, Jim Henderson, Kenneth Seasholes, Barbara Tellman, Gary Woodard, Kyle Carpenter, Chris Hudson, and Souad Sherif, *Water in the Tucson Area: Seeking Sustainability*, Water Resources Research Center, Issue Paper #20, summer 1999.

62. Nancy Laney, *Desert Water: From Ancient Aquifers to Modern Demands* (Tucson: Arizona-Sonora Desert Museum Press, 1998).

63. Tony Davis, "Gains Seen On Area's Water Goals: Groundwater table up in urban core, but levels falling in other spots," *Arizona Daily Star*, January 30, 2012, p. A1; article's corrected Tucson-Area Water-Table Levels map, Arizona Daily Star, 1 February 2012, p. A6.

64. Michael F. Logan, *The Lessening Stream: An Environmental History of the Santa Cruz River* (Tucson: University of Arizona Press, 2002).

65. Ibid.

66. Arizona Department of Environmental Quality, *Superfund Programs Section Site and Program Information—C 02-04*, July 2002.

67. Paul Green and Kendall Kroesen, "Water-Energy Nexus: Would You Pay for Clean Air?," Tucson Audubon Society's *Vermilion Flycatcher*, July-September 2011, Vol. 56, No. 3.

68. "Possible Climate Shift Could Worsen Water Deficit in the Southwest," Science Daily, 16 February 2000, http://www.sciencedaily.com/releases/2000/02/000216052551.htm (accessed 1 February 2013).

69. American Rivers, "Colorado River 'Most Endangered'," April 14, 2004 press release. www.americanrivers.org/site/News2?abbr=AMR_&page=NewsArticle&id=6699

70. Charles Bergman, *Red Delta: Fighting for Life at the End of the Colorado River* (Golden CO: Fulcrum, 2002).

71. Marc Reisner, *Cadillac Desert* (New York: Penguin Books, 1986).

72. Dr. Jim Riley, Associate Professor Soil, Water and Environmental Science Department, University of Arizona, personal communication, 16 June 2005.

73. Kirk Vincent and Laurie Wirt, "Urban Runoff—Lessen the Strain on Public Works by Using That Water at Home," *The Arizona Hydrological Society Newsletter*, v. 10, 1993, pp. 1–3.

74. David Confer, email to author, 18 April 2012.

75. Ibid.

76. Beorn Courtney, P.E. of the Headwaters Corporation, "Rainwater and Snowmelt Harvesting in Colorado," http://attachments.wetpaintserv.us/ByFprEQ1NM7kdOAKWDADlg==562514 (accessed 2 April 2013).

77. Ibid.

78. Ibid.

79. Beorn Courtney, P.E. of the Headwaters Corporation, email correspondence, 11 April 2013.

80. Matt Corrion, "Collecting Rainwater Still Illegal in Much of Colorado," *Lot Lines Blog*, http://www.lot-lines.com/collecting-rainwater-still-illegal-in-much-of-colorado (accessed 4 April 2013).

81. Frank Ramberg, Research Scientist, Department of Entomology, University of Arizona, personal communication, 24 January 2005.

CHAPTER 1

1. PELUM Association, "Water Harvesting: Some General Principles and Methods for Areas of Intensive Use and Dryland Cropping." PELUM Association, Box CY301, Causeway, Harare, Zimbabwe, July 1995.

2. Ibid.

3. Anil Agarwal, Sunita Narain, and Indira Khurana, *Making Water Everybody's Business* (New Delhi: Centre for Science and Environment, 2001).

4. Ibid.

5. Ibid.

6. Ibid.

7. PELUM Association, "Water Harvesting."

8. Ibid.

9. Bill Mollison, *Introduction to Permaculture* (Tyalgum, Tasmania: Tagari Publications, 1988).

10. Ibid.

11. Ibid.

12. Ibid.

13. Ibid.

CHAPTER 2

1. Ben Haggard, Sol y Sombra Foundation, *Drylands Watershed Restoration: Introductory Workshop Activities*, (Santa Fe: Center for the Study of Community, 1994).

2. Ibid.

3. U.S. Environmental Protection Agency (EPA), "What is a Watershed?" http://water.epa.gov/type/watersheds/whatis.cfm (accessed 30 January 2013).

4. Figures determined from calculations from Pima County, Arizona, Cooperative Extension, Water Resources Center, Low 4 Program, "How to Develop a Drip Irrigation Schedule" handout.

5. Art Ludwig, *Create an Oasis with Greywater* (Santa Barbara CA: Oasis Design, 2007).

6. "Lapse rate," http://en.wikipedia.org/wiki/Lapse_rate (accessed 30 January 2013).

7. David Boehnlein of Bullock's Permaculture Homestead (www.PermaculturePortal.com), email to author, 21 January 2011.

8. "Windpumps," http://en.wikipedia.org/wiki/Windpump (accessed 30 January 2013).

9. Ian Woofenden and Mick Sagrillo, "Is Wind Electricity Right for You?" *Home Power* 143, June/July 2011.

10. Frank Walker, Tucson Water Quality Lab, email, 23 May 2012.

Box 2.9. Estimates of Discharge Volumes from Household Evaporative Coolers, Air Conditioners, and Reverse-Osmosis Filters

a. Roy Otterbein, "Installing and Maintaining Evaporative Coolers," *Home Energy* magazine online, May/June 1996, http://www.homeenergy.org/show/article/year/1996/magazine/97/id/1211 (accessed 31 January 2013).

b. Martin M. Karpiscak, Thomas M. Babcock, Glenn W. France, Jeffrey Zauderer, Susan B. Hopf, and Kennith E. Foster, "Evaporative Cooler Water Use in Phoenix," *Journal AWWA*, Volume 90, Issue 4, April 1998.

c. Bill Hoffman, Austin Water Utility Commerical and Industrial Water Conservation Program Coordinator, personal communication, 27 July 2006.

d. Bill Hoffman, "Combining Storm and Rainwater Harvesting at Commercial Sites," First American Rainwater Harvesting Conference, 21–23 August 2003, Proceedings.

e. Art Ludwig, *Create an Oasis with Greywater*, (Santa Barbara CA: Oasis Design, 2006).

Figure 2.7. One-Page Place Assessment: Tucson, Arizona

1. University of Arizona station (#28815), www.wrcc.dri.edu (accessed 22 March 2010).

2. Tucson Magnetic Observatory station (#28800), www.wrcc.dri.edu (accessed 22 March 2010).

3. www.esrl.noaa.gov/gmd/grad/solcalc (accessed 13 June 2010).

4. Ibid.

5. ftp://ftp-fc.sc.egov.usda.gov/AZ/NRI/prevailing_winds.pdf (accessed 2 February 2013).

6. www.myforecast.com/bin/climate.m?city=10954 (accessed 11 March 2012).

7. www.wrcc.dri.edu/htmlfiles/westevap.final.html (accessed 2 February 2013).

8. Michelle Breckner, Service Climatologist, WRCC, personal communication (telephone), 23 April 2010.

9. www.Census.gov (accessed 12 May 2012).

10. Tucson Water, *Homeowner's Guide to Using Water Wisely*, rev. 2009, www.tucsonaz.gov/water/docs/homeowner.pdf (accessed 31 January 2013).

11. Arizona GroundWater Monitoring Site Hydrograph, gisweb.azwater.gov/gwsi/Detail.aspx (accessed 31 January 2013). Well: Local ID D-14-13 13CBC, Site ID 321227110574801, Registry ID 619923, Latitude 32° 12' 38.5", Longitude 110° 58' 33.4", Altitude 2368', Water Use–Public Supply, Drill date 3/1/1946.

12. Demand and Supply Assessment DRAFT, Tucson Active Management Area, 28 May 2010, Arizona Dept of Water Resources, www.azwater.gov/AzDWR/WaterManagement/Assessments/documents/FINALTAMAASSESSMENT.pdf (accessed 2 February 2013).

13. Tony Davis, "Gains Seen on Area's Water Goals," *Arizona Daily Star*, Monday, 30 January 2012, p. A1.

14. Bruce Plenk, City of Tucson Solar Energy Coordinator, 2007 data, email to author, 22 March 2010.

15. "Priority Vulnerable Species in Pima County," pima.gov/cmo/sdcp/species/fsheets/vuln/vuln.html (accessed 31 January 2013).

16. Lobos of the Southwest, Mapping the Lobos Range, mexicanwolves.org/index.php/wolf-country (accessed 8 January 2013).

17. Jonathan DuHamel, "Jaguar Sighted Near Tucson," *Tucson Citizen*, tucsoncitizen.com/wryheat/2011/11/21/jaguar-sighted-near-tucson (accessed 8 January 2013).

18. Doug Kreutz, "Series Reminds: Once Grizzlies Roamed Nearby," Arizona Daily Star, 30 January 2012, azstarnet.com/mobi/news/article_357f3ef5-74e4-5d65-839c-87248b982688.html (accessed 1/8/2013).

CHAPTER 3

1. Training Course on Water Harvesting, FAO Land and Water Digital Media Series #26, 2003.

2. Michelle Matthews and Marco Barrantes of La Loma Development, personal communication to author, 3 August 2012.

3. Jimmy L. Tipton, *Water Requirements of Landscape Trees: Final Report* (Phoenix: Arizona Department of Water Resources, 1997).

4. Jeff Lowenfels and Wayne Lewis, *Teaming with Microbes: The Organic Gardener's Guide to the Soil Food Web* (Portland: Timber Press, 2010).

CHAPTER 4

1. Ann Phillips, ed., *City of Tucson Water Harvesting Guidance Manual*, City of Tucson, Department of Transportation, Stormwater Section, June 2003.; rev. 2005 available at http://cms3.tucsonaz.gov/files/transportation/2006WaterHarvesting.pdf (accessed 31 January 2013).

2. Jonathan Hammond, Marshall Hunt, Richard Cramer, and Loren Neubauer, *A Strategy for Energy Conservation: Proposed Energy Conservation and Solar Utilization Ordinance for the City of Davis, California*, City of Davis, August 1974.

3. Jeff Saunders, personal communication to author, 10 February 2012.

4. Ed Mazria, *The Passive Solar Energy Book: A Complete Guide to Passive Solar Home, Greenhouse and Building Design.* (Emmaus PA: Rodale Press, 1979).

5. U.S. Department of Energy, "Passive Solar Design: Increase Energy Efficiency and Comfort in Homes by Incorporating Passive Solar Design Features," Technology Fact Sheet, apps1.eere.energy.gov/buildings/publications/pdfs/building_america/29236.pdf (accessed 31 January 2013).

6. David A. Bainbridge and Ken Haggard, *Passive Solar Architecture: Heating, Cooling, Ventilation, Daylighting, and More Using Natural Flows* (White River Jct VT: Chelsea Green Publishing, 2011)

7. Richard Heede and Staff of Rocky Mountain Institute, *HOMEmade Money*, (Snowmass CO: Rocky Mountain Institute, 1995).

8. John Rosenow, "Every Tree Is Part of the Global Forest," *Arizona Urban and Community Forestry*, Vol. 5, No. 1 (March 1999), p. 3.

9. Martin M. Karpiscak, Thomas M. Babcock, Glenn W. France, Jeffrey Zauderer, Susan B. Hopf, and Kenneth E. Foster, *Evaporative Cooler Water Use Within the City of Phoenix: Final Report*, Arizona Department of Water Resources, Phoenix Active Management Area, April 1995.

10. Plant water use figures determined from calculations from Pima County, Arizona, Cooperative Extension, Water Resources Center, Low 4 Program, "How to Develop a Drip Irrigation Schedule" handout.

11. Richard Heede and Staff of Rocky Mountain Institute, *HOMEmade Money*, (Snowmass CO: Rocky Mountain Institute, 1995).

12. James R. Simpson and E. Gregory McPherson, "Potential of Tree Shade for Reducing Residential Energy Use in California," *Journal of Arboriculture* 22(1), 1996.

13. Judy Corbett and Michael Corbett, *Designing Sustainable Communities: Learning from Village Homes* (Washington DC: Island Press, 2000).

14. Robert Kourik, *Designing and Maintaining Your Edible Landscape Naturally* (Santa Rosa CA: Metamorphic Press, 1986).

15. Living Building Challenge 2.1, https://ilbi.org/lbc/standard (accessed 4 February 2013).

16. Ralph L. Knowles, *Ritual House* (Washington DC: Island Press 2006).

17. Sara C.Bronin, "Solar Rights," *Boston University Law Review*, Vol. 89, 2009, p. 1217.

18. Corbett and Corbett, *Designing Sustainable Communities*.

19. Milagro Co-Housing Declaration of Covenants, Conditions, and Restrictions.

20. "Recognizing Skin Cancer," Boston University, 1997, www.bu.edu/cme/modules/2002/skincancer02/content/04-malig.html.

21. Barbara Mobely, Alabama Cooperative Extension System, "Extension Launches Skin Cancer Inititiative," www.aces.edu/dept/extcomm/newspaper/may20a03.html (accessed 31 January 2013).

22. Janis Keating, "Trees: The Oldest New Thing in Stormwater Treatment?" *Stormwater*, March 2002; available at http://www.gscience.org/uploads%5Cresources%5C894%5Cforestry-2012-envirothon-resource-2-of-3-trees-oldest-new-thing.pdf (accessed 31 January 2013).

23. Hammond et al., *Strategy for Energy Conservation*.

24. Corbett and Corbett, *Designing Sustainable Communities*.

25. Ibid.

26. William James, "Green Roads: Research Into Permeable Pavers," *Stormwater*, March 2002, www.forester.net/sw_0203_green.html.

CHAPTER 5

1. Amy Vickers, *Handbook of Water Use and Conservation* (Amherst MA: WaterPlow Press, 2001).

2. Keoleian, G.A., S. Blanchard, and P. Reppe, "Life-Cycle Energy, Costs, and Strategies for Improving a Single-Family House," *Journal of Industrial Ecology* 4 (2000), pp. 135-136.

3. Richard Heede and Staff of Rocky Mountain Institute, *HOMEmade Money*, (Snowmass CO: Rocky Mountain Institute, 1995).

4. Florida Solar Energy Center, "Comparative Evaluation of the Impact of Roofing Systems on Residential Cooling Energy Demand in Florida," http://www.fsec.ucf.edu/en/publications/html/FSEC-CR-1220-00-es/index.htm (accessed 23 March 2013).

5. The Living Building Challenge User's Guide, Prerequisite Five–Materials Red List, Cascadia Region Green Buidling Council, 2007, https://ilbi.org.

6. Feed-in tariff, Wikipedia, http://en.wikipedia.org/wiki/Feed-in_tariff, (accessed 3 March 2013).

7. "Municipal Solid Waste (MSW): Basic Facts, Almanac of Policy Issues," http://www.policyalmanac.org/environment/archive/epa_municipal_solid_waste.shtml (accessed 31 January 2013)

8. Karyn Maier, Demand Media, National Geographic, "Household Recycling Personal Waste Facts," http://greenliving.nationalgeographic.com/household-recycling-personal-waste-3234.html (accessed 31 January 2013).

9. M.A. Pavao-Zuckerman and C. Sookhdeo, "Soil Ecological Knowledge Promotes the Ecosystem Services of Green Infrastructure," *Watershed Science Bulletin*, in review, 2012.

10. Watershed Management Group, "Soil's Role in Processing Pollutants for Air and Water: Case Studies of Green Infrastructure and Carbon Sequestration with Dr. Mitchell Pavao Zuckerman," webinar: 24 January 2013, http://watershedmg.org/webinars/soils

11. Alan Sundermeier, Randall Reeder, and Rattan Lal, Soil Carbon Sequestration Fundamentals, Ohio State University Extension Fact Sheet, AEX-510-05, http://ohioline.osu.edu/aex-fact/0510.html (accessed 2 February 2013).

12. Pavao-Zuckerman and Sookhdeo, "Soil Ecological Knowledge."

13. SDCP: Sonoran Desert Conservation Plan, "Priority Vulnerable Species in Pima County," Sonoran Desert Conservation Plan, 2002, http://pima.gov/cmo/sdcp/species/fsheets/vuln/vuln.html (accessed 31 January 2013).

14. Charline Profiri and Wade C. Sherbrooke, "The Mystery of the Wet Lizards," http://www.highlightskids.com/science-stories/mystery-wet-lizards (accessed 1 February 2013).

APPENDIX 5

1. David A. Bainbridge and Ken Haggard, *Passive Solar Architecture: Heating, Cooling, Ventilation*, Daylighting, and More Using Natural Flows (White River Jct VT: Chelsea Green Publishing, 2011).

2. "Lapse_rate," http://en.wikipedia.org/wiki/Lapse_rate, (accessed 1 February 2013).

3. http://www.onthesnow.com/news/a/15157/ask-a-weatherman-how-does-elevation-affect-temperature (accessed 1 February 2013).

APPENDIX 7

1. Pasqualetti, Martin and Scott Kelley, "The Water Costs of Electricity in Arizona," Project Fact Sheet, Arizona Water Institute, 2007, http://www.circleofblue.org/waternews/wp-content/uploads/2010/08/AZ-Solar-Water-Fact-Sheet.pdf (accessed 1 February 2013).

2. Charles R. Landau, "Optimal Tilt of Solar Panels," 2012, www.macslab.com/optsolar.html (accessed 1 February 2013

3. ITACA, "Solar Panel Angles for Various Latitudes," http://www.itacanet.org/solar-panel-angles-for-various-latitudes/ (accessed 1 February 2013).

4. Bill Yanda, and Rick Fisher, *The Food and Heat Producing Solar Greenhouse* (Santa Fe: John Muir Publications, 1980).

APPENDIX 8

1. G.Z. Brown, and Mark DeKay, *Sun, Wind, and Light: Architectural Design Strategies* (New York: John Wiley & Sons, 2001).

2. Ibid.

3. Aermotor sales staff, personal communication to author, 20 February 2012.

4. http://www.aermotorwindmill.com/Sales/CommonQuestions.asp (accessed 20 February 2012).

5. Ian Woofenden and Mick Sagrillo, "Is Wind Electricity Right for You?" Home Power 143, June & July 2011.

6. William Kamkwamba and Bryan Mealer, *The Boy Who Harnessed the Wind: Creating Currents of Electricity and Hope* (New York: William Morrow, 2010).

7. David A. Bainbridge and Ken Haggard, *Passive Solar Architecture: Heating, Cooling, Ventilation, Daylighting, and More Using Natural Flows* (White River Jct VT: Chelsea Green Publishing, 2011), p 117.

8. Brown and DeKay, *Sun, Wind, and Light.*

9. Bainbridge and Haggard, *Passive Solar Architecture.*

10. Ibid.

11. Ibid.

12. Ibid.

13. Vernon Quam, Bruce Wight, and Harvey Hirning. "Farmstead Windbreak," F-1055, May 1993, reviewed and reprinted April 1996, www.ag.ndsu.edu/pubs/plantsci/trees/f1055w.htm (accessed 3 August 2012).

14. Bainbridge and Haggard, Passive Solar Architecture.

15. N.P. Woodruff, "Shelterbelt and Surface Barrier Effects," Agricultural Experiment Station, Manhattan, Kansas, Technical Bulletin 77, December 1954, www.weru.ksu.edu/new_weru/publications/Andrew_pdf/507.pdf (accessed 1 February 2013).

APPENDIX 9

1. "Climate Change," http://en.wikipedia.org/wiki/Climate_change (accessed 27 January 2013).

2. Skeptical Science website, "Is There a Scientific Consensus on Global Warming?", http://www.skepticalscience.com/global-warming-scientific-consensus-basic.htm (accessed 27 January 2013).

3. http://www.chasingice.com/learn/is-climate-change-man-made/ (accessed 2 February 2013).

4. http://co2now.org/ (accessed 29 December 2012).

5. James Hansen et al. "Target Atmospheric CO2: Where Should Humanity Aim?," *Open Atmospheric Science Journal*, 2008, Vol. 2, pp. 217-31.

6. http://www.350.org/about/science (accessed 2 February 2013).

7. Anthony Leiserowitz, interview, "Making People Care About Climate Change," segment of Bill Moyers & Company Show 152: Ending the Silence on Climate Change, 4 January 2013.

8. Bill McKibben, "Global Warming's Terrifying New Math: Three Simple Numbers That Add Up to Global catastrophe—and That Make Clear Who the Real Enemy Is," *Rolling Stone*, 19 July 2012, http://www.rollingstone.com/politics/news/global-warmings-terrifying-new-math-20120719 (accessed 2 February 2013).

9. Ibid.

10. Ibid.

11. Ibid.

12. Federal Highway Administration's Highway Statistics 2006 Table VM-1 (PDF) found on EPA's Climate Change–Greenhouse Gas Emissions page (accessed 28 December 2011).

13. "Emission Facts: Average Annual Emissions and Fuel Consumption for Passenger Cars and Light Trucks" found on EPA's Consumer Information page, http://www.epa.gov/otaq/consumer/f00013.htm (accessed 28 December 2011).

14. David J. Nowak, Robert E. Hoehn III, Daniel Crane, Lorraine Weller, and Antonio Davila, *Assessing Urban Forest Effects and Values: Los Angeles' Urban Forest*, USDA Forest Service Resource Bulletin NRS-47, March 2011, http://www.nrs.fs.fed.us/pubs/rb/rb_nrs47.pdf (accessed 2 February 2013). *Note: this reference gives figures for carbon (C) sequestered by trees, not carbon dioxide. A molecule of carbon dioxide weighs 3.66 times more than an atom of carbon (C). We converted the carbon (C) figures from the reference to carbon dioxide.*

15. Ibid.

16. Carey King and Michael Webber, Center for International Energy and Environmental Policy, Jackson School of Geosciences, University of Texas-Austin, "Water Intensity of Transportation," *Environmental Science and Technology*, Vol. 42, No. 21, 2008, pp. 7866-72.

17. Water Footprint Calculator Methodology and Tips, Energy and Transportation, http://environment.nationalgeographic.com/environment/freshwater/water-calculator-methodology/#page=2 (accessed 2 February 2013).

18. 2006-2010 statistic for average number of people per AZ household, http://quickfacts.census.gov/qfd/states/00000.html (accessed 15 February 2012).

19. Water Costs of Energy (WCE) chart.

20. Unisource Energy Corporation, Facts At Your Fingertips 2011, https://www.tep.com/doc/uns-facts-2011.pdf (accessed 11 February 2012).

21. Percentage of grid-tied household electrical consumption and production based on numbers provided by Kevin Koch, owner of Technicians for Sustainability (TFS), Tucson 13 February 2012.

22. Kevin Koch, owner of Technicians for Sustainability and installer of the PV solar system, personal communication to author, 13 February 2012.

23. 2006–2010 statistic for average number of people per U.S. household, http://quickfacts.census.gov/qfd/states/00000.html (accessed 15 January 2012).

24. Energy Costs of Water (ECW) chart.

25. Bruce Prior, Hydrologist, Tucson Water, personal communication, 14 February 2012.

26. Western Resource Advocates and Environmental Defense Fund, "Protecting the Lifeline of the West: How Climate and Clean Energy Policies Can Safeguard Water," 2010,

p. 15, www.westernresourceadvocates.org/water/lifeline/lifeline.pdf (accessed 2 February 2012).

27. Christopher Scott et al., "Water and Energy Sustainability with Rapid Growth and Climate Change in the Arizona-Sonora Border Region, a Report to the Arizona Water Institute," 2008, http://wsp.arizona.edu/sites/wsp.arizona.edu/files/Scott%20final%20report%2008.pdf (accessed 2 February 2013). Percentages calculated with data from figure 7 on page 10 of the report.

28. Ibid.

Water Costs of Energy (WCE) chart

1. www.eia.gov/cneaf/electricity/esr/table5.html (accessed 2 February 2013).

2. data.worldbank.org/indicator/EG.USE.ELEC.KH.PC (accessed 17 August 2011).

3. Martin Pasqualetti and Scott Kelley, "The Water Costs of Electricity in Arizona," Project Fact Sheet, Arizona Water Institute, 2007, http://www.circleofblue.org/waternews/wp-content/uploads/2010/08/AZ-Solar-Water-Fact-Sheet.pdf (accessed 1 February 2013).

4. "CO2 Emissions & Water Use Rates for U.S. Electrical Generation," naturesvoice-ourchoice.org/images/pdf/co2_and_water_use.pdf (accessed 2 February 2013).

5. U.S. Department of Energy, Report to Congress, "Energy Demands on Water Resources," www.sandia.gov/energy-water/docs/121-RptToCongress-EWwEIAcomments-FINAL.pdf (accessed 2 February 2013).

6. Western Resource Advocates and Environmental Defense Fund, "Protecting the Lifeline of the West: How Climate and Clean Energy Policies Can Safeguard Water," 2010, p 11, www.westernresourceadvocates.org/water/lifeline/lifeline.pdf (accessed 2 February 2013).

7. Water, Energy, CO2 for All eGRID Subregions spreadsheet, via email from Brandy Lellou, Nature's Voice–Our Choice, 27 July 2011.

8. Ashlynn S. Stillwell et al., "Energy-Water Nexus in Texas," www.edf.org/documents/9479_Energy-WaterNexusinTexasApr2009.pdf (accessed 2 February 2013).

9. www.epa.gov/cleanenergy/energy-and-you/affect/water-resource.html (accessed 2 February 2013).

10. www2.intota.com/experts.asp?strSearchType=all&strQuery=wet+cooling+tower (accessed 18 February 2012).

11. "Concentrated solar power," http://en.wikipedia.org/wiki/Concentrated_solar_power (accessed 2 February 2013).

12. "Combined cycle," http://en.wikipedia.org/wiki/Combined_cycle (accessed 2 February 2013).

Energy Costs of Water (ECW) Chart

1. 98 gpcd (USGS, see ECW reference #7) X 30.4 days/month X 2.59 people/household (2006-2010, census.gov) = 7,716 gallons/household/month.

2. Ashlynn S. Stillwell et al., "Energy-Water Nexus in Texas," www.edf.org/documents/9479_Energy-WaterNexusinTexasApr2009.pdf (accessed 2 February 2013).

3. Bevan Griffiths-Sattenspiel and Wendy Wilson, "The Carbon Footprint of Water," River Network, May 2009, http://www.rivernetwork.org/sites/default/files/The%20Carbon%20Footprint%20of%20Water-River%20Network-2009.pdf (accessed 2 February 2013).

4. U.S. Department of Energy, Report to Congress, "Energy Demands on Water Resources," www.sandia.gov/energy-water/docs/121-RptToCongress-EWwEIAcomments-FINAL.pdf (accessed 2 February 2013).

5. Tamin Younos and Kimberly E. Tulou, "Energy Needs, Consumption, and Sources," *Journal of Contemporary Water Research and Education*, December 2005.

6. LADWP's Draft 2010 Urban Water Management Plan, January 2011, from ladwp.com/ladwp/cms/ladwp013956.pdf (accessed 28 July 2011).

7. pubs.usgs.gov/circ/1344/pdf/c1344.pdf, pp 4-5, 19 (accessed 2 November 2011).

8. Total national water use ÷ population. Food and Agriculture Organization of the United Nations (FAO) AQUASTAT database at www.fao.org/nr/water/aquastat/data/query/index.html (accessed 17 August 2011). In 2007, 1,583 m3/U.S. person/yr x 264.17 gal/m3 ÷ 365 days/year = 1,146 gpcd.

9. Total national water use ÷ population. Chartsbin.com/view/1455 (accessed 17 August 2011). 506 m3/U.S. person/yr x 264.17 gal/m3 ÷ 365 days/year = 366 gpcd.

10. HomeDepot.com (accessed 5 November 2012).

11. Mark Ragel, Water Harvesting International, via email, 7 November 2011.

12. Jeremiah Kidd, San Isidro Permaculture, via email, 8 November 2011.

13. Paul James, BestHomeWaterSavers.com, via email, 28 October 2011.

14. F.L. Burton et al., *Water and Wastewater Industries: Characteristics and DSM Opportunities*, Report TR-102015s (Palo Alto: Electric Power Research Institute (EPRI), 1993).

15. Cec.net/tips/brochures/heatcool.pdf (accessed 18 January 2012).

16. Brad Lancaster, *Rainwater Harvesting for Dylands and Beyond, Vol. 2*, (Tucson: RainSource Press, 2008), p. 301

17. Brandy Lellou, Nature's Voice–Our Choice, via email, 18 January 2012.

18. Christopher Scott et al., "Water and Energy Sustainability with Rapid Growth and Climate Change in the Arizona-Sonora Border Region, a Report to the Arizona Water Institute," 2008, http://wsp.arizona.edu/sites/wsp.arizona.edu/files/Scott%20final%20report%2008.pdf (accessed 2 February 2013).

19. Western Resource Advocates and Environmental Defense Fund, "Protecting the Lifeline of the West: How Climate and Clean Energy Policies Can Safeguard Water," 2010, p. 15, www.westernresourceadvocates.org/water/lifeline/lifeline.pdf (accessed 2 February 2012).

20. "Brackish water," http://en.wikipedia.org/wiki/Brackish_water (accessed 3 February 2013).

Carbon Costs of Energy (CCE) Chart

1. epa.gov/cleanenergy/energy-and-you/affect/air-emissions.html (accessed 3 November 2011).

2. Emissions calculated with data from eGRID2010V1_1_year07_PLANT.xls at epa.gov/cleanenergy/energy-resources/egrid/index.html (accessed 4 November 2011), per Art Diem, U.S. EPA, via phone, 4 November 2011.

3. eGRID2010 Year 07 Summary Tables, http://www.epa.gov/cleanenergy/energy-resources/egrid/index.html (accessed 2 February 2013).

4. www.eia.doe.gov/cneaf/electricity/esr/table5.html (accessed 2 February 2013).

5. data.worldbank.org/indicator/EG.USE.ELEC.KH.PC (accessed 21 August 2011).

6. Alan W. Hodges and Mohammad Rahmani, "Fuel Sources and CO2 Emissions by Electric Power Plants in the United States," 2009, edis.ifas.ufl.edu/fe796 (accessed 2 February 2013).

7. U.S. Energy Information Administration; ChartsBin statistics collector team, "Current Worldwide Carbon Dioxide Emissions Per Person," chartsbin.com/view/1519 (accessed 8 November 2011).

8. Nuclear Energy Institute website, www.nei.org/resourcesandstats/nuclear_statistics/nuclearwasteamountsandonsitestorage (accessed 9 December 2011).

9. "Concentrating Solar Power," gigatonthrowdown.org/files/Gigaton_ConSolPow.pdf (accessed 2 February 2013).

10. Brandy Lellou, Nature's Voice–Our Choice, via email, 19 January 2012.

11. eGrid2007TechnicalSupportDocument.pdf,), epa.gov/cleanenergy/energy-resources/egrid/index.html (accessed 2 February 2013).

12. "Concentrated solar power," http://en.wikipedia.org/wiki/Concentrated_solar_power (accessed 2 February 2013).

13. "Photovoltaics," http://en.wikipedia.org/wiki/Photovoltaics (accessed 2 February 2013).

Table A9.1

1. 2006-2010 statistic, http://quickfacts.census.gov/qfd/states/00000.html (accessed 15 February 2012).

2. http://www.eia.gov/cneaf/electricity/esr/table5.html (accessed 21 August 2012).

3. Based on mean of low- and high-end values from Water Costs of Energy (WCE) of 0.605 gallons (2.29 liters) /kWh.

4. Based on average of 1.293 lb (0.587 kg) /kWh from Carbon Costs of Energy (CCE).

5. http://www.eia.gov/emeu/recs/recs2001/enduse2001/enduse2001.html (accessed 21 August 2012).

6. http://www.designrecycleinc.com/led%20comp%20chart.html (accessed 21 August 2012).

7. EPA's LightingCalculator.xlsx, http://www.energystar.gov/ia/business/bulk_purchasing/bpsavings_calc/LightingCalculator.xlsx?6ed4-1491&6ed4-1491 (downloaded 21 August 2012).

8. Karpiscak, M., Babcock, T., France, G., Zauderer, J., Hopf, S., Foster, K., Evaporative cooler water use within the City of Phoenix: Final Report, Arizona Department of Water Resources, Phoenix Active Management Area, April 1995.

GLOSSARY

1. David Rosgen, *The Cross-Vane, W-Wier and J-Hook Vane Structures: Their Description, Design and Application for Stream Stablization and River Restoration* (Fort Collins, CO: Wildland Hydrology, 2001); http://www.wildlandhydrology.com/assets/cross-vane.pdf (accesssed 2 February 2013).

Glossary

Active system. A system requiring a person, switch, or valve to actively turn it on/open or off/closed. The more active a system (perhaps its water is also pumped), the more energy it requires.

Aerobic. A condition which supports organisms that thrive only in the presence of oxygen.

Algae. Microscopic plants which contain chlorophyll and live in water. Algae can impart tastes and odors to stored water.

Alluvial fan. Fan-shaped depositional landforms created when sediment-laden tributary channels widen and their slope is significantly reduced, such as occurs when channels intersect broad gently sloping valley bottoms.

Altitude angle. Sometimes called Elevation angle. The height of the sun in degrees above a true horizon—not affected by mountains, buildings, or other topography.

Anaerobic. A condition, which supports organisms that thrive only in the absence of oxygen.

Angle of repose. The maximum angle or steepness of slope at which a pile of unconsolidated material can remain stable.

Angular open-graded gravel. Aggregate whose particles are angular in shape so their flat faces interlock with each other to resist rotating and shifting, and having a narrow range of particle sizes and open void spaces that improve interlocking between the particles, while maintaining good porosity.

Annual. Plant that takes 1 year or less to go through its entire life cycle: germination of the seed, vegetative growth, flowering, and seed production, after which it dies.

Aquifer. Subterranean layers of sedimentary particles (sand, gravel, and rocks) laid down over geologic time, in which water fills the tiny spaces between the particles.

Azimuth angle. The distance of the sun in degrees east or west of True South in the northern hemisphere, or the distance in degrees east or west of True North in the southern hemisphere.

Backwater valve. A sewer drain pipe valve consisting of a flap that opens with water flowing out, but that otherwise remains closed to prevent the inflow of backed-up sewage. It is commonly made from ABS plastic, typically used with gravity-fed sewer drains, and is available from better-stocked plumbing suppliers. I install backwater valves at the ends of cistern overflow pipes to keep insects, critters, and sunlight out of the stored water.

Baffle. A low deflecting structure placed on one side of an unhealthily narrowed, deepened, or straightened channel to nudge the water flow into the opposite bank. It is designed to create lateral erosion on the opposite bank to widen the channel; to harvest some of the resulting sediment on the baffle, increasing vegetated in-channel floodplain; and to induce a more gradually-sloped, meandering channel flow.

Bankfull. Elevation of normal full channel water flows just before they'd begin to spill over the channel's banks onto the floodplain.

Baseflow. Sustained, dry-weather runoff or water flow. It is streamflow, which results from precipitation that has infiltrated into the soil, then eventually moves through the soil to the stream channel.

Berm 'n basin. A water-harvesting earthwork laid perpendicular to land slope, consisting of an excavated basin and a raised berm located just downslope of the basin.

Bicycle boulevard. A bicycle- and pedestrian-friendly, low/slow-traffic (and typically tree-lined) residential street that connects neighborhoods with safe bike/pedestrian crossings at major intersections.

Biocompatible. A material whose decomposition creates products that are beneficial for, or at least not harmful to, the environment in which it is deposited.

Blackwater. Wastewater from toilets (some regulators consider kitchen sink wastewater as blackwater too), that has higher levels of solids and coliform bacteria than greywater sources.

Boomerang berm. Semicircular berm open to, and harvesting, incoming runoff from upslope.

Branched drain greywater system. System of pipes, valves, or "double L" or "Y" fittings that "branch" or split a gravity-fed flow of greywater to as many as sixteen outlets within mulched basins distributed in a landscape.

Carbon offset. A reduction in emissions of carbon dioxide or other climate-altering gases made in order to compensate for or to offset an emission made elsewhere.

Catchment surface. Surface from which runoff is captured within earthworks or a cistern for beneficial on-site use.

Central Arizona Project (CAP). A multi-billion dollar canal project, which diverts water from the Colorado River and pumps it 2,400 feet (731 m) uphill and over 336 miles (540 km) through the desert to reach farms and the cities of Phoenix and Tucson, Arizona.

Channel flow. The concentrated distribution of runoff within distinct channels or drainages. Look for nick points, rills, gullies, bank cutting, different sediment sizes, vegetation growing within channels, and exposed roots to assess the force of the flow and the health of the channel.

Channelization. Constricting and straightening water flow by sealing and smoothing the banks and sometimes the bed of a waterway, often with concrete. It can be compared to a shotgun barrel for water. Channelization increases the velocity of water flow through and downstream of the channelized area, reducing infiltration of water into the soil and sometimes deepening the channel.

Check dam. A low, leaky barrier placed perpendicular to the flow of water within a drainage to slow the water's flow, infiltrate more water into the soil, and hold soil and organic matter higher in the watershed. Unlike a one-rock dam, it can be more than one course of rock tall if built atop bedrock, which can endure the erosive force of the water falling over the dam. Otherwise a stabilizing spillway and/or pool must be created.

Cistern. A tank used to store rainwater.

Climate change. A significant and lasting change in weather patterns, conditions, and distribution over periods ranging from decades to millions of years. Climate change is caused by factors that include oceanic processes (such as oceanic circulation), variations in radiation from the sun received by Earth, plate tectonics and volcanic eruptions, and human-induced alterations of the natural world.

Clock time. Also called *mean solar time* or *mean time*, is time based on the motion of the mean sun (an imaginary sun moving uniformly along the celestial equator), resulting in equal 24-hour days throughout the year. If days were measured by the actual movement of the Sun (solar time), they would vary slightly in length at different times of the year due to differences in Earth's orbital speed and other factors. Mean time is the standard clock time throughout most of the world.

Cold frame. A low enclosure capped with a transparent surface such as glass under which plants are grown in cold seasons. The glass can be raised for ventilation on warmer days as needed.

Combined sewer. Sewer that contains sewage, household wastewater, and rain runoff from streets, yards, and driveways.

Commodify. To turn a natural resource into a limited-access commodity to be bought, sold, and hoarded.

Commons. Cultural and natural resources such as land, water, air, ecosystems, and a habitable Earth that provide the ecological basis of life and whose sustainability and equitable allocation depends on cooperation among its community members. These resources are held in common, not owned privately.

Communify. To work together to enhance a natural resource and the related community by managing the sustainable, fair use and equal accessibility of the resource.

Community. Represents all living and interacting organisms in an ecosystem, including people, other animals, plants, fungi, and bacteria.

Compost. A soil amendment made from decomposed organic matter. The act of composting speeds up the decomposition of the organic matter, while retaining more nutrients by keeping the compost pile moist (in a pit, in the shade, covered in mulch), lightly aerated, and by balancing the amount of carbon material (dry woody material like straw or sawdust) to nitrogen-rich material (green plant material, fresh manure, urine), which also prevents odors.

Composting toilet. A waterless toilet in which dry carbon-rich material such as straw or sawdust is added to its aerobic composting chamber to help decompose (without objectionable odor) nitrogen-rich human feces and urine into high-grade fertilizer.

Contour berm. A berm 'n basin constructed along a contour line.

Contour line. A level line perpendicular to the slope of the land.

Crossover riffle. Also called a "meander crossover," this more level-bottomed section of a watercourse between bends is where larger sediment tends to drop out of the water's flow.

Cross-vane. A grade-control structure that focuses stream power toward the center of a channel, and maintains a permanent scour pool below the structure. (The pool of water diffuses the force of water falling over the structure.) This structure is built with footer rocks/boulders installed in the channel bed to a depth up to three times the height of the cross-vane above the bed. Strategy developed by Dr. David Rosgen.[1]

Culvert. A drainage pipe made to transport water beneath a roadway. Metal culverts can be made into above-ground cisterns.

Daylighting pipe. Outletting a pipe into the open air.

Daylighting a waterway. Uncovering and revegetating a previously piped or buried waterway to recreate a natural, living watercourse.

Degenerative. A type of investment that starts to break down as soon as it is used, typically serves only one function, consumes more resources than it produces, and which degrades the health of its surroundings and/or the world.

Detention/Retention basin. A structure that decreases stormwater flow from a site by temporarily holding the runoff on site. This is not a water-harvesting structure unless the held water is beneficially utilized on site (irrigating vegetation for example).

Diversion swale. A gently sloping drainageway that moves water slowly downslope across a landscape, while simultaneously allowing much of it to infiltrate into the soil.

Double-glazed windows. Also called "double-paned windows," these windows have two panes of glass separated by an air- or other gas-filled space to reduce excessive heat and cold transfer through the window.

Downcutting. Erosion deepening a channel by removing material from the channel's bed.

Drip irrigation. An irrigation strategy applying water via an emitter to the root zone of a plant at a rate slow enough (usually less than 3 gallons [11 liters] per hour) to allow the soil to absorb it without runoff.

Dry farm. A type of farming only using local rainfall and runoff for irrigation.

Dryland. Areas of the world where potential average yearly moisture loss (evapotranspiration) exceeds average yearly moisture gain (precipitation).

Dry-stacked retaining wall. A naturally porous wall of stone, brick, or salvaged concrete laid "dry" without mortar, and maintaining a batter or lean of 5 to 15° into the slope to help counter the weight of upslope earth.

Dry-system downspout. A rainwater conveyance pipe maintaining a continuous slope so that all water drains from it, leaving it dry between rains. A dry system downspout does not collect sludge, nor is it prone to freezing damage.

Dumpster. An exterior trash container from which opportunistic scavengers can salvage discarded resources.

Earthworks. Simple structures and strategies that change the topography and surface of the soil to speed the infiltration of water; augment soil moisture and nutrient availability with more organic matter; and maximize the density and diversity of vegetation utilizing that water to generate more resources, while swelling the gains of water percolation and nutrient cultivation.

Embodied energy. The sum of all the energy required to produce goods or services, considered as if that energy was incorporated or "embodied" in the product itself.

Ephemeral water flow. Water that only flows seasonally or during and just after storms.

Erosion. Wearing away of soil and rock by gravity, wind, and water, intensified by human land-clearing practices.

Evaporation. The change of water from a liquid to a gas.

Evapotranspiration. The combined measurement of water loss to evaporation and transpiration through the pores of vegetation.

Feed-in tariff. A payment from the government or utility provider to a small-scale renewable energy producer for the energy it produces.

Ferrocement. Metal-reinforced cement.

First flush system. A device or length of capped pipe that diverts the dirtiest or foulest first flush of water running off a catchment surface away from a cistern.

Flow-through system. A water-harvesting earthwork through which surface water flows (in one side and out the other when full) with much of the flow infiltrating the soil of the earthwork before exiting. This differs from a calmer "backwater" earthwork where the inflow and overflow point are the same resulting in no more water flowing into the system when full, and typically very little to no water flowing out of the system.

French drain. A trench or basin filled with porous materials such as gravel or mulch that have ample air spaces between them, allowing water to infiltrate quickly into the drain and percolate into the root zone of the surrounding soil, while creating a stable surface you can walk on.

Gabion. A check dam in which the rocks are encased in a wrapping of wire fencing or a wire basket that holds everything together—sort of a rock burrito in a wire tortilla.

Gabion basket. A rectangular wire basket made to contain many rocks, forming a check dam across a drainage.

Galvalume. A protective coating (typically over steel) primarily composed of aluminum, zinc, and silicone.

Generative. A type of investment that starts to break down as soon as it is used, typically serves multiple functions, produces or generates more resources than it consumes, and conserves other resources.

Gnomon. A vertical shaft casting a shadow used as an indicator of the sun's position and time.

Gpcd. Gallons (consumed or used) per capita (person) per day.

Gravity-fed. Powered, pressurized, or transported solely by the natural force of gravity.

Greywater. Wastewater originating from a clothes washer, bathtub, shower, drinking fountain, or sink that can be safely reused to irrigate a landscape.

Greywater harvesting. The practice of safely directing the greywater generated at a site to the root zone of perennial plants in the yard where it can help grow beautiful and productive landscapes.

Greywater stub out. See *stub out*.

Grid-tied solar PV system. A solar photovoltaic (PV) power system that is connected to a utility's electrical grid, and into which the solar system feeds surplus power on sunny days, and from which the system draws power at night and on cloudy days.

Groundwater. Water that has naturally infiltrated into and is stored within an underground aquifer.

Guild. A harmonious assembly of living species such as plants, animals, and people and non-living elements such as rocks or buildings that perform better through their cooperative interrelationships than they would as individuals.

Gully. A large erosive drainage or arroyo.

Hammock effect. A stabilizing fabric of woven roots (growing from the trees along the banks of a water channel) over which the channel bed rests and the water flows.

Hard water. A characteristic of water containing dissolved calcium and magnesium, which is responsible for most scale formation in water heaters and pipes.

Hardscape. Hard paving material such as concrete sidewalks, asphalt streets, and brick patios.

Headcut. The growing upstream edge of an erosive gully or rill.

Hydrologic cycle. The continual movement of water between the earth and the atmosphere through precipitation, infiltration into and release from living systems, evaporation, evapotranspiration, and precipitation again.

Hydroperiod. Period of time water is available in the soil for growth of associated plants.

Hydroseeding. A planting method that sprays a slurry of mulch and seed over the ground.

Impervious. A non-permeable solid surface.

Imprinter roller. An imprinting tool fabricated from a 10- to 20-foot (300- to 600-cm) long, 20- or 24-inch (50 to 60-cm) diameter smooth roller with 10-inch (25-cm) lengths of 6- X 6-inch (15-cm) to 8- X 8-inch (20-cm) angle iron welded onto the roller in a pattern of staggered star rings.

Imprinting. A water-harvesting technique used to accelerate the revegetation of disturbed or denuded land with annual precipitation from 3 to 14 inches (76 mm to 356 mm), by creating numerous small, well-formed depressions in the soil that collect seed, rainwater, sediment, and plant litter, and provide sheltered microclimates for germinating seed and establishing seedlings.

Induced meandering. A method of transforming incised (and often unnaturally straightened) water channels by striving to understand and assist a stream's natural evolution using the power of floods to shape/heal the channel and banks over time (often by forming a calmer, more meandering flow).

Infiltration. The movement of water from the land's surface into the soil.

Infiltration basin. A landscaped level-bottomed, relatively shallow depression dug into the earth that collects, infiltrates, and utilizes the rain that falls within it, the runoff draining into it from the surrounding area, and potentially household greywater too.

Infiltration chamber. An empty, bottomless subsurface plastic chamber into which greywater is released, reducing direct human or animal contact with the greywater, and reducing the risk of roots growing into the greywater pipe.

Integrated design. A very efficient design methodology that provides on-site needs (e.g., water, shelter, food, aesthetics) with on-site elements (e.g., stormwater runoff, greywater, cooling shade, warming sun, vegetation) by assessing all on-site resources, and placing and designing all new elements so they build on these existing resources and help divert, deflect, or convert the site's challenges into still more resources. Integrated design saves resources (e.g.,

energy, water, money), while enhancing the function and sustainability of a site.

Jandy valve. My preferred, fully adjustable, three-way diversion valve for a household greywater distribution system. Available from pool and spa suppliers.

Kilowatt-hour (kWh). When you buy electricity you are charged by the kilowatt-hour. When you use 1,000 watts for 1 hour, that's a kilowatt-hour. If a 100 watt light bulb is on for one hour per day for 30 days, the energy used is 100 W Å~ 30 h = 3,000 Wh = 3 kWh. In 2005, the American average monthly residential electricity consumption was 938 kWh, according to the Energy Information Administration.

Land subsidence. The settling or sinking of land that results from the compaction of the sedimentary layers of an aquifer. This occurs when groundwater is withdrawn from the pore spaces of these sedimentary layers faster than the water can be naturally replaced.

Leeward. The side of an object sheltered from the wind, or the direction the wind is blowing to.

Low-water-use vegetation. Vegetation that can subsist on natural rainfall alone.

Lpcd. Liters (consumed or used) per capita (person) per day.

Media luna. See *sheet flow spreader*.

Mesquite pods. The naturally sweet and nutritious edible seed pods/fruit of the mesquite (*Prosopis spp.*) tree. The soft chewable pods are good sources of calcium, manganese, potassium, iron, and zinc, while the hard seeds within are good sources of protein.

Microclimate. A more temperate or extreme localized climate created by the shelter or exposure of adjacent landscape features or buildings.

Microorganisms. Plants or animals of microscopic size.

Mulch. A porous layer of organic matter or rock on the surface of the soil (not mixed into the soil) that increases the porosity and fertility of the underlying soil, while reducing soil moisture loss to evaporation.

Municipal water use. Use of "city water" that you pay for, and which often is piped and pumped long distances. The source is typically surface water (such as from reservoirs or rivers) or groundwater pumped from aquifers.

Native vegetation. Vegetation indigenous to a 25-mile (40-km) radius of a site and found within 500 feet (150 m) of the site's elevation. Some sites may require defining native with a larger radius to bring in more plant diversity, but the smaller the radius the more likely the plants can thrive within the climatic constraints of the site.

Natural recharge. The rate at which water naturally fills or replenishes an aquifer.

Net and pan system. A modified series of boomerang berms connected directly to one another, concentrating harvested runoff at multiple points in the landscape. A completed system looks like a "net" of berms draped over a hillside with "pans" or basins inside each segment of the "net."

Nonpoint-source pollution. Pollutants from many diffuse sources. Nonpoint-source pollution is caused by stormwater or snowmelt moving over the ground. The runoff picks up and carries away natural and human-made pollutants, finally depositing them into lakes, rivers, wetlands, coastal waters, and even underground sources of drinking water.

Nonpotable water. Water that is not safe for human consumption without adequate filtration and/or treatment.

Oasis zone. The area around gathering spots like patios, front porches, and paths within 30 feet (9 m) of a home where water resources such as roof runoff and household greywater are readily available to support a greater density and/or diversity of vegetation.

One-rock dam. A low grade-control structure or "dam" placed across a small drainage, and, unlike a check dam, built just one rock high (no stacking of rock) with rock placed in three to five parallel rows packed tightly together, through which stabilizing and sediment-harvesting vegetation can easily grow.

On-site water budget. The amount of rainwater falling on, and runoff running onto a site, minus the amount of water lost to runoff. Sometimes greywater generated and reused on site is also included, but unlike the rainwater, this is not necessarily a sustainable water source if pumped or trucked in from off site.

Open-pollinated. Non-hybrid plants produced by transferring pollen from two parents from the same variety, which in turn produce offspring just like the parent plants. Heirloom vegetables are open-pollinated varieties passed down from generation to generation.

Organic. Non-genetically modified life grown or raised without synthetic fertilizers, pesticides, or sewer sludge in such a way that the fertility of associated soil improves with time.

Organic groundcover. Natural materials that break down and improve the soil such as dead plant material, manure, or compost.

Orientation. How a building or plantings are oriented in relationship to the winter's noonday sun, and the angles of the rising and setting sun year round. Buildings with their longer wall and windows facing the winter noonday sun, and shorter walls and fewer windows facing the rising and setting sun are much easier to passively heat and cool than a building of the opposite orientation.

Overflow. The planned and stabilized exit route for excess water from a water-harvesting earthwork or tank.

Overflow water. Excess water exceeding the storage capacity of a water-harvesting earthwork or tank.

P-trap. Drain pipe in the shape of the letter P used to prevent sewer gas from entering a building by keeping a water seal in the bend of the pipe.

PAN evaporation. A measurement of the combined effects of temperature, humidity, rainfall, drought dispersion, solar radiation, and wind taken by measuring the rate of evaporation from an exposed pan of standing water.

Parts per million (ppm) and parts per billion (ppb). Used to quantify amounts of pollutants and/or trace minerals in water, etc.

Passive system. A system that passively or freely works without the need of any person, switch, or valve. The more passive a system (perhaps its water simply moves with gravity), the less energy it requires.

Pathogen. An organism that may cause disease.

Peak surge. The highest short-term volume of expected water flow.

Percolation. The downward movement of water infiltrating the soil.

Perennial. Plant that lives longer than two years.

Perennial water flow. Water that continually flows year round, year after year.

Permaculture. A methodology of integrated, sustainable design based on natural systems.

Permeable paving. A broad term for water-harvesting techniques that use porous hardscape/paving materials to enable water to pass through the pavement and infiltrate into soil, passively irrigating adjoining plantings, dissipating the heat of the sun, reducing soil compaction, allowing tree roots beneath the paving to breathe, filtering pollutants, and decreasing the need for expensive drainage infrastructure.

pH. The measure of acidity or alkalinity ranging from 1 to 14. Below 7 is increasingly acid, 7 is neutral, and above 7 is increasingly alkaline.

Place assessment. The distillation of an exploration of various characteristics (such as watershed, climate, sun patterns, geology, wildlife, cultures) that make a place unique.

Point bar. A deposit of particles (such as sand, gravel, rock, and boulders) on the inside bend of a meandering waterway.

Potable water. Water that is safe for human consumption and can be used for the greatest variety of uses.

Prevailing wind. The most frequent or common wind that blows predominantly from a single general direction.

Radiant heat loss. The transfer/loss of heat energy through space. No physical contact (as would be the case with *conduction*) is involved.

Rainhead. An angled downspout screen placed below a roof gutter. The rainhead is designed to direct most debris off the screen to the ground below, while allowing water through and down the downspout.

Ramada. An outdoor shade structure under which it is comfortable to gather.

Regenerative. A type of investment that can repair, reproduce, recreate, or regenerate itself; typically serves multiple functions; produces more resources than it consumes; and improves the health of its surroundings and the world.

Renewable. A resource that can be replaced in a short period of time. Renewable does not necessarily mean sustainable, for instance the transport of "renewable" Colorado River water pumped to Tucson and Phoenix, Arizona consumes huge amounts of resources, while its over-allocation has so depleted the river's flow, that vast tracks of the Colorado river delta and the culture of the indigenous people of the area have been destroyed.

Reservoir. A structure for storing water. It may be open or covered.

Retaining wall. A structure that holds back a slope, preventing erosion.

Revolving community loan fund. A source of money a community puts together to fund projects through loans that enhance the lives of individual community members in a way that simultaneously enhances the health of the community (for example construction of a composting toilet is financed, which results in improved health of those that use it, while also improving water quality and quantity, soil fertility, and thus the health of the larger community). The fund gets its name from the revolving aspect of loan repayment, where the central fund is replenished as individual projects pay back their loans, creating the opportunity to issue other loans to new projects.

Rill. A tiny erosive drainage in which loose soil has washed away. It is very common on eroding slopes where roadways have been cut into hillsides or on bare dirt driveways and roads that run downslope.

Rock-lined plunge pool. Also called a *Zuni bowl*, it is a headcut-control structure constructed of a rock-mulch rundown leading into a rock-lined plunge pool(s) that dissipates the force of water falling over the headcut's pour-over, while hydrating the soil to establish stabilizing vegetation.

Rock-mulch rundown. An erosion-controlling, one-rock high layer of rock mulch used to armor a sloped, low-energy water flow location.

Rolling dip. An excavated cross drain designed to divert water off the road surface and consisting of two main features: 1) a broadly angled dip drain with a cross slope of 4-8% (steep enough to flush accumulating sediment off the road and drain) and 2) a reverse grade mound (rollout) used to reinforce the drainage channel and prevent water flowing down road's wheel ruts. Length of dip and length of rollout are about equal to the length of the average vehicle traveling over them for smooth travel over structure.

Runoff. Water that flows off a surface when more rain falls than the surface can absorb.

Runoff coefficient. The average percentage of rainwater that runs off a type of surface. For example, a rooftop with a runoff coefficient of 0.95 indicates that 95% of the rain falling on that roof will run off.

Runon. Runoff water that runs onto a site.

Saturated soil. Soil in which the pore space is completely filled with water.

Sediment. Soil, sand, and minerals washed from land into water or lower reaches of land, usually after rain. Excessive sediment can destroy fish-nesting areas; clog animal habitats, French drains, and porous pavement; and obscure waters so that sunlight does not reach aquatic plants.

Sediment traps. Native grasses, a basin/pool, or other strategies that slowdown the flow of water, enabling the sediments carried in the water to drop out of the flow. Such traps can be placed at the inlet of a water-harvesting earthwork, and made accessible for easy sediment removal.

Sewer. A pipe used to transport sewage elsewhere.

Sheet flow. The relatively even distribution of runoff water over the land surface, following the slope of the land downward, but not focused into distinct channels. Sheet flow has most likely occurred after a large rainfall if you don't see distinct channels in an area of sloping bare dirt.

Sheet flow spreader. Also called a *media luna* (half moon), it is a level-topped, one-rock high, crescent-shaped rock mulch structure (where the ends of the crescent point uphill) used on relatively flat to alluvial-fan-shaped ground to disperse erosive channelized flow and reestablish sheet flow where it once occurred. It is laid on contour.

Site assessment. The exploration of various site-specific characteristics (such as slope and water flows; solar aspect; microclimates; paths and uses of people, wildlife, and/or livestock; and vegetation) to better understand what is happening in order to better aid the site in reaching its potential.

Slope. A measurable steepness indicating a change in elevation from one point to another.

Soft water. Water containing little or no dissolved calcium and magnesium.

Solar access. Maintaining full winter sun exposure to winter sunfacing windows, solar water heaters, solar photovoltaic panels, solar ovens, and winter gardens.

Solar arc. A number of shading elements such as trees, cisterns, trellises, covered porches, and overhangs laid out in the shape of an arc or semi-circle open to the winter sun, and deflecting the rising and setting summer sun from any objects, such as a home or garden placed within the arc.

Solar hot air collector. A black-interior backed box capped with a transparent surface such as glass facing the winter sun, which traps heat like a greenhouse, and then vents that heated air into an adjoining building.

Solar noon. The time of the day that divides the daylight hours for that day exactly in half. To determine solar noon, calculate the length of the day from the time of sunset and sunrise and divide by two. Solar noon is the time of day the sun is highest in the sky.

Solar time. A way of telling time based on the Sun's position in the sky. It can be crudely measured by a sundial, and is also called *sundial time* or *apparent solar time*. Solar time is not the same as *clock time*.

Spillway. A planned and stabilized route for overflow water.

Sponge. Living sponge-like condition of mulched and vegetated soil, a rain garden, and/or a healthy watershed, whereby water rapidly infiltrates, then cycles through (and regenerates) the soil and its associated life (plants, microorganisms, etc), which greatly reduces the rate of the water's loss (via evaporation, runoff, erosion, or pollution) from the system.

Step pools. A stepped series of drop offs and pools within a water channel. The force of the water falling over the drop off is diffused by the pool of water into which the falling water drops. These can occur naturally or be constructed.

Stormwater. Rainwater once it has landed on a ground-surface.

Stub out (greywater). Plumbing connections installed in new construction that allow easy and inexpensive future access to a greywater stream in order to redirect the greywater to irrigate plantings in the landscape.

Subsoil. The naturally compacted soil found beneath less compacted, more organic-matter-rich topsoil.

Subwatershed. A smaller watershed within, and making up part of a larger watershed.

Sun path. Our perception (looking from a given point on Earth) of how and where the sun moves through the sky on a given day, and how that path changes throughout the year, affecting shadows, temperature, length of day, light, and more.

Sun & Shade trap. An area having a more comfortable and moderate microclimate due the site being open to the winter's rising and noonday sun, while shaded from the afternoon sun—primarily in summer.

Superfund site. A site contaminated with hazardous substances, and approved for use of superfund trust money to fund cleanup efforts of the contamination.

Supplemental water. An auxiliary source of water meant to augment natural on-site rainfall resources.

Surface water. Water that flows on the surface of the land, such as water flowing in creeks and rivers.

Sustainable. A condition in which biodiversity and renewability of ecosystems, cultures, and natural resource production and quality are maintained over time.

Terrace. Sometimes called a bench, it is a relatively flat "shelf" of soil built parallel to the contour of a slope. The terrace reduces the steepness of a section of a slope, reducing run-off and erosion, while increasing infiltration. Terraces can be built with or without a retaining wall depending on the steepness of the slope.

Three-way diverter valve. A valve used to direct or divert water flow in one of two directions. Found at pool and spa supply stores, these valves can be incorporated into greywater plumbing to allow the user to send the greywater to the landscape or sewer as they please.

Tinaja. A desert water hole naturally carved into bedrock.

Topsoil. The upper layer of soil containing most of the organic matter and fertility.

Total dissolved solids (TDS). Measures the quantity of materials dissolved in one gallon or liter of water. Technically, these are the dry residues that remain after the water has been heated to 180°C.

Toxic. Any substance able to cause injury to living organisms when eaten, absorbed through the skin, or inhaled into the lungs.

Transpiration. The loss of moisture from plants to the air via the stomata within their leaves.

Trombe wall. A winter sun-facing masonry wall separated from the outdoors by a glass or other transparent membrane, and an air space. It is designed to absorb the sun's radiant heat during sunny days and release it into a building's interior.

True North. North according to the Earth's axis and thus the direction toward the geographic North Pole, not magnetic north, which can vary from place to place and over time due to local magnetic anomalies. The North Star (Polaris) is located very close to true north.

True South. South according to the Earth's axis and thus the direction toward the geographic South Pole. The opposite direction of True North.

Vent. A screened opening installed above the elevation of a closed water tank's inlet pipe, which allows air to flow in and out of the tank, permitting the smooth inflow and outflow of water, and preventing a vacuum-caused implosion when large quantities of water are quickly drawn from the tank.

Virtual water footprint. The total volume of freshwater that is used to produce the commodities, goods, and services consumed by a person, company, or nation.

Wastewater. Water used by humans and considered a "waste" needing to be disposed of. Creating such a thing is the real waste.

Water softener. A device that replaces calcium and magnesium ions from hard water with sodium ions. Without the calcium and magnesium the water becomes "soft," but the added sodium or salt is not good for plants or soil. So softened water is not good for use with greywater systems, and the softener backwash is even worse since it has an even higher salt content.

Water table. The upper limit of a body of ground-water.

Watershed. The total area of a landscape draining or contributing water to a particular site or drainage.

Well. A human-made hole in the earth from which groundwater is withdrawn.

Wet-system downspout. A rainwater conveyance pipe having a section of its run below the level of both its input point and its output point, resulting in water (and sometimes sludge) always being in the low run of pipe (unless manually drained between rains). Note: For gravity-fed water to properly move through this system, the inlet point of the pipe must be higher than the outlet point.

Windward. The side or direction from which the wind is blowing.

Winter-sun side. For those living in the northern hemisphere, the south-facing side of buildings, walls, and trees is the "winter-sun side" and the north-facing side is the "winter-shade side." This is because the winter sun stays in the southern sky all day.

Zuni bowl. See *rock-lined plunge pool*.

Index

NOTES

In the Arid Southwest water scarcity has taught generations upon generation of desert dwellers to respect their environment and to cherish the most precious natural resource of this area, WATER. These teachings have and continue to come directly from the interconnectedness of land, people, and culture, and although modernity encourages us to forget, the environment constantly reminds us. For this reason, we must be diligent in gathering and sharing knowledge that will create new paths that support a just, equitable, and sustainable future, that addresses root problems, creates resilience and restores a balance between people and their environment. *Rainwater Harvesting for Drylands and Beyond* is a great book helping do just that.

— Luis A. Perales, M.S., Community Organizer;
Tierra Y Libertad Organization; www.facebook.com/TierraYLibertadOrganization

"Brad's Rainwater Harvesting books may be the most important books I have ever read, as an architect, permaculturist, and homeowner. The amazingly simple, natural, and beautiful strategies presented are much-needed tools for solving an astounding array of water-connected problems, from energy use and water scarcity to food production and biodiverse landscapes. Brad reveals the importance of an integrated, wholistic approach to dealing with water on a site, and he reminds us of the great rewards of treating rainwater as the valuable resource that it is."

— Gayle Borst, Architect and Executive Director of Design~Build~Live, Austin, Texas

"Our modern society is afflicted with a severe case of hydrological illiteracy. Brad's *Rainwater Harvesting for Drylands and Beyond*, in three volumes, is the antidote needed to mitigate this epidemic of cerebral imperviousness impacting the collective head-waters of our cultural ego-system! Pragmatic practitioners of "waterspread" restoration for arid lands and beyond, will find a wealth of accessible and practical information in these books. The Conservation Hydrology mantra of Slow It - Spread It - Sink It has never been better articulated in as clear and concise terms for the homeowner, ranch owner, subdivision developer or city stormwater engineer. I will require this book for all my Basins of Relations community watershed students and feel it should be so as well for all land managers and land use planners. This book is sure to become a classic for all people who believe in a future based on rehydration instead of dehydration and for that I say, Bravo Brad!"

— Brock Dolman, WATER Institute Director, Occidental Arts and Ecology Center

"What a wonderful, enthusiastic book. Brad Lancaster lives what he preaches—a water-careful lifestyle that is all about more life. Brad is a worthy teacher—his love and deep respect for water shines through on every page."

— Ben Haggard, author, artist, regenerative systems practitioner and teacher

"Brad Lancaster presents the first of three volumes of his vision on what might be called Eco-hydrology. It is not just a book about harvesting rainfall, although there are many practical ideas on how to make use of the water that falls on your land. It is a guide designed to help the reader see what Brad sees when looking at a city lot or homesite. He has set as his goal to train you to see your land and the environment in which it is set in a new way, as a natural resource to be managed in harmony with your living there. While water harvesting is central to living a new paradigm it is only part of a broader vision designed to enrich your quality of life while enhancing the surrounding environment."

— James J. Riley, Ph.D., Soil, Water and the Environmental Science Department,
The University of Arizona

"This book is the best of both worlds, highly user friendly and highly informational. Its practical and straightforward nature is an antidote to inaction from overwhelm and confusion. It will definitely be an essential resource for our work at the Institute of Permaculture Education for Children and facilitate bringing Permaculture and Water Harvesting into schools."

— Patty Parks-Wasserman,
Director and Founder of Institute of Permaculture Education for Children

"Like small acorns that grow into mighty oaks, Ben Franklin's succinct and wise words are perhaps more valuable today: 'Waste not; want not...A penny saved is a penny earned.' The anticipation of rain and its eventual harvest and storage for nurturing the native habitat, our source of food, the quality of our air and water, and visual delight for our senses is a natural model for us to mimic. Brad Lancaster's pioneering series of books, Rainwater Harvesting for Drylands and Beyond, shows us how we can mimic the way nature works, as it immediately provides the resources necessary to support a world of efficient and effective use of water that helps create abundance in all our lives."

— Dr. Wayne Moody, American Institute of Certified Planners (AICP), Planner

"Brad Lancaster's *Rainwater Harvesting for Drylands and Beyond, Vol. 1*, is a fantastic resource for all of us dryland dwellers! Lancaster's enthusiasm and commitment to more sustainable and rewarding dryland lifestyles bubbles out of every paragraph, and his writing is personable and easily accessible, even though packed with information. This combination makes rainwater harvesting seem not only absolutely necessary if we want to survive for long in drylands, but easy and fun as well!"

— David A. Cleveland, Prof., Environmental Studies; University of California Santa Barbara;
Co-Director, Center for People, Food & Environment; co-author of *Food from Dryland Gardens*

"Brad's book is amazing. He has made it so readable and the photos and diagrams and even cartoons make it easy to follow and very practical, yet the research and figures are there to read which make it a very scholarly book as well. His passion for the subject shines through in a very non-preachy way. It is a remarkable achievement. It is so relevant for places all over the world. It is clear that water is going to be one of the most important issues for the 21st century and he tackles it in a way that everybody can play their part."

— Paul Hallowes, Santa Barbara, California

"Brad Lancaster is one of those rare individuals who combines a practical ability to design and implement common-sense solutions to rainwater management issues with a clear ecological and political vision of the importance of doing so. In *Rainwater Harvesting for Drylands and Beyond* Brad shows us how to use rainwater around our homes and in our communities so that our human-created landscapes reflect the abundance of nature. As we move from assumptions of scarcity to participation in abundance, our lives and our communities can be transformed."

— David Confer, Ph.D., environmental engineer and sustainable design and development consultant

"On a water-world such as ours, Brad's book should be a required study for all human beings. Scholarly, forthright and, above all, practical, this work delivers critical knowledge to those thirsty for a positive relationship with water. Thank you Brad for your friendly presentation of such a complex and important component of global sustainability!"

— Paul A. Branson, Earthwise Technologies Ecological Restoration

see www.HarvestingRainwater.com for more testimonials and book reviews

The series continues...

Rainwater Harvesting for Drylands and Beyond, Volume 2

Water-Harvesting Earthworks

Earthworks are one of the easiest, least expensive, and most effective ways of passively harvesting and conserving multiple sources of water in the soil. Associated vegetation then pumps the harvested water back out in the form of beauty, food, shelter, wildlife habitat, timber, and passive heating and cooling strategies, while controlling erosion, increasing soil fertility, reducing downstream flooding, and improving water and air quality.

Building on the information presented in Volume 1, this award-winning book shows you how to select, place, size, construct, and plant your chosen water-harvesting earthworks. It presents detailed how-to information and variations of all the earthworks, including chapters on mulch, vegetation, and greywater recycling so you can customize the techniques to the unique requirements of your site. Info on how to create cheap and simple tools to read slope and water flow are also included along with sample plant guides.

Real life stories and examples permeate the book, including:

- How homesteading grandmothers are restoring their land with simple earthworks made on their daily walks
- How curb cuts and infiltration basins redirect street runoff to passively irrigate flourishing shade trees planted along the street
- How check dams have helped create springs and perennial flows in once-dry creeks
- How infiltration basins or rain gardens are creating thriving rain-fed gardens
- How backyard greywater laundromats are turning "wastewater" into a resource growing food, beauty, and shade that builds community, and more.
- More than 450 illustrations and photographs.

For more details and ordering info see www.HarvestingRainwater.com.

Rainwater Harvesting for Drylands and Beyond, Volume 3

Roof Catchment and Cistern Systems

Cisterns harvesting runoff from rooftop catchments have the potential to harvest the highest quality rainwater runoff on site, and allow for the greatest range of potential uses for that water. Water uses range from irrigation to fire protection, bathing, and washing, as well as potable water for drinking.

Building on the information presented in Volume 1, this book shows you how to select, size, design, build, or buy and install cistern systems harvesting roof run-off. Guiding principles unique to such systems are presented along with numerous tank options, and design strategies that will improve water quality, save you money, reduce maintenance, and expand the ways you can use your tanks and the water harvested within.

Additional information includes:

- How to size gutters and downspouts
- How to turn a drain pipe into a storage tank—how to build culvert cisterns
- How to retrofit ferrocement septic tanks for cistern use
- How to create a gravity-fed drip irrigation system for use with above-ground tanks
- How to create a pump-fed drip irrigation system for use with below-ground tanks
- System setup, maintenance tips, and water treatment options
- More than 60 illustrations and photographs

For more details, publication dates, and ordering info see www.HarvestingRainwater.com.

Rainwater Harvesting for Drylands and Beyond, Volume 1 in Spanish, Arabic, and electronic editions

For more details, publication dates, and ordering info see www.HarvestingRainwater.com.

Front inside cover figure captions:

All images below appear again in the book to illustrate case studies or concepts.
Figure numbers correspond to where in the book they can be found.

Fig. 3.11A. Before sponge installation. Rainfall and roof runoff collected in newly constructed infiltration basins minutes after large summer storm. The basins have not yet been mulched or planted. Milagro Cohousing, Tucson, Arizona. Credit: Natalie Hill

Fig. 3.11B. After sponge installation. Same basins mulched and vegetated. Basins are designed to infiltrate water quickly so there are no problems with mosquitoes or anaerobic soils. These basins, with their spongey mulch and soil-burrowing plant roots infiltrate all stormwater within 20 minutes. Plants are irrigated solely with harvested rainwater and recycled household "wastewater."

Fig. 3.26A. One-rock check dam before rains, Red Windmill Draw, Malapai Ranch, near Douglas, Arizona. One-rock dams are just one rock high and placed in three to five parallel rows packed tightly together across the drainage. Reproduced with permission from *An Introduction to Erosion Control* by Bill Zeedyk and Jan-Willem Jansens. Photo credit: Van Clothier of Stream Dynamics

Fig. 3.26B. One-rock check dam two months later, after summer rains. New vegetation has grown upstream of the stabilizing dam. Because the structure is only one rock high, vegetation will easily grow between the rock to further stabilize the structure and slow and infiltrate water flow. Reproduced with permission from *An Introduction to Erosion Control* by Bill Zeedyk and Jan-Willem Jansens. Photo credit: Van Clothier of Stream Dynamics

Fig. 5.10A. Before the 1996 planting of rain and trees in public right-of-way adjoining Lancaster property with asphalt driveway freshly removed, Tucson, Arizona, 1994

Fig.5.10B. After planting of rain and trees a tree-lined footpath now revives the once sterile right-of-way, 2006. Landscape is irrigated solely with harvested rainfall and street runoff.

Back inside cover figure captions:

Fig. 5.3. Late winter garden grown entirely from rainfall and roof runoff harvested in 1,300-gallon (4,920-liter) tank at Lancaster household receiving 11 inches (280 mm) annual rainfall

Fig. 5.8. Rainwater-irrigated food grown and processed at Lancaster household

Fig. 3.5. Mulched and planted terraces stabilized with dry-stacked salvaged concrete retaining walls designed and installed by La Loma Development Company in Pasadena, California

Fig. 3.18. A small yard with hardscape kept permeable by installing recycled sidewalk chunks with ample gaps for water infiltration, Amado residence, Tucson, Arizona. Designed by Blue Agave Landscape Design

Fig. 3.35. A 10,000-gallon (38,000-liter) steel tank/mural collecting roof runoff for irrigation, Children's Museum, Santa Fe, New Mexico

Fig. 3.41. A 7,000-gallon (26,460-liter) poured-in-place concrete tank faced with stone doubles as a patio and collects roof runoff for gravity-fed irrigation of citrus trees and other plants below. Note how the tank's mass creates a warmer microclimate for the frost-sensitive citrus trees on the south-facing side of this northern hemisphere tank. Tucson, Arizona